VOLUME FOUR HUNDRED AND FIFTY

METHODS IN ENZYMOLOGY

Fluorescence Spectroscopy

METHODS IN ENZYMOLOGY

Editors-in-Chief

JOHN N. ABELSON AND MELVIN I. SIMON

*Division of Biology
California Institute of Technology
Pasadena, California*

Founding Editors

SIDNEY P. COLOWICK AND NATHAN O. KAPLAN

VOLUME FOUR HUNDRED AND FIFTY

METHODS IN ENZYMOLOGY

Fluorescence Spectroscopy

EDITED BY

LUDWIG BRAND
Johns Hopkins University
Department of Biology
Baltimore, MD, USA

MICHAEL L. JOHNSON
University of Virginia Health Sciences Center
Department of Pharmacology
Jordan Hall, Charlottesville, VA, USA

ELSEVIER

AMSTERDAM • BOSTON • HEIDELBERG • LONDON
NEW YORK • OXFORD • PARIS • SAN DIEGO
SAN FRANCISCO • SINGAPORE • SYDNEY • TOKYO
Academic Press is an imprint of Elsevier

Academic Press is an imprint of Elsevier
525 B Street, Suite 1900, San Diego, California 92101-4495, USA
30 Corporate Drive, Suite 400, Burlington, MA 01803, USA
32 Jamestown Road, London NW1 7BY, UK

Copyright © 2008, Elsevier Inc. All Rights Reserved.

No part of this publication may be reproduced or transmitted in any form or by any means, electronic or mechanical, including photocopy, recording, or any information storage and retrieval system, without permission in writing from the Publisher.

The appearance of the code at the bottom of the first page of a chapter in this book indicates the Publisher's consent that copies of the chapter may be made for personal or internal use of specific clients. This consent is given on the condition, however, that the copier pay the stated per copy fee through the Copyright Clearance Center, Inc. (www.copyright.com), for copying beyond that permitted by Sections 107 or 108 of the U.S. Copyright Law. This consent does not extend to other kinds of copying, such as copying for general distribution, for advertising or promotional purposes, for creating new collective works, or for resale. Copy fees for pre-2008 chapters are as shown on the title pages. If no fee code appears on the title page, the copy fee is the same as for current chapters. 0076-6879/2008 $35.00

Permissions may be sought directly from Elsevier's Science & Technology Rights Department in Oxford, UK: phone: (+44) 1865 843830, fax: (+44) 1865 853333, E-mail: permissions@elsevier.com. You may also complete your request on-line via the Elsevier homepage (http://elsevier.com), by selecting "Support & Contact" then "Copyright and Permission" and then "Obtaining Permissions."

For information on all Elsevier Academic Press publications visit our Web site at www.elsevierdirect.com

ISBN-13: 978-0-12-374586-6

PRINTED IN THE UNITED STATES OF AMERICA
08 09 10 11   9 8 7 6 5 4 3 2 1

Working together to grow
libraries in developing countries

www.elsevier.com | www.bookaid.org | www.sabre.org

ELSEVIER   BOOK AID International   Sabre Foundation

Contents

Contributors	xi
Preface	xv
Volumes in Series	xvii

1. A Method in Enzymology for Measuring Hydrolytic Activities in Live Cell Environments   1
Beverly Z. Packard and Akira Komoriya

1.	Protease Substrate Design	2
2.	Protease Specificity	5
3.	Cell-Permeable Fluorogenic Probes	7
4.	Conclusions	16
	References	17

2. Heterogeneity of Fluorescence Determined by the Method of Area-Normalized Time-Resolved Emission Spectroscopy   21
N. Periasamy

1.	Introduction	22
2.	TRES and TRANES Methods	24
3.	TRES and TRANES Spectra of Simple Cases	25
4.	Physical Significance of TRANES and Isoemissive Point	29
5.	TRES and TRANES of Fluorophores in Microheterogeneous Media	30
6.	Fluorescence in Microheterogeneous and Biological Media: Special Cases	32
	Acknowledgment	33
	References	33

3. Multiparametric Probing of Microenvironment with Solvatochromic Fluorescent Dyes   37
Andrey S. Klymchenko and Alexander P. Demchenko

1.	"Universal" and "Specific" Noncovalent Interactions	39
2.	The Methodology of Multiparametric Approach with Application of 3-Hydroxyflavone Dyes	40
3.	Correlations of Spectroscopic Data with Solvatochromic Variables	43

4. Algorithm for Multiparametric Probing Based on Parameters of Absorption and Dual Emission	47
5. Application of Multiparametric Probing	49
6. Limitations of the Multiparametric Approach	51
7. Concluding Remarks	55
References	56

4. Site-Selective Red-Edge Effects — 59

Alexander P. Demchenko

1. Introduction	60
2. Molecular Disorder and the Origin of Red-Edge Effects	61
3. The Principle of Photoselection	62
4. Ground-State Heterogeneity	65
5. The Magnitude of Red-Edge Excitation Fluorescence Shift and Its Connection with Dielectric Relaxations	67
6. Red-Edge Effect with High Resolution in Time	71
7. Red-Edge Effects in Excited-State Reactions	72
8. Fluorescent Probes for Optimal Observation of Red-Edge Effects	74
9. Peculiarities of the Red-Edge Effects of Indole and Tryptophan	75
10. Conclusions	76
References	77

5. Fluorescence Approaches to Quantifying Biomolecular Interactions — 79

Catherine A. Royer and Suzanne F. Scarlata

1. Introduction	80
2. Fluorescence Observables	80
3. Designing a Fluorescence Experiment	85
4. Conclusions	103
References	103

6. Forster Resonance Energy Transfer Measurements of Transmembrane Helix Dimerization Energetics — 107

Mikhail Merzlyakov and Kalina Hristova

1. Introduction	108
2. Challenges in Quantitative Measurements of Interactions Between TM Helices in Bilayers	108
3. Bilayer Platforms for Measuring TM Helix Interactions Using FRET	110
4. FRET Due to Random Colocalization of Donors and Acceptors (Proximity FRET)	114

5.	FRET Efficiencies and Energetics of TM Helix Dimerization	115
6.	Biological Insights from FRET Measurements	124
7.	Conclusion	125
	References	125

7. Application of Single-Molecule Spectroscopy in Studying Enzyme Kinetics and Mechanism — 129

Jue Shi, Joseph Dertouzos, Ari Gafni, and Duncan Steel

1.	Introduction	130
2.	Experimental Considerations	134
3.	Sample Preparation	136
4.	Instrumentation	138
5.	Data Analysis of Single-Molecule Trajectories	141
6.	Additional Considerations	151
7.	Summary	154
	References	155

8. Ultrafast Fluorescence Spectroscopy via Upconversion: Applications to Biophysics — 159

Jianhua Xu and Jay R. Knutson

1.	Introduction	160
2.	Basic Concepts	162
3.	Upconversion Spectrophotofluorometer and Experimental Considerations	168
4.	Ultrafast Photophysics of Single Tryptophan, Peptides, Proteins, and Nucleic Acids	171
5.	Summary and Future Directions	179
	References	179

9. Use of Fluorescence Resonance Energy Transfer (FRET) in Studying Protein-induced DNA Bending — 185

Anatoly I. Dragan and Peter L. Privalov

1.	Introduction	186
2.	Preparation of Labeled DNA Duplexes	187
3.	Fluorescence Resonance Energy Transfer	189
4.	FRET in Studying Large Protein-Induced DNA Bends	191
5.	Dependence of the Protein-induced DNA Bend on the Forces Involved in Binding	194
6.	FRET in Studying Small Protein-Induced DNA Bends	194
7.	Conclusions	198

Acknowledgment	198
References	199

10. Fluorescent Pteridine Probes for Nucleic Acid Analysis — 201
Mary E. Hawkins

1. Introduction	202
2. Pteridine Analog Characteristics	205
3. Procedures for Oligonucleotide Synthesis with Pteridine Analogs	208
4. Characterization of Pteridine-Containing Sequences	210
5. Applications	211
6. Summary	227
Acknowledgments	228
References	228

11. Single-Molecule Fluorescence Methods for the Analysis of RNA Folding and Ribonucleoprotein Assembly — 233
Goran Pljevaljčić and David P. Millar

1. Introduction	233
2. Labeling Methods	236
3. Single-Molecule Fluorescence Detection Methods	238
4. Diffusion Single-Pair FRET for RNA Based Systems	238
5. Total Internal Reflection Fluorescence (TIRF) for RNA Based Systems	240
6. Application 1: Folding of the Hairpin Ribozyme	243
7. Application 2: Assembly of the Rev–RRE Complex	247
Acknowledgments	249
References	249

12. Using Fluorophore-Labeled Oligonucleotides to Measure Affinities of Protein–DNA Interactions — 253
Brian J. Anderson, Chris Larkin, Kip Guja, and Joel F. Schildbach

1. Introduction	254
2. Definitions of Fluorescence Anisotropy and Intensity	254
3. Advantages of Fluorescence Measurements	256
4. Disadvantages of Fluorescence Measurements	257
5. Designing the Oligonucleotide	258
6. Designing the Experiment	266
7. Competition Assays for Determining Specificity	269
8. Conclusions	270
Acknowledgments	270
References	270

13. Identifying Small Pulsatile Signals within Noisy Data: A Fluorescence Application   273
Michael L. Johnson, Leon S. Farhy, Paula P. Veldhuis, and Joseph R. Lakowicz

   1. Introduction   274
   2. Methods   274
   3. Results   280
   4. Discussion   284
   Acknowledgments   286
   References   286

14. Determination of Zinc Using Carbonic Anhydrase-Based Fluorescence Biosensors   287
Rebecca Bozym, Tamiika K. Hurst, Nissa Westerberg, Andrea Stoddard, Carol A. Fierke, Christopher J. Frederickson, and Richard B. Thompson

   1. Introduction   288
   2. Principles of CA-Based Zinc Sensing   288
   3. Transducing Zinc Binding as a Fluorescence Change   290
   4. "Free" Versus Bound Zinc Ion: Speciation   295
   5. Metal Ion Buffers   297
   6. Kinetics   301
   7. Applications: Ratiometric Determination of Free Zinc in Solution   302
   8. Preparation of Apocarbonic Anhydrase   303
   9. Intracellular Sensing with TAT Tag   304
   10. Intracellular Sensing with an Expressible CA Sensor   305
   References   307

15. Instrumentation for Fluorescence-Based Fiber Optic Biosensors   311
Richard B. Thompson, Hui-Hui Zeng, Daniel Ohnemus, Bryan McCranor, Michele Cramer, and James Moffett

   1. Introduction and Rationale for Fluorescence-Based Fiber Optic Sensors   312
   2. Basic Principles of Fiber Optics for Fluorescence Sensors   313
   3. Optics and Mechanics of Fluorescence-Based Fiber Optic Sensors   316
   4. Mounting and Alignment of the Instrument   319
   5. Standards for Ratiometric and Lifetime-Based Fiber Optic Fluorescence Sensors   323
   6. Construction of Fiber Optic Probes   326
   7. Zn^{2+} Probe   329
   8. Cu^{2+} Probe   330
   9. Use of Fiber Optic Probes for Discrete Samples   332

10. Operating Issues: Noise, Background, Thermal Drift, and
    Mode Hopping   333
11. Field and Shipboard Use   334
Acknowledgments   335
References   336

Author Index   *339*
Subject Index   *353*

Contributors

Brian J. Anderson
George Washington University School of Medicine, Washington, District of Columbia, and Department of Biology, Johns Hopkins University, Baltimore, Maryland

Rebecca Bozym
Cellumen, Inc., 3180 William Pitt Way, Pittsburgh, Pennsylvania, and Department of Biochemistry and Molecular Biology, University of Maryland School of Medicine, Baltimore, Maryland

Michele Cramer
Department of Biochemistry and Molecular Biology, University of Maryland School of Medicine, Baltimore, Maryland

Alexander P. Demchenko
A.V. Palladin Institute of Biochemistry, Kiev 252030, Ukraine

Joseph Dertouzos
Department of Physics and EECS, University of Michigan, Ann Arbor, Michigan

Anatoly I. Dragan
The Institute of Fluorescence, University of Maryland Biotechnology Institute, Columbus Center, Baltimore, Maryland

Leon S. Farhy
Department of Medicine (Endocrine Division), University of Virginia Health System, Charlottesville, Virginia

Carol A. Fierke
Department of Chemistry, University of Michigan, Ann Arbor, Michigan

Christopher J. Frederickson
NeuroBioTex, Inc., Galveston, Texas

Ari Gafni
Department of Biological Chemistry, University of Michigan, Ann Arbor, Michigan, and Biophysics Research Division, University of Michigan, Ann Arbor, Michigan

Kip Guja
School of Medicine at Stony Brook University Medical Center, Stony Brook, New York, and Department of Biology, Johns Hopkins University, Baltimore, Maryland

Mary E. Hawkins
National Institutes of Health, National Cancer Institute, Bethesda, MD 20892

Kalina Hristova
Department of Materials Science and Engineering, Johns Hopkins University, Baltimore, Maryland 21218

Tamiika K. Hurst
Department of Chemistry, University of Michigan, Ann Arbor, Michigan

Michael L. Johnson
Department of Medicine (Endocrine Division), University of Virginia Health System, Charlottesville, Virginia, and Department of Pharmacology, University of Virginia Health System, Charlottesville, Virginia

Andrey S. Klymchenko
A.V. Palladin Institute of Biochemistry, Kiev 01030, Ukraine, and Lab. de Pharmacologie et Physicochimie, UMR 7034 du CNRS, Faculté de Pharmacie, Université Louis Pasteur, 67401 Illkirch, France

Jay R. Knutson
Optical Spectroscopy Section, Laboratory of Molecular Biophysics, National Heart, Lung and Blood Institute, National Institutes of Health, Bethesda, Maryland 20892-1412

Akira Komoriya
OncoImmunin, Inc., Gaithersburg, Maryland 20877

Joseph R. Lakowicz
Department of Biochemistry, University of Maryland at Baltimore, Baltimore, Maryland

Chris Larkin
Food and Drug Administration, Rockville, Maryland, and Department of Biology, Johns Hopkins University, Baltimore, Maryland

Bryan McCranor
Department of Biochemistry and Molecular Biology, University of Maryland School of Medicine, Baltimore, Maryland

Mikhail Merzlyakov
Department of Materials Science and Engineering, Johns Hopkins University, Baltimore, Maryland 21218

David P. Millar
Department of Molecular Biology, The Scripps Research Institute, La Jolla, California

James Moffett
Department of Biological Sciences, University of Southern California, Los Angeles, California

Daniel Ohnemus
Woods Hole Oceanographic Institution, Woods Hole, Massachusetts

Beverly Z. Packard
OncoImmunin, Inc., Gaithersburg, Maryland 20877

N. Periasamy
Department of Chemical Sciences, Tata Institute of Fundamental Research, Homi Bhabha Road, Mumbai 400005, India

Goran Pljevaljčić
Department of Molecular Biology, The Scripps Research Institute, La Jolla, California

Peter L. Privalov
The Institute of Fluorescence, University of Maryland Biotechnology Institute, Columbus Center, Baltimore, Maryland

Catherine A. Royer
Centre de Biochimie Structurale, 29 rue de Navacelles, 34090 Montpellier Cedex, France

Suzanne F. Scarlata
Department of Physiology and Biophysics, School of Medicine, State University of New York at Stony Brook, Stony Brook, New York

Joel F. Schildbach
Department of Biology, Johns Hopkins University, Baltimore, Maryland

Jue Shi
Biophysics Research Division, University of Michigan, Ann Arbor, Michigan, and Department of Physics, Hong Kong, Baptist University, Hong Kong

Duncan Steel
Department of Physics and EECS, University of Michigan, Ann Arbor, Michigan, and Biophysics Research Division, University of Michigan, Ann Arbor, Michigan

Andrea Stoddard
Department of Chemistry, University of Michigan, Ann Arbor, Michigan

Richard B. Thompson
Department of Biochemistry and Molecular Biology, University of Maryland School of Medicine, Baltimore, MD 21201

Paula P. Veldhuis
Department of Medicine (Endocrine Division), University of Virginia Health System, Charlottesville, Virginia, and Department of Pharmacology, University of Virginia Health System, Charlottesville, Virginia

Nissa Westerberg
U.S. Patent and Trademark Office, Alexandria, Virginia, and Department of Biochemistry and Molecular Biology, University of Maryland School of Medicine, Baltimore, Maryland

Jianhua Xu
Optical Spectroscopy Section, Laboratory of Molecular Biophysics, National Heart, Lung and Blood Institute, National Institutes of Health, Bethesda, Maryland 20892-1412

Hui-Hui Zeng
Department of Biochemistry and Molecular Biology, University of Maryland School of Medicine, Baltimore, Maryland

Preface

Methods in Enzymology (Fluorescence 2008)

Light is a relatively non-invasive probe. As ever more theoretical and experimental details emerge regarding the behavior of molecules and atoms in the excited state, fluorescence spectroscopy continues to be a method of choice to study the behavior of biological macromolecules, membranes and living cells. Some of the topics covered in volume 278 of *Methods in Enzymology* are revisited and new areas of interest are covered.

The chapter by Packard and Komoriya on intracellular protease activity is an example of the application of novel photophysics to studies in the living cell. Merzelyakov and Hristova describe how Förster type resonance energy transfer (FRET) can be utilized to measure dimerization energetics of transmembrane helices. An important feature of fluorescence is that there are several quantities that can be measured. Advantageous applications of the "multiparametric" approach is evident in the chapters by Klymchenko and Demchenko on solvatochromic dyes and Demchenko on site-selective red edge effects and also in the contribution by Periasamy on heterogeneity of fluorescence. During the past few years there has been an explosion of interest in the use of fluorescence spectroscopy to study single molecules. The chapter by Shi, Dertouzos, Gafni, and Steel discusses the application of single molecule spectroscopy in studying enzyme kinetics and mechanism. The single molecule approach is also discussed by Pljevaljcic and Millar in studies of RNA folding and ribonucleoprotein assembly.

Royer and Scarlata describe fluorescence approaches, including fluorescence correlation spectroscopy and fluorescence anisotropy, for quantifying bio-molecular interactions. Hawkins describes several pteridine probes for nucleic acid analysis. Anderson and Schildbach discuss the use of fluophore-labeled oligonucleotides to measure affinities of protein–DNA interactions. Dragan and Privalov describe the use of FRET in studying protein-induced DNA bending.

Two chapters describe biosensors for inorganic ions. Bozym, Hurst, Westerberg, Stoddard, Fierce, and Thompson discuss the determination of zinc using carbonic anhydrase-based fluorescence biosensors. Thompson, Zeng, Ohnemus, and Moffett describe the determination of free metal ions in solution using fiber optic biosensors.

The development of instrumentation and procedures for data analysis involved in fluorescence lifetime measurements had a revolutionary effect on fluorescence spectroscopy, allowing direct studies of a variety of photophysical processes. These studies have been limited by the fact that the time resolution of both pulse and phase instruments are no better than fractions of a nanosecond. Xu and Knutson describe ultrafast florescence spectroscopy via upconversion. This methodology which works well in the picosecond time domain can now be used for tryptophan lifetime studies and is likely to provide new insight in studies of protein fluorescence. Finally Johnson, Fahy, Veldhuis, and Lakowicz describe a fluorescence application of a numerical technique for identifying small pulsatile signals within noisy data.

It is clear that applications of fluorescence spectroscopy to studies of biological macromolecules and cells are proceeding at a rapid rate. Advances are described in this volume.

LUDWIG BRAND
MICHAEL L. JOHNSON

METHODS IN ENZYMOLOGY

VOLUME I. Preparation and Assay of Enzymes
Edited by SIDNEY P. COLOWICK AND NATHAN O. KAPLAN

VOLUME II. Preparation and Assay of Enzymes
Edited by SIDNEY P. COLOWICK AND NATHAN O. KAPLAN

VOLUME III. Preparation and Assay of Substrates
Edited by SIDNEY P. COLOWICK AND NATHAN O. KAPLAN

VOLUME IV. Special Techniques for the Enzymologist
Edited by SIDNEY P. COLOWICK AND NATHAN O. KAPLAN

VOLUME V. Preparation and Assay of Enzymes
Edited by SIDNEY P. COLOWICK AND NATHAN O. KAPLAN

VOLUME VI. Preparation and Assay of Enzymes *(Continued)*
Preparation and Assay of Substrates
Special Techniques
Edited by SIDNEY P. COLOWICK AND NATHAN O. KAPLAN

VOLUME VII. Cumulative Subject Index
Edited by SIDNEY P. COLOWICK AND NATHAN O. KAPLAN

VOLUME VIII. Complex Carbohydrates
Edited by ELIZABETH F. NEUFELD AND VICTOR GINSBURG

VOLUME IX. Carbohydrate Metabolism
Edited by WILLIS A. WOOD

VOLUME X. Oxidation and Phosphorylation
Edited by RONALD W. ESTABROOK AND MAYNARD E. PULLMAN

VOLUME XI. Enzyme Structure
Edited by C. H. W. HIRS

VOLUME XII. Nucleic Acids (Parts A and B)
Edited by LAWRENCE GROSSMAN AND KIVIE MOLDAVE

VOLUME XIII. Citric Acid Cycle
Edited by J. M. LOWENSTEIN

VOLUME XIV. Lipids
Edited by J. M. LOWENSTEIN

VOLUME XV. Steroids and Terpenoids
Edited by RAYMOND B. CLAYTON

VOLUME XVI. Fast Reactions
Edited by KENNETH KUSTIN

VOLUME XVII. Metabolism of Amino Acids and Amines (Parts A and B)
Edited by HERBERT TABOR AND CELIA WHITE TABOR

VOLUME XVIII. Vitamins and Coenzymes (Parts A, B, and C)
Edited by DONALD B. MCCORMICK AND LEMUEL D. WRIGHT

VOLUME XIX. Proteolytic Enzymes
Edited by GERTRUDE E. PERLMANN AND LASZLO LORAND

VOLUME XX. Nucleic Acids and Protein Synthesis (Part C)
Edited by KIVIE MOLDAVE AND LAWRENCE GROSSMAN

VOLUME XXI. Nucleic Acids (Part D)
Edited by LAWRENCE GROSSMAN AND KIVIE MOLDAVE

VOLUME XXII. Enzyme Purification and Related Techniques
Edited by WILLIAM B. JAKOBY

VOLUME XXIII. Photosynthesis (Part A)
Edited by ANTHONY SAN PIETRO

VOLUME XXIV. Photosynthesis and Nitrogen Fixation (Part B)
Edited by ANTHONY SAN PIETRO

VOLUME XXV. Enzyme Structure (Part B)
Edited by C. H. W. HIRS AND SERGE N. TIMASHEFF

VOLUME XXVI. Enzyme Structure (Part C)
Edited by C. H. W. HIRS AND SERGE N. TIMASHEFF

VOLUME XXVII. Enzyme Structure (Part D)
Edited by C. H. W. HIRS AND SERGE N. TIMASHEFF

VOLUME XXVIII. Complex Carbohydrates (Part B)
Edited by VICTOR GINSBURG

VOLUME XXIX. Nucleic Acids and Protein Synthesis (Part E)
Edited by LAWRENCE GROSSMAN AND KIVIE MOLDAVE

VOLUME XXX. Nucleic Acids and Protein Synthesis (Part F)
Edited by KIVIE MOLDAVE AND LAWRENCE GROSSMAN

VOLUME XXXI. Biomembranes (Part A)
Edited by SIDNEY FLEISCHER AND LESTER PACKER

VOLUME XXXII. Biomembranes (Part B)
Edited by SIDNEY FLEISCHER AND LESTER PACKER

VOLUME XXXIII. Cumulative Subject Index Volumes I–XXX
Edited by MARTHA G. DENNIS AND EDWARD A. DENNIS

VOLUME XXXIV. Affinity Techniques (Enzyme Purification: Part B)
Edited by WILLIAM B. JAKOBY AND MEIR WILCHEK

VOLUME XXXV. Lipids (Part B)
Edited by JOHN M. LOWENSTEIN

VOLUME XXXVI. Hormone Action (Part A: Steroid Hormones)
Edited by BERT W. O'MALLEY AND JOEL G. HARDMAN

VOLUME XXXVII. Hormone Action (Part B: Peptide Hormones)
Edited by BERT W. O'MALLEY AND JOEL G. HARDMAN

VOLUME XXXVIII. Hormone Action (Part C: Cyclic Nucleotides)
Edited by JOEL G. HARDMAN AND BERT W. O'MALLEY

VOLUME XXXIX. Hormone Action (Part D: Isolated Cells, Tissues, and Organ Systems)
Edited by JOEL G. HARDMAN AND BERT W. O'MALLEY

VOLUME XL. Hormone Action (Part E: Nuclear Structure and Function)
Edited by BERT W. O'MALLEY AND JOEL G. HARDMAN

VOLUME XLI. Carbohydrate Metabolism (Part B)
Edited by W. A. WOOD

VOLUME XLII. Carbohydrate Metabolism (Part C)
Edited by W. A. WOOD

VOLUME XLIII. Antibiotics
Edited by JOHN H. HASH

VOLUME XLIV. Immobilized Enzymes
Edited by KLAUS MOSBACH

VOLUME XLV. Proteolytic Enzymes (Part B)
Edited by LASZLO LORAND

VOLUME XLVI. Affinity Labeling
Edited by WILLIAM B. JAKOBY AND MEIR WILCHEK

VOLUME XLVII. Enzyme Structure (Part E)
Edited by C. H. W. HIRS AND SERGE N. TIMASHEFF

VOLUME XLVIII. Enzyme Structure (Part F)
Edited by C. H. W. HIRS AND SERGE N. TIMASHEFF

VOLUME XLIX. Enzyme Structure (Part G)
Edited by C. H. W. HIRS AND SERGE N. TIMASHEFF

VOLUME L. Complex Carbohydrates (Part C)
Edited by VICTOR GINSBURG

VOLUME LI. Purine and Pyrimidine Nucleotide Metabolism
Edited by PATRICIA A. HOFFEE AND MARY ELLEN JONES

VOLUME LII. Biomembranes (Part C: Biological Oxidations)
Edited by SIDNEY FLEISCHER AND LESTER PACKER

VOLUME LIII. Biomembranes (Part D: Biological Oxidations)
Edited by SIDNEY FLEISCHER AND LESTER PACKER

VOLUME LIV. Biomembranes (Part E: Biological Oxidations)
Edited by SIDNEY FLEISCHER AND LESTER PACKER

VOLUME LV. Biomembranes (Part F: Bioenergetics)
Edited by SIDNEY FLEISCHER AND LESTER PACKER

VOLUME LVI. Biomembranes (Part G: Bioenergetics)
Edited by SIDNEY FLEISCHER AND LESTER PACKER

VOLUME LVII. Bioluminescence and Chemiluminescence
Edited by MARLENE A. DELUCA

VOLUME LVIII. Cell Culture
Edited by WILLIAM B. JAKOBY AND IRA PASTAN

VOLUME LIX. Nucleic Acids and Protein Synthesis (Part G)
Edited by KIVIE MOLDAVE AND LAWRENCE GROSSMAN

VOLUME LX. Nucleic Acids and Protein Synthesis (Part H)
Edited by KIVIE MOLDAVE AND LAWRENCE GROSSMAN

VOLUME 61. Enzyme Structure (Part H)
Edited by C. H. W. HIRS AND SERGE N. TIMASHEFF

VOLUME 62. Vitamins and Coenzymes (Part D)
Edited by DONALD B. MCCORMICK AND LEMUEL D. WRIGHT

VOLUME 63. Enzyme Kinetics and Mechanism (Part A: Initial Rate and Inhibitor Methods)
Edited by DANIEL L. PURICH

VOLUME 64. Enzyme Kinetics and Mechanism
(Part B: Isotopic Probes and Complex Enzyme Systems)
Edited by DANIEL L. PURICH

VOLUME 65. Nucleic Acids (Part I)
Edited by LAWRENCE GROSSMAN AND KIVIE MOLDAVE

VOLUME 66. Vitamins and Coenzymes (Part E)
Edited by DONALD B. MCCORMICK AND LEMUEL D. WRIGHT

VOLUME 67. Vitamins and Coenzymes (Part F)
Edited by DONALD B. MCCORMICK AND LEMUEL D. WRIGHT

VOLUME 68. Recombinant DNA
Edited by RAY WU

VOLUME 69. Photosynthesis and Nitrogen Fixation (Part C)
Edited by ANTHONY SAN PIETRO

VOLUME 70. Immunochemical Techniques (Part A)
Edited by HELEN VAN VUNAKIS AND JOHN J. LANGONE

VOLUME 71. Lipids (Part C)
Edited by JOHN M. LOWENSTEIN

VOLUME 72. Lipids (Part D)
Edited by JOHN M. LOWENSTEIN

VOLUME 73. Immunochemical Techniques (Part B)
Edited by JOHN J. LANGONE AND HELEN VAN VUNAKIS

VOLUME 74. Immunochemical Techniques (Part C)
Edited by JOHN J. LANGONE AND HELEN VAN VUNAKIS

VOLUME 75. Cumulative Subject Index Volumes XXXI, XXXII, XXXIV–LX
Edited by EDWARD A. DENNIS AND MARTHA G. DENNIS

VOLUME 76. Hemoglobins
Edited by ERALDO ANTONINI, LUIGI ROSSI-BERNARDI, AND EMILIA CHIANCONE

VOLUME 77. Detoxication and Drug Metabolism
Edited by WILLIAM B. JAKOBY

VOLUME 78. Interferons (Part A)
Edited by SIDNEY PESTKA

VOLUME 79. Interferons (Part B)
Edited by SIDNEY PESTKA

VOLUME 80. Proteolytic Enzymes (Part C)
Edited by LASZLO LORAND

VOLUME 81. Biomembranes (Part H: Visual Pigments and Purple Membranes, I)
Edited by LESTER PACKER

VOLUME 82. Structural and Contractile Proteins (Part A: Extracellular Matrix)
Edited by LEON W. CUNNINGHAM AND DIXIE W. FREDERIKSEN

VOLUME 83. Complex Carbohydrates (Part D)
Edited by VICTOR GINSBURG

VOLUME 84. Immunochemical Techniques (Part D: Selected Immunoassays)
Edited by JOHN J. LANGONE AND HELEN VAN VUNAKIS

VOLUME 85. Structural and Contractile Proteins (Part B: The Contractile Apparatus and the Cytoskeleton)
Edited by DIXIE W. FREDERIKSEN AND LEON W. CUNNINGHAM

VOLUME 86. Prostaglandins and Arachidonate Metabolites
Edited by WILLIAM E. M. LANDS AND WILLIAM L. SMITH

VOLUME 87. Enzyme Kinetics and Mechanism (Part C: Intermediates, Stereo-chemistry, and Rate Studies)
Edited by DANIEL L. PURICH

VOLUME 88. Biomembranes (Part I: Visual Pigments and Purple Membranes, II)
Edited by LESTER PACKER

VOLUME 89. Carbohydrate Metabolism (Part D)
Edited by WILLIS A. WOOD

VOLUME 90. Carbohydrate Metabolism (Part E)
Edited by WILLIS A. WOOD

VOLUME 91. Enzyme Structure (Part I)
Edited by C. H. W. HIRS AND SERGE N. TIMASHEFF

VOLUME 92. Immunochemical Techniques (Part E: Monoclonal Antibodies and General Immunoassay Methods)
Edited by JOHN J. LANGONE AND HELEN VAN VUNAKIS

VOLUME 93. Immunochemical Techniques (Part F: Conventional Antibodies, Fc Receptors, and Cytotoxicity)
Edited by JOHN J. LANGONE AND HELEN VAN VUNAKIS

VOLUME 94. Polyamines
Edited by HERBERT TABOR AND CELIA WHITE TABOR

VOLUME 95. Cumulative Subject Index Volumes 61–74, 76–80
Edited by EDWARD A. DENNIS AND MARTHA G. DENNIS

VOLUME 96. Biomembranes [Part J: Membrane Biogenesis: Assembly and Targeting (General Methods; Eukaryotes)]
Edited by SIDNEY FLEISCHER AND BECCA FLEISCHER

VOLUME 97. Biomembranes [Part K: Membrane Biogenesis: Assembly and Targeting (Prokaryotes, Mitochondria, and Chloroplasts)]
Edited by SIDNEY FLEISCHER AND BECCA FLEISCHER

VOLUME 98. Biomembranes (Part L: Membrane Biogenesis: Processing and Recycling)
Edited by SIDNEY FLEISCHER AND BECCA FLEISCHER

VOLUME 99. Hormone Action (Part F: Protein Kinases)
Edited by JACKIE D. CORBIN AND JOEL G. HARDMAN

VOLUME 100. Recombinant DNA (Part B)
Edited by RAY WU, LAWRENCE GROSSMAN, AND KIVIE MOLDAVE

VOLUME 101. Recombinant DNA (Part C)
Edited by RAY WU, LAWRENCE GROSSMAN, AND KIVIE MOLDAVE

VOLUME 102. Hormone Action (Part G: Calmodulin and Calcium-Binding Proteins)
Edited by ANTHONY R. MEANS AND BERT W. O'MALLEY

VOLUME 103. Hormone Action (Part H: Neuroendocrine Peptides)
Edited by P. MICHAEL CONN

VOLUME 104. Enzyme Purification and Related Techniques (Part C)
Edited by WILLIAM B. JAKOBY

VOLUME 105. Oxygen Radicals in Biological Systems
Edited by LESTER PACKER

VOLUME 106. Posttranslational Modifications (Part A)
Edited by FINN WOLD AND KIVIE MOLDAVE

VOLUME 107. Posttranslational Modifications (Part B)
Edited by FINN WOLD AND KIVIE MOLDAVE

VOLUME 108. Immunochemical Techniques (Part G: Separation and Characterization of Lymphoid Cells)
Edited by GIOVANNI DI SABATO, JOHN J. LANGONE, AND HELEN VAN VUNAKIS

VOLUME 109. Hormone Action (Part I: Peptide Hormones)
Edited by LUTZ BIRNBAUMER AND BERT W. O'MALLEY

VOLUME 110. Steroids and Isoprenoids (Part A)
Edited by JOHN H. LAW AND HANS C. RILLING

VOLUME 111. Steroids and Isoprenoids (Part B)
Edited by JOHN H. LAW AND HANS C. RILLING

VOLUME 112. Drug and Enzyme Targeting (Part A)
Edited by KENNETH J. WIDDER AND RALPH GREEN

VOLUME 113. Glutamate, Glutamine, Glutathione, and Related Compounds
Edited by ALTON MEISTER

VOLUME 114. Diffraction Methods for Biological Macromolecules (Part A)
Edited by HAROLD W. WYCKOFF, C. H. W. HIRS, AND SERGE N. TIMASHEFF

VOLUME 115. Diffraction Methods for Biological Macromolecules (Part B)
Edited by HAROLD W. WYCKOFF, C. H. W. HIRS, AND SERGE N. TIMASHEFF

VOLUME 116. Immunochemical Techniques (Part H: Effectors and Mediators of Lymphoid Cell Functions)
Edited by GIOVANNI DI SABATO, JOHN J. LANGONE, AND HELEN VAN VUNAKIS

VOLUME 117. Enzyme Structure (Part J)
Edited by C. H. W. HIRS AND SERGE N. TIMASHEFF

VOLUME 118. Plant Molecular Biology
Edited by ARTHUR WEISSBACH AND HERBERT WEISSBACH

VOLUME 119. Interferons (Part C)
Edited by SIDNEY PESTKA

VOLUME 120. Cumulative Subject Index Volumes 81–94, 96–101

VOLUME 121. Immunochemical Techniques (Part I: Hybridoma Technology and Monoclonal Antibodies)
Edited by JOHN J. LANGONE AND HELEN VAN VUNAKIS

VOLUME 122. Vitamins and Coenzymes (Part G)
Edited by FRANK CHYTIL AND DONALD B. MCCORMICK

VOLUME 123. Vitamins and Coenzymes (Part H)
Edited by FRANK CHYTIL AND DONALD B. MCCORMICK

VOLUME 124. Hormone Action (Part J: Neuroendocrine Peptides)
Edited by P. MICHAEL CONN

VOLUME 125. Biomembranes (Part M: Transport in Bacteria, Mitochondria, and Chloroplasts: General Approaches and Transport Systems)
Edited by SIDNEY FLEISCHER AND BECCA FLEISCHER

VOLUME 126. Biomembranes (Part N: Transport in Bacteria, Mitochondria, and Chloroplasts: Protonmotive Force)
Edited by SIDNEY FLEISCHER AND BECCA FLEISCHER

VOLUME 127. Biomembranes (Part O: Protons and Water: Structure and Translocation)
Edited by LESTER PACKER

VOLUME 128. Plasma Lipoproteins (Part A: Preparation, Structure, and Molecular Biology)
Edited by JERE P. SEGREST AND JOHN J. ALBERS

VOLUME 129. Plasma Lipoproteins (Part B: Characterization, Cell Biology, and Metabolism)
Edited by JOHN J. ALBERS AND JERE P. SEGREST

VOLUME 130. Enzyme Structure (Part K)
Edited by C. H. W. HIRS AND SERGE N. TIMASHEFF

VOLUME 131. Enzyme Structure (Part L)
Edited by C. H. W. HIRS AND SERGE N. TIMASHEFF

VOLUME 132. Immunochemical Techniques (Part J: Phagocytosis and Cell-Mediated Cytotoxicity)
Edited by GIOVANNI DI SABATO AND JOHANNES EVERSE

VOLUME 133. Bioluminescence and Chemiluminescence (Part B)
Edited by MARLENE DELUCA AND WILLIAM D. MCELROY

VOLUME 134. Structural and Contractile Proteins (Part C: The Contractile Apparatus and the Cytoskeleton)
Edited by RICHARD B. VALLEE

VOLUME 135. Immobilized Enzymes and Cells (Part B)
Edited by KLAUS MOSBACH

VOLUME 136. Immobilized Enzymes and Cells (Part C)
Edited by KLAUS MOSBACH

VOLUME 137. Immobilized Enzymes and Cells (Part D)
Edited by KLAUS MOSBACH

VOLUME 138. Complex Carbohydrates (Part E)
Edited by VICTOR GINSBURG

VOLUME 139. Cellular Regulators (Part A: Calcium- and Calmodulin-Binding Proteins)
Edited by ANTHONY R. MEANS AND P. MICHAEL CONN

VOLUME 140. Cumulative Subject Index Volumes 102–119, 121–134

VOLUME 141. Cellular Regulators (Part B: Calcium and Lipids)
Edited by P. MICHAEL CONN AND ANTHONY R. MEANS

VOLUME 142. Metabolism of Aromatic Amino Acids and Amines
Edited by SEYMOUR KAUFMAN

VOLUME 143. Sulfur and Sulfur Amino Acids
Edited by WILLIAM B. JAKOBY AND OWEN GRIFFITH

VOLUME 144. Structural and Contractile Proteins (Part D: Extracellular Matrix)
Edited by LEON W. CUNNINGHAM

VOLUME 145. Structural and Contractile Proteins (Part E: Extracellular Matrix)
Edited by LEON W. CUNNINGHAM

VOLUME 146. Peptide Growth Factors (Part A)
Edited by DAVID BARNES AND DAVID A. SIRBASKU

VOLUME 147. Peptide Growth Factors (Part B)
Edited by DAVID BARNES AND DAVID A. SIRBASKU

VOLUME 148. Plant Cell Membranes
Edited by LESTER PACKER AND ROLAND DOUCE

VOLUME 149. Drug and Enzyme Targeting (Part B)
Edited by RALPH GREEN AND KENNETH J. WIDDER

VOLUME 150. Immunochemical Techniques (Part K: *In Vitro* Models of B and T Cell Functions and Lymphoid Cell Receptors)
Edited by GIOVANNI DI SABATO

VOLUME 151. Molecular Genetics of Mammalian Cells
Edited by MICHAEL M. GOTTESMAN

VOLUME 152. Guide to Molecular Cloning Techniques
Edited by SHELBY L. BERGER AND ALAN R. KIMMEL

VOLUME 153. Recombinant DNA (Part D)
Edited by RAY WU AND LAWRENCE GROSSMAN

VOLUME 154. Recombinant DNA (Part E)
Edited by RAY WU AND LAWRENCE GROSSMAN

VOLUME 155. Recombinant DNA (Part F)
Edited by RAY WU

VOLUME 156. Biomembranes (Part P: ATP-Driven Pumps and Related Transport: The Na, K-Pump)
Edited by SIDNEY FLEISCHER AND BECCA FLEISCHER

VOLUME 157. Biomembranes (Part Q: ATP-Driven Pumps and Related Transport: Calcium, Proton, and Potassium Pumps)
Edited by SIDNEY FLEISCHER AND BECCA FLEISCHER

VOLUME 158. Metalloproteins (Part A)
Edited by JAMES F. RIORDAN AND BERT L. VALLEE

VOLUME 159. Initiation and Termination of Cyclic Nucleotide Action
Edited by JACKIE D. CORBIN AND ROGER A. JOHNSON

VOLUME 160. Biomass (Part A: Cellulose and Hemicellulose)
Edited by WILLIS A. WOOD AND SCOTT T. KELLOGG

VOLUME 161. Biomass (Part B: Lignin, Pectin, and Chitin)
Edited by WILLIS A. WOOD AND SCOTT T. KELLOGG

VOLUME 162. Immunochemical Techniques (Part L: Chemotaxis and Inflammation)
Edited by GIOVANNI DI SABATO

VOLUME 163. Immunochemical Techniques (Part M: Chemotaxis and Inflammation)
Edited by GIOVANNI DI SABATO

VOLUME 164. Ribosomes
Edited by HARRY F. NOLLER, JR., AND KIVIE MOLDAVE

VOLUME 165. Microbial Toxins: Tools for Enzymology
Edited by SIDNEY HARSHMAN

VOLUME 166. Branched-Chain Amino Acids
Edited by ROBERT HARRIS AND JOHN R. SOKATCH

VOLUME 167. Cyanobacteria
Edited by LESTER PACKER AND ALEXANDER N. GLAZER

VOLUME 168. Hormone Action (Part K: Neuroendocrine Peptides)
Edited by P. MICHAEL CONN

VOLUME 169. Platelets: Receptors, Adhesion, Secretion (Part A)
Edited by JACEK HAWIGER

VOLUME 170. Nucleosomes
Edited by PAUL M. WASSARMAN AND ROGER D. KORNBERG

VOLUME 171. Biomembranes (Part R: Transport Theory: Cells and Model Membranes)
Edited by SIDNEY FLEISCHER AND BECCA FLEISCHER

VOLUME 172. Biomembranes (Part S: Transport: Membrane Isolation and Characterization)
Edited by SIDNEY FLEISCHER AND BECCA FLEISCHER

VOLUME 173. Biomembranes [Part T: Cellular and Subcellular Transport: Eukaryotic (Nonepithelial) Cells]
Edited by SIDNEY FLEISCHER AND BECCA FLEISCHER

VOLUME 174. Biomembranes [Part U: Cellular and Subcellular Transport: Eukaryotic (Nonepithelial) Cells]
Edited by SIDNEY FLEISCHER AND BECCA FLEISCHER

VOLUME 175. Cumulative Subject Index Volumes 135–139, 141–167

VOLUME 176. Nuclear Magnetic Resonance (Part A: Spectral Techniques and Dynamics)
Edited by NORMAN J. OPPENHEIMER AND THOMAS L. JAMES

VOLUME 177. Nuclear Magnetic Resonance (Part B: Structure and Mechanism)
Edited by NORMAN J. OPPENHEIMER AND THOMAS L. JAMES

VOLUME 178. Antibodies, Antigens, and Molecular Mimicry
Edited by JOHN J. LANGONE

VOLUME 179. Complex Carbohydrates (Part F)
Edited by VICTOR GINSBURG

VOLUME 180. RNA Processing (Part A: General Methods)
Edited by JAMES E. DAHLBERG AND JOHN N. ABELSON

VOLUME 181. RNA Processing (Part B: Specific Methods)
Edited by JAMES E. DAHLBERG AND JOHN N. ABELSON

VOLUME 182. Guide to Protein Purification
Edited by MURRAY P. DEUTSCHER

VOLUME 183. Molecular Evolution: Computer Analysis of Protein and Nucleic Acid Sequences
Edited by RUSSELL F. DOOLITTLE

VOLUME 184. Avidin-Biotin Technology
Edited by MEIR WILCHEK AND EDWARD A. BAYER

VOLUME 185. Gene Expression Technology
Edited by DAVID V. GOEDDEL

VOLUME 186. Oxygen Radicals in Biological Systems (Part B: Oxygen Radicals and Antioxidants)
Edited by LESTER PACKER AND ALEXANDER N. GLAZER

VOLUME 187. Arachidonate Related Lipid Mediators
Edited by ROBERT C. MURPHY AND FRANK A. FITZPATRICK

VOLUME 188. Hydrocarbons and Methylotrophy
Edited by MARY E. LIDSTROM

VOLUME 189. Retinoids (Part A: Molecular and Metabolic Aspects)
Edited by LESTER PACKER

VOLUME 190. Retinoids (Part B: Cell Differentiation and Clinical Applications)
Edited by LESTER PACKER

VOLUME 191. Biomembranes (Part V: Cellular and Subcellular Transport: Epithelial Cells)
Edited by SIDNEY FLEISCHER AND BECCA FLEISCHER

VOLUME 192. Biomembranes (Part W: Cellular and Subcellular Transport: Epithelial Cells)
Edited by SIDNEY FLEISCHER AND BECCA FLEISCHER

VOLUME 193. Mass Spectrometry
Edited by JAMES A. MCCLOSKEY

VOLUME 194. Guide to Yeast Genetics and Molecular Biology
Edited by CHRISTINE GUTHRIE AND GERALD R. FINK

VOLUME 195. Adenylyl Cyclase, G Proteins, and Guanylyl Cyclase
Edited by ROGER A. JOHNSON AND JACKIE D. CORBIN

VOLUME 196. Molecular Motors and the Cytoskeleton
Edited by RICHARD B. VALLEE

VOLUME 197. Phospholipases
Edited by EDWARD A. DENNIS

VOLUME 198. Peptide Growth Factors (Part C)
Edited by DAVID BARNES, J. P. MATHER, AND GORDON H. SATO

VOLUME 199. Cumulative Subject Index Volumes 168–174, 176–194

VOLUME 200. Protein Phosphorylation (Part A: Protein Kinases: Assays, Purification, Antibodies, Functional Analysis, Cloning, and Expression)
Edited by TONY HUNTER AND BARTHOLOMEW M. SEFTON

VOLUME 201. Protein Phosphorylation (Part B: Analysis of Protein Phosphorylation, Protein Kinase Inhibitors, and Protein Phosphatases)
Edited by TONY HUNTER AND BARTHOLOMEW M. SEFTON

VOLUME 202. Molecular Design and Modeling: Concepts and Applications (Part A: Proteins, Peptides, and Enzymes)
Edited by JOHN J. LANGONE

VOLUME 203. Molecular Design and Modeling: Concepts and Applications (Part B: Antibodies and Antigens, Nucleic Acids, Polysaccharides, and Drugs)
Edited by JOHN J. LANGONE

VOLUME 204. Bacterial Genetic Systems
Edited by JEFFREY H. MILLER

VOLUME 205. Metallobiochemistry (Part B: Metallothionein and Related Molecules)
Edited by JAMES F. RIORDAN AND BERT L. VALLEE

VOLUME 206. Cytochrome P450
Edited by MICHAEL R. WATERMAN AND ERIC F. JOHNSON

VOLUME 207. Ion Channels
Edited by BERNARDO RUDY AND LINDA E. IVERSON

VOLUME 208. Protein–DNA Interactions
Edited by ROBERT T. SAUER

VOLUME 209. Phospholipid Biosynthesis
Edited by EDWARD A. DENNIS AND DENNIS E. VANCE

VOLUME 210. Numerical Computer Methods
Edited by LUDWIG BRAND AND MICHAEL L. JOHNSON

VOLUME 211. DNA Structures (Part A: Synthesis and Physical Analysis of DNA)
Edited by DAVID M. J. LILLEY AND JAMES E. DAHLBERG

VOLUME 212. DNA Structures (Part B: Chemical and Electrophoretic Analysis of DNA)
Edited by DAVID M. J. LILLEY AND JAMES E. DAHLBERG

VOLUME 213. Carotenoids (Part A: Chemistry, Separation, Quantitation, and Antioxidation)
Edited by LESTER PACKER

VOLUME 214. Carotenoids (Part B: Metabolism, Genetics, and Biosynthesis)
Edited by LESTER PACKER

VOLUME 215. Platelets: Receptors, Adhesion, Secretion (Part B)
Edited by JACEK J. HAWIGER

VOLUME 216. Recombinant DNA (Part G)
Edited by RAY WU

VOLUME 217. Recombinant DNA (Part H)
Edited by RAY WU

VOLUME 218. Recombinant DNA (Part I)
Edited by RAY WU

VOLUME 219. Reconstitution of Intracellular Transport
Edited by JAMES E. ROTHMAN

VOLUME 220. Membrane Fusion Techniques (Part A)
Edited by NEJAT DÜZGÜNEŞ

VOLUME 221. Membrane Fusion Techniques (Part B)
Edited by NEJAT DÜZGÜNEŞ

VOLUME 222. Proteolytic Enzymes in Coagulation, Fibrinolysis, and Complement Activation (Part A: Mammalian Blood Coagulation Factors and Inhibitors)
Edited by LASZLO LORAND AND KENNETH G. MANN

VOLUME 223. Proteolytic Enzymes in Coagulation, Fibrinolysis, and Complement Activation (Part B: Complement Activation, Fibrinolysis, and Nonmammalian Blood Coagulation Factors)
Edited by LASZLO LORAND AND KENNETH G. MANN

VOLUME 224. Molecular Evolution: Producing the Biochemical Data
Edited by ELIZABETH ANNE ZIMMER, THOMAS J. WHITE, REBECCA L. CANN, AND ALLAN C. WILSON

VOLUME 225. Guide to Techniques in Mouse Development
Edited by PAUL M. WASSARMAN AND MELVIN L. DEPAMPHILIS

VOLUME 226. Metallobiochemistry (Part C: Spectroscopic and Physical Methods for Probing Metal Ion Environments in Metalloenzymes and Metalloproteins)
Edited by JAMES F. RIORDAN AND BERT L. VALLEE

VOLUME 227. Metallobiochemistry (Part D: Physical and Spectroscopic Methods for Probing Metal Ion Environments in Metalloproteins)
Edited by JAMES F. RIORDAN AND BERT L. VALLEE

VOLUME 228. Aqueous Two-Phase Systems
Edited by HARRY WALTER AND GÖTE JOHANSSON

VOLUME 229. Cumulative Subject Index Volumes 195–198, 200–227

VOLUME 230. Guide to Techniques in Glycobiology
Edited by WILLIAM J. LENNARZ AND GERALD W. HART

VOLUME 231. Hemoglobins (Part B: Biochemical and Analytical Methods)
Edited by JOHANNES EVERSE, KIM D. VANDEGRIFF, AND ROBERT M. WINSLOW

VOLUME 232. Hemoglobins (Part C: Biophysical Methods)
Edited by JOHANNES EVERSE, KIM D. VANDEGRIFF, AND ROBERT M. WINSLOW

VOLUME 233. Oxygen Radicals in Biological Systems (Part C)
Edited by LESTER PACKER

VOLUME 234. Oxygen Radicals in Biological Systems (Part D)
Edited by LESTER PACKER

VOLUME 235. Bacterial Pathogenesis (Part A: Identification and Regulation of Virulence Factors)
Edited by VIRGINIA L. CLARK AND PATRIK M. BAVOIL

VOLUME 236. Bacterial Pathogenesis (Part B: Integration of Pathogenic Bacteria with Host Cells)
Edited by VIRGINIA L. CLARK AND PATRIK M. BAVOIL

VOLUME 237. Heterotrimeric G Proteins
Edited by RAVI IYENGAR

VOLUME 238. Heterotrimeric G-Protein Effectors
Edited by RAVI IYENGAR

VOLUME 239. Nuclear Magnetic Resonance (Part C)
Edited by THOMAS L. JAMES AND NORMAN J. OPPENHEIMER

VOLUME 240. Numerical Computer Methods (Part B)
Edited by MICHAEL L. JOHNSON AND LUDWIG BRAND

VOLUME 241. Retroviral Proteases
Edited by LAWRENCE C. KUO AND JULES A. SHAFER

VOLUME 242. Neoglycoconjugates (Part A)
Edited by Y. C. LEE AND REIKO T. LEE

VOLUME 243. Inorganic Microbial Sulfur Metabolism
Edited by HARRY D. PECK, JR., AND JEAN LEGALL

VOLUME 244. Proteolytic Enzymes: Serine and Cysteine Peptidases
Edited by ALAN J. BARRETT

VOLUME 245. Extracellular Matrix Components
Edited by E. RUOSLAHTI AND E. ENGVALL

VOLUME 246. Biochemical Spectroscopy
Edited by KENNETH SAUER

VOLUME 247. Neoglycoconjugates (Part B: Biomedical Applications)
Edited by Y. C. LEE AND REIKO T. LEE

VOLUME 248. Proteolytic Enzymes: Aspartic and Metallo Peptidases
Edited by ALAN J. BARRETT

VOLUME 249. Enzyme Kinetics and Mechanism (Part D: Developments in Enzyme Dynamics)
Edited by DANIEL L. PURICH

VOLUME 250. Lipid Modifications of Proteins
Edited by PATRICK J. CASEY AND JANICE E. BUSS

VOLUME 251. Biothiols (Part A: Monothiols and Dithiols, Protein Thiols, and Thiyl Radicals)
Edited by LESTER PACKER

VOLUME 252. Biothiols (Part B: Glutathione and Thioredoxin; Thiols in Signal Transduction and Gene Regulation)
Edited by LESTER PACKER

VOLUME 253. Adhesion of Microbial Pathogens
Edited by RON J. DOYLE AND ITZHAK OFEK

VOLUME 254. Oncogene Techniques
Edited by PETER K. VOGT AND INDER M. VERMA

VOLUME 255. Small GTPases and Their Regulators (Part A: Ras Family)
Edited by W. E. BALCH, CHANNING J. DER, AND ALAN HALL

VOLUME 256. Small GTPases and Their Regulators (Part B: Rho Family)
Edited by W. E. BALCH, CHANNING J. DER, AND ALAN HALL

VOLUME 257. Small GTPases and Their Regulators (Part C: Proteins Involved in Transport)
Edited by W. E. BALCH, CHANNING J. DER, AND ALAN HALL

VOLUME 258. Redox-Active Amino Acids in Biology
Edited by JUDITH P. KLINMAN

VOLUME 259. Energetics of Biological Macromolecules
Edited by MICHAEL L. JOHNSON AND GARY K. ACKERS

VOLUME 260. Mitochondrial Biogenesis and Genetics (Part A)
Edited by GIUSEPPE M. ATTARDI AND ANNE CHOMYN

VOLUME 261. Nuclear Magnetic Resonance and Nucleic Acids
Edited by THOMAS L. JAMES

VOLUME 262. DNA Replication
Edited by JUDITH L. CAMPBELL

VOLUME 263. Plasma Lipoproteins (Part C: Quantitation)
Edited by WILLIAM A. BRADLEY, SANDRA H. GIANTURCO, AND JERE P. SEGREST

VOLUME 264. Mitochondrial Biogenesis and Genetics (Part B)
Edited by GIUSEPPE M. ATTARDI AND ANNE CHOMYN

VOLUME 265. Cumulative Subject Index Volumes 228, 230–262

VOLUME 266. Computer Methods for Macromolecular Sequence Analysis
Edited by RUSSELL F. DOOLITTLE

VOLUME 267. Combinatorial Chemistry
Edited by JOHN N. ABELSON

VOLUME 268. Nitric Oxide (Part A: Sources and Detection of NO; NO Synthase)
Edited by LESTER PACKER

VOLUME 269. Nitric Oxide (Part B: Physiological and Pathological Processes)
Edited by LESTER PACKER

VOLUME 270. High Resolution Separation and Analysis of Biological Macromolecules (Part A: Fundamentals)
Edited by BARRY L. KARGER AND WILLIAM S. HANCOCK

VOLUME 271. High Resolution Separation and Analysis of Biological Macromolecules (Part B: Applications)
Edited by BARRY L. KARGER AND WILLIAM S. HANCOCK

VOLUME 272. Cytochrome P450 (Part B)
Edited by ERIC F. JOHNSON AND MICHAEL R. WATERMAN

VOLUME 273. RNA Polymerase and Associated Factors (Part A)
Edited by SANKAR ADHYA

VOLUME 274. RNA Polymerase and Associated Factors (Part B)
Edited by SANKAR ADHYA

VOLUME 275. Viral Polymerases and Related Proteins
Edited by LAWRENCE C. KUO, DAVID B. OLSEN, AND STEVEN S. CARROLL

VOLUME 276. Macromolecular Crystallography (Part A)
Edited by CHARLES W. CARTER, JR., AND ROBERT M. SWEET

VOLUME 277. Macromolecular Crystallography (Part B)
Edited by CHARLES W. CARTER, JR., AND ROBERT M. SWEET

VOLUME 278. Fluorescence Spectroscopy
Edited by LUDWIG BRAND AND MICHAEL L. JOHNSON

VOLUME 279. Vitamins and Coenzymes (Part I)
Edited by DONALD B. MCCORMICK, JOHN W. SUTTIE, AND CONRAD WAGNER

VOLUME 280. Vitamins and Coenzymes (Part J)
Edited by DONALD B. MCCORMICK, JOHN W. SUTTIE, AND CONRAD WAGNER

VOLUME 281. Vitamins and Coenzymes (Part K)
Edited by DONALD B. MCCORMICK, JOHN W. SUTTIE, AND CONRAD WAGNER

VOLUME 282. Vitamins and Coenzymes (Part L)
Edited by DONALD B. MCCORMICK, JOHN W. SUTTIE, AND CONRAD WAGNER

VOLUME 283. Cell Cycle Control
Edited by WILLIAM G. DUNPHY

VOLUME 284. Lipases (Part A: Biotechnology)
Edited by BYRON RUBIN AND EDWARD A. DENNIS

VOLUME 285. Cumulative Subject Index Volumes 263, 264, 266–284, 286–289

VOLUME 286. Lipases (Part B: Enzyme Characterization and Utilization)
Edited by BYRON RUBIN AND EDWARD A. DENNIS

VOLUME 287. Chemokines
Edited by RICHARD HORUK

VOLUME 288. Chemokine Receptors
Edited by RICHARD HORUK

VOLUME 289. Solid Phase Peptide Synthesis
Edited by GREGG B. FIELDS

VOLUME 290. Molecular Chaperones
Edited by GEORGE H. LORIMER AND THOMAS BALDWIN

VOLUME 291. Caged Compounds
Edited by GERARD MARRIOTT

VOLUME 292. ABC Transporters: Biochemical, Cellular, and Molecular Aspects
Edited by SURESH V. AMBUDKAR AND MICHAEL M. GOTTESMAN

VOLUME 293. Ion Channels (Part B)
Edited by P. MICHAEL CONN

VOLUME 294. Ion Channels (Part C)
Edited by P. MICHAEL CONN

VOLUME 295. Energetics of Biological Macromolecules (Part B)
Edited by GARY K. ACKERS AND MICHAEL L. JOHNSON

VOLUME 296. Neurotransmitter Transporters
Edited by SUSAN G. AMARA

VOLUME 297. Photosynthesis: Molecular Biology of Energy Capture
Edited by LEE MCINTOSH

VOLUME 298. Molecular Motors and the Cytoskeleton (Part B)
Edited by RICHARD B. VALLEE

VOLUME 299. Oxidants and Antioxidants (Part A)
Edited by LESTER PACKER

VOLUME 300. Oxidants and Antioxidants (Part B)
Edited by LESTER PACKER

VOLUME 301. Nitric Oxide: Biological and Antioxidant Activities (Part C)
Edited by LESTER PACKER

VOLUME 302. Green Fluorescent Protein
Edited by P. MICHAEL CONN

VOLUME 303. cDNA Preparation and Display
Edited by SHERMAN M. WEISSMAN

VOLUME 304. Chromatin
Edited by PAUL M. WASSARMAN AND ALAN P. WOLFFE

VOLUME 305. Bioluminescence and Chemiluminescence (Part C)
Edited by THOMAS O. BALDWIN AND MIRIAM M. ZIEGLER

VOLUME 306. Expression of Recombinant Genes in Eukaryotic Systems
Edited by JOSEPH C. GLORIOSO AND MARTIN C. SCHMIDT

VOLUME 307. Confocal Microscopy
Edited by P. MICHAEL CONN

VOLUME 308. Enzyme Kinetics and Mechanism (Part E: Energetics of Enzyme Catalysis)
Edited by DANIEL L. PURICH AND VERN L. SCHRAMM

VOLUME 309. Amyloid, Prions, and Other Protein Aggregates
Edited by RONALD WETZEL

VOLUME 310. Biofilms
Edited by RON J. DOYLE

VOLUME 311. Sphingolipid Metabolism and Cell Signaling (Part A)
Edited by ALFRED H. MERRILL, JR., AND YUSUF A. HANNUN

VOLUME 312. Sphingolipid Metabolism and Cell Signaling (Part B)
Edited by ALFRED H. MERRILL, JR., AND YUSUF A. HANNUN

VOLUME 313. Antisense Technology (Part A: General Methods, Methods of Delivery, and RNA Studies)
Edited by M. IAN PHILLIPS

VOLUME 314. Antisense Technology (Part B: Applications)
Edited by M. IAN PHILLIPS

VOLUME 315. Vertebrate Phototransduction and the Visual Cycle (Part A)
Edited by KRZYSZTOF PALCZEWSKI

VOLUME 316. Vertebrate Phototransduction and the Visual Cycle (Part B)
Edited by KRZYSZTOF PALCZEWSKI

VOLUME 317. RNA–Ligand Interactions (Part A: Structural Biology Methods)
Edited by DANIEL W. CELANDER AND JOHN N. ABELSON

VOLUME 318. RNA–Ligand Interactions (Part B: Molecular Biology Methods)
Edited by DANIEL W. CELANDER AND JOHN N. ABELSON

VOLUME 319. Singlet Oxygen, UV-A, and Ozone
Edited by LESTER PACKER AND HELMUT SIES

VOLUME 320. Cumulative Subject Index Volumes 290–319

VOLUME 321. Numerical Computer Methods (Part C)
Edited by MICHAEL L. JOHNSON AND LUDWIG BRAND

VOLUME 322. Apoptosis
Edited by JOHN C. REED

VOLUME 323. Energetics of Biological Macromolecules (Part C)
Edited by MICHAEL L. JOHNSON AND GARY K. ACKERS

VOLUME 324. Branched-Chain Amino Acids (Part B)
Edited by ROBERT A. HARRIS AND JOHN R. SOKATCH

VOLUME 325. Regulators and Effectors of Small GTPases (Part D: Rho Family)
Edited by W. E. BALCH, CHANNING J. DER, AND ALAN HALL

VOLUME 326. Applications of Chimeric Genes and Hybrid Proteins (Part A: Gene Expression and Protein Purification)
Edited by JEREMY THORNER, SCOTT D. EMR, AND JOHN N. ABELSON

VOLUME 327. Applications of Chimeric Genes and Hybrid Proteins (Part B: Cell Biology and Physiology)
Edited by JEREMY THORNER, SCOTT D. EMR, AND JOHN N. ABELSON

VOLUME 328. Applications of Chimeric Genes and Hybrid Proteins (Part C: Protein–Protein Interactions and Genomics)
Edited by JEREMY THORNER, SCOTT D. EMR, AND JOHN N. ABELSON

VOLUME 329. Regulators and Effectors of Small GTPases (Part E: GTPases Involved in Vesicular Traffic)
Edited by W. E. BALCH, CHANNING J. DER, AND ALAN HALL

VOLUME 330. Hyperthermophilic Enzymes (Part A)
Edited by MICHAEL W. W. ADAMS AND ROBERT M. KELLY

VOLUME 331. Hyperthermophilic Enzymes (Part B)
Edited by MICHAEL W. W. ADAMS AND ROBERT M. KELLY

VOLUME 332. Regulators and Effectors of Small GTPases (Part F: Ras Family I)
Edited by W. E. BALCH, CHANNING J. DER, AND ALAN HALL

VOLUME 333. Regulators and Effectors of Small GTPases (Part G: Ras Family II)
Edited by W. E. BALCH, CHANNING J. DER, AND ALAN HALL

VOLUME 334. Hyperthermophilic Enzymes (Part C)
Edited by MICHAEL W. W. ADAMS AND ROBERT M. KELLY

VOLUME 335. Flavonoids and Other Polyphenols
Edited by LESTER PACKER

VOLUME 336. Microbial Growth in Biofilms (Part A: Developmental and Molecular Biological Aspects)
Edited by RON J. DOYLE

VOLUME 337. Microbial Growth in Biofilms (Part B: Special Environments and Physicochemical Aspects)
Edited by RON J. DOYLE

VOLUME 338. Nuclear Magnetic Resonance of Biological Macromolecules (Part A)
Edited by THOMAS L. JAMES, VOLKER DÖTSCH, AND ULI SCHMITZ

VOLUME 339. Nuclear Magnetic Resonance of Biological Macromolecules (Part B)
Edited by THOMAS L. JAMES, VOLKER DÖTSCH, AND ULI SCHMITZ

VOLUME 340. Drug–Nucleic Acid Interactions
Edited by JONATHAN B. CHAIRES AND MICHAEL J. WARING

VOLUME 341. Ribonucleases (Part A)
Edited by ALLEN W. NICHOLSON

VOLUME 342. Ribonucleases (Part B)
Edited by ALLEN W. NICHOLSON

VOLUME 343. G Protein Pathways (Part A: Receptors)
Edited by RAVI IYENGAR AND JOHN D. HILDEBRANDT

VOLUME 344. G Protein Pathways (Part B: G Proteins and Their Regulators)
Edited by RAVI IYENGAR AND JOHN D. HILDEBRANDT

VOLUME 345. G Protein Pathways (Part C: Effector Mechanisms)
Edited by RAVI IYENGAR AND JOHN D. HILDEBRANDT

VOLUME 346. Gene Therapy Methods
Edited by M. IAN PHILLIPS

VOLUME 347. Protein Sensors and Reactive Oxygen Species (Part A: Selenoproteins and Thioredoxin)
Edited by HELMUT SIES AND LESTER PACKER

VOLUME 348. Protein Sensors and Reactive Oxygen Species (Part B: Thiol Enzymes and Proteins)
Edited by HELMUT SIES AND LESTER PACKER

VOLUME 349. Superoxide Dismutase
Edited by LESTER PACKER

VOLUME 350. Guide to Yeast Genetics and Molecular and Cell Biology (Part B)
Edited by CHRISTINE GUTHRIE AND GERALD R. FINK

VOLUME 351. Guide to Yeast Genetics and Molecular and Cell Biology (Part C)
Edited by CHRISTINE GUTHRIE AND GERALD R. FINK

VOLUME 352. Redox Cell Biology and Genetics (Part A)
Edited by CHANDAN K. SEN AND LESTER PACKER

VOLUME 353. Redox Cell Biology and Genetics (Part B)
Edited by CHANDAN K. SEN AND LESTER PACKER

VOLUME 354. Enzyme Kinetics and Mechanisms (Part F: Detection and Characterization of Enzyme Reaction Intermediates)
Edited by DANIEL L. PURICH

VOLUME 355. Cumulative Subject Index Volumes 321–354

VOLUME 356. Laser Capture Microscopy and Microdissection
Edited by P. MICHAEL CONN

VOLUME 357. Cytochrome P450, Part C
Edited by ERIC F. JOHNSON AND MICHAEL R. WATERMAN

VOLUME 358. Bacterial Pathogenesis (Part C: Identification, Regulation, and Function of Virulence Factors)
Edited by VIRGINIA L. CLARK AND PATRIK M. BAVOIL

VOLUME 359. Nitric Oxide (Part D)
Edited by ENRIQUE CADENAS AND LESTER PACKER

VOLUME 360. Biophotonics (Part A)
Edited by GERARD MARRIOTT AND IAN PARKER

VOLUME 361. Biophotonics (Part B)
Edited by GERARD MARRIOTT AND IAN PARKER

VOLUME 362. Recognition of Carbohydrates in Biological Systems (Part A)
Edited by YUAN C. LEE AND REIKO T. LEE

VOLUME 363. Recognition of Carbohydrates in Biological Systems (Part B)
Edited by YUAN C. LEE AND REIKO T. LEE

VOLUME 364. Nuclear Receptors
Edited by DAVID W. RUSSELL AND DAVID J. MANGELSDORF

VOLUME 365. Differentiation of Embryonic Stem Cells
Edited by PAUL M. WASSAUMAN AND GORDON M. KELLER

VOLUME 366. Protein Phosphatases
Edited by SUSANNE KLUMPP AND JOSEF KRIEGLSTEIN

VOLUME 367. Liposomes (Part A)
Edited by NEJAT DÜZGÜNEŞ

VOLUME 368. Macromolecular Crystallography (Part C)
Edited by CHARLES W. CARTER, JR., AND ROBERT M. SWEET

VOLUME 369. Combinational Chemistry (Part B)
Edited by GUILLERMO A. MORALES AND BARRY A. BUNIN

VOLUME 370. RNA Polymerases and Associated Factors (Part C)
Edited by SANKAR L. ADHYA AND SUSAN GARGES

VOLUME 371. RNA Polymerases and Associated Factors (Part D)
Edited by SANKAR L. ADHYA AND SUSAN GARGES

VOLUME 372. Liposomes (Part B)
Edited by NEJAT DÜZGÜNEŞ

VOLUME 373. Liposomes (Part C)
Edited by NEJAT DÜZGÜNEŞ

VOLUME 374. Macromolecular Crystallography (Part D)
Edited by CHARLES W. CARTER, JR., AND ROBERT W. SWEET

VOLUME 375. Chromatin and Chromatin Remodeling Enzymes (Part A)
Edited by C. DAVID ALLIS AND CARL WU

VOLUME 376. Chromatin and Chromatin Remodeling Enzymes (Part B)
Edited by C. DAVID ALLIS AND CARL WU

VOLUME 377. Chromatin and Chromatin Remodeling Enzymes (Part C)
Edited by C. DAVID ALLIS AND CARL WU

VOLUME 378. Quinones and Quinone Enzymes (Part A)
Edited by HELMUT SIES AND LESTER PACKER

VOLUME 379. Energetics of Biological Macromolecules (Part D)
Edited by JO M. HOLT, MICHAEL L. JOHNSON, AND GARY K. ACKERS

VOLUME 380. Energetics of Biological Macromolecules (Part E)
Edited by JO M. HOLT, MICHAEL L. JOHNSON, AND GARY K. ACKERS

VOLUME 381. Oxygen Sensing
Edited by CHANDAN K. SEN AND GREGG L. SEMENZA

VOLUME 382. Quinones and Quinone Enzymes (Part B)
Edited by HELMUT SIES AND LESTER PACKER

VOLUME 383. Numerical Computer Methods (Part D)
Edited by LUDWIG BRAND AND MICHAEL L. JOHNSON

VOLUME 384. Numerical Computer Methods (Part E)
Edited by LUDWIG BRAND AND MICHAEL L. JOHNSON

VOLUME 385. Imaging in Biological Research (Part A)
Edited by P. MICHAEL CONN

VOLUME 386. Imaging in Biological Research (Part B)
Edited by P. MICHAEL CONN

VOLUME 387. Liposomes (Part D)
Edited by NEJAT DÜZGÜNEŞ

VOLUME 388. Protein Engineering
Edited by DAN E. ROBERTSON AND JOSEPH P. NOEL

VOLUME 389. Regulators of G-Protein Signaling (Part A)
Edited by DAVID P. SIDEROVSKI

VOLUME 390. Regulators of G-Protein Signaling (Part B)
Edited by DAVID P. SIDEROVSKI

VOLUME 391. Liposomes (Part E)
Edited by NEJAT DÜZGÜNEŞ

VOLUME 392. RNA Interference
Edited by ENGELKE ROSSI

VOLUME 393. Circadian Rhythms
Edited by MICHAEL W. YOUNG

VOLUME 394. Nuclear Magnetic Resonance of Biological Macromolecules (Part C)
Edited by THOMAS L. JAMES

VOLUME 395. Producing the Biochemical Data (Part B)
Edited by ELIZABETH A. ZIMMER AND ERIC H. ROALSON

VOLUME 396. Nitric Oxide (Part E)
Edited by LESTER PACKER AND ENRIQUE CADENAS

VOLUME 397. Environmental Microbiology
Edited by JARED R. LEADBETTER

VOLUME 398. Ubiquitin and Protein Degradation (Part A)
Edited by RAYMOND J. DESHAIES

VOLUME 399. Ubiquitin and Protein Degradation (Part B)
Edited by RAYMOND J. DESHAIES

VOLUME 400. Phase II Conjugation Enzymes and Transport Systems
Edited by HELMUT SIES AND LESTER PACKER

VOLUME 401. Glutathione Transferases and Gamma Glutamyl Transpeptidases
Edited by HELMUT SIES AND LESTER PACKER

VOLUME 402. Biological Mass Spectrometry
Edited by A. L. BURLINGAME

VOLUME 403. GTPases Regulating Membrane Targeting and Fusion
Edited by WILLIAM E. BALCH, CHANNING J. DER, AND ALAN HALL

VOLUME 404. GTPases Regulating Membrane Dynamics
Edited by WILLIAM E. BALCH, CHANNING J. DER, AND ALAN HALL

VOLUME 405. Mass Spectrometry: Modified Proteins and Glycoconjugates
Edited by A. L. BURLINGAME

VOLUME 406. Regulators and Effectors of Small GTPases: Rho Family
Edited by WILLIAM E. BALCH, CHANNING J. DER, AND ALAN HALL

VOLUME 407. Regulators and Effectors of Small GTPases: Ras Family
Edited by WILLIAM E. BALCH, CHANNING J. DER, AND ALAN HALL

VOLUME 408. DNA Repair (Part A)
Edited by JUDITH L. CAMPBELL AND PAUL MODRICH

VOLUME 409. DNA Repair (Part B)
Edited by JUDITH L. CAMPBELL AND PAUL MODRICH

VOLUME 410. DNA Microarrays (Part A: Array Platforms and Web-Bench Protocols)
Edited by ALAN KIMMEL AND BRIAN OLIVER

VOLUME 411. DNA Microarrays (Part B: Databases and Statistics)
Edited by ALAN KIMMEL AND BRIAN OLIVER

VOLUME 412. Amyloid, Prions, and Other Protein Aggregates (Part B)
Edited by INDU KHETERPAL AND RONALD WETZEL

VOLUME 413. Amyloid, Prions, and Other Protein Aggregates (Part C)
Edited by INDU KHETERPAL AND RONALD WETZEL

VOLUME 414. Measuring Biological Responses with Automated Microscopy
Edited by JAMES INGLESE

VOLUME 415. Glycobiology
Edited by MINORU FUKUDA

VOLUME 416. Glycomics
Edited by MINORU FUKUDA

VOLUME 417. Functional Glycomics
Edited by MINORU FUKUDA

VOLUME 418. Embryonic Stem Cells
Edited by IRINA KLIMANSKAYA AND ROBERT LANZA

VOLUME 419. Adult Stem Cells
Edited by IRINA KLIMANSKAYA AND ROBERT LANZA

VOLUME 420. Stem Cell Tools and Other Experimental Protocols
Edited by IRINA KLIMANSKAYA AND ROBERT LANZA

VOLUME 421. Advanced Bacterial Genetics: Use of Transposons and Phage for Genomic Engineering
Edited by KELLY T. HUGHES

VOLUME 422. Two-Component Signaling Systems, Part A
Edited by MELVIN I. SIMON, BRIAN R. CRANE, AND ALEXANDRINE CRANE

VOLUME 423. Two-Component Signaling Systems, Part B
Edited by MELVIN I. SIMON, BRIAN R. CRANE, AND ALEXANDRINE CRANE

VOLUME 424. RNA Editing
Edited by JONATHA M. GOTT

VOLUME 425. RNA Modification
Edited by JONATHA M. GOTT

VOLUME 426. Integrins
Edited by DAVID CHERESH

VOLUME 427. MicroRNA Methods
Edited by JOHN J. ROSSI

VOLUME 428. Osmosensing and Osmosignaling
Edited by HELMUT SIES AND DIETER HAUSSINGER

VOLUME 429. Translation Initiation: Extract Systems and Molecular Genetics
Edited by JON LORSCH

VOLUME 430. Translation Initiation: Reconstituted Systems and Biophysical Methods
Edited by JON LORSCH

VOLUME 431. Translation Initiation: Cell Biology, High-Throughput and Chemical-Based Approaches
Edited by JON LORSCH

VOLUME 432. Lipidomics and Bioactive Lipids: Mass-Spectrometry–Based Lipid Analysis
Edited by H. ALEX BROWN

VOLUME 433. Lipidomics and Bioactive Lipids: Specialized Analytical Methods and Lipids in Disease
Edited by H. ALEX BROWN

VOLUME 434. Lipidomics and Bioactive Lipids: Lipids and Cell Signaling
Edited by H. ALEX BROWN

VOLUME 435. Oxygen Biology and Hypoxia
Edited by HELMUT SIES AND BERNHARD BRÜNE

VOLUME 436. Globins and Other Nitric Oxide-Reactive Protiens (Part A)
Edited by ROBERT K. POOLE

VOLUME 437. Globins and Other Nitric Oxide-Reactive Protiens (Part B)
Edited by ROBERT K. POOLE

VOLUME 438. Small GTPases in Disease (Part A)
Edited by WILLIAM E. BALCH, CHANNING J. DER, AND ALAN HALL

VOLUME 439. Small GTPases in Disease (Part B)
Edited by WILLIAM E. BALCH, CHANNING J. DER, AND ALAN HALL

VOLUME 440. Nitric Oxide, Part F Oxidative and Nitrosative Stress in Redox Regulation of Cell Signaling
Edited by ENRIQUE CADENAS AND LESTER PACKER

VOLUME 441. Nitric Oxide, Part G Oxidative and Nitrosative Stress in Redox Regulation of Cell Signaling
Edited by ENRIQUE CADENAS AND LESTER PACKER

VOLUME 442. Programmed Cell Death, General Principles for Studying Cell Death (Part A)
Edited by ROYA KHOSRAVI-FAR, ZAHRA ZAKERI, RICHARD A. LOCKSHIN, AND MAURO PIACENTINI

VOLUME 443. Angiogenesis: *In Vitro* Systems
Edited by DAVID A. CHERESH

VOLUME 444. Angiogenesis: *In Vivo* Systems (Part A)
Edited by DAVID A. CHERESH

VOLUME 445. Angiogenesis: *In Vivo* Systems (Part B)
Edited by DAVID A. CHERESH

VOLUME 446. Programmed Cell Death, The Biology and Therapeutic Implications of Cell Death (Part B)
Edited by ROYA KHOSRAVI-FAR, ZAHRA ZAKERI, RICHARD A. LOCKSHIN, AND MAURO PIACENTINI

VOLUME 447. RNA Turnover in Prokaryotes, Archae and Organelles
Edited by LYNNE E. MAQUAT AND CECILIA M. ARRAIANO

VOLUME 448. RNA Turnover in Eukaryotes: Nucleases, Pathways and Anaylsis of mRNA Decay
Edited by LYNNE E. MAQUAT AND MEGERDITCH KILEDJIAN

VOLUME 449. RNA Turnover in Eukaryotes: Analysis of Specialized and Quality Control RNA Decay Pathways
Edited by LYNNE E. MAQUAT AND MEGERDITCH KILEDJIAN

VOLUME 450. Fluorescence Spectroscopy
Edited by LUDWING BRAND AND MICHAEL JOHNSON

CHAPTER ONE

A Method in Enzymology for Measuring Hydrolytic Activities in Live Cell Environments

Beverly Z. Packard *and* Akira Komoriya

Contents

1. Protease Substrate Design	2
2. Protease Specificity	5
3. Cell-Permeable Fluorogenic Probes	7
3.1. Toxicity study of cell-permeable IHED substrates	8
3.2. Cellular organization: Ordering of the activation of proteases in live cells	8
3.3. Quantification and visualization of cell-mediated cytotoxicity	12
3.4. Measurement of an early event in inflammation: Activation of caspase 1 in macrophages	13
3.5. Measurement of migration and invasiveness in metastatic cancer cells	14
3.6. Promise of new protease substrates: Measurement of viral protease activities in infected cells and tissues	15
3.7. Measurement of nuclease activity and hybridization with cell-permeable fluorogenic oligonucleotides inside live cells	15
4. Conclusions	16
References	17

Abstract

The capability of determining the physiologic role(s) of cellular enzymes requires probes with access to all intracellular and extracellular environments. Importantly, reporter molecules must be able to cross not only the plasma membrane but also enter organelles inside live cells without disturbing the physiologic integrity of the system under study. Additionally, each enzyme must recognize a probe by the same linear and conformational characteristics as it would a physiologic substrate or inhibitor. This chapter focuses on the design

OncoImmunin, Inc., Gaithersburg, Maryland 20877

and use of cell- and tissue-permeable fluorogenic protease substrates. Their applications, which are far-reaching, include measurements for apoptosis, cytotoxicity, inflammation, cancer metastasis, and viral infections such as HIV. Recently, substitution of amino acids with nucleotides in the probe backbone has allowed measurements of nuclease activities and hybridization of oligonucleotides inside live cells and an example thereof is presented.

Abbreviations

CTLs	cytotoxic T-lymphocytes
ICE	interleukin-1 converting enzyme
IHED	intramolecular H-type excitonic dimer
IL-1$_\beta$	interleukin-1 type beta
LPS	lipopolysaccharide
PEC	peritoneal exudate cells
PI	propidium iodide

1. Protease Substrate Design

With the objective of measuring protease activities in native environments, specifically inside live cells, requirements for reporter molecules, that is, substrates, include specificity, selectivity, and sensitivity. Additional requisites such as substrates' cell permeability and ability to access all subcellular compartments must be considered and determined. As a starting point, structural elements found in physiological macromolecular protease substrates and inhibitors are an excellent guide. The basic design elements of a macromolecular substrate or inhibitor that confer high protease specificity begin with recognition by a protease of a substrate's or inhibitor's amino acid sequence; the latter includes amino acids from both amino and carboxyl sides of the recognition/cleavage site where the amino acids are designated as P_{1-n} and P'_{1-n}, respectively (Fig. 1.1). This is easily seen by reviewing the

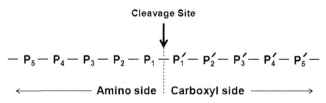

Figure 1.1 Nomenclature for substrates. Positions of amino acids in target recognition/cleavage sites.

binding cleft of a given protease's crystal structure. If one locates the catalytic residues within this pocket, one finds their position to be near the middle of the cleft rather than at an extreme end. Thus, protease substrate recognition extends *across* the cleavage site. It is clear that an optimal substrate design must include the entire substrate's/inhibitor's protease recognition sequence, not just one side, for example, the amino terminal half, as is the case with many commonly used substrates and inhibitors (Bell *et al.*, 2003; Richardson, 2002).

Typical conformation(s) for these protease binding residues can be found in the crystal structures of the macromolecular family of serine protease inhibitors known as SERPINs (Silverman *et al.*, 2001) (*note*: some SERPINs have no known inhibitory capability but do contain SERPIN structural elements). Here, the commonly found conformations for protease recognition domains of substrates or inhibitors are found to be a turn or loop conformation in the domain connecting two well-defined secondary structures, for example, β-sheets or α-helices. For example in α_1-antitrypsin, nature's most effective neutrophil elastase inhibitor, the protease cleavage site extends out from the remaining folded structure making this domain extremely accessible for docking with elastase's catalytic substrate binding cleft (Song *et al.*, 1995). Thus, a substrate that mimicked this conformation well would have a stable loop or horseshoe-like bent conformation. In contrast a substrate without a stable folded conformation would have a larger positive free energy contribution to the overall free energy of substrate binding. (This latter unfavorable free energy contribution to the Gibbs free energy of substrate binding is derived largely from the unfavorable conformational entropy change associated with the transition from the conformationally unrestrained free unbound state to the bound constrained state.) Moreover, a protease substrate consisting of a linear short tetrapeptide where the P'_1 residue position is occupied by a chromophore/fluorophore (*vide infra*) inherently would have an even larger unfavorable conformational entropy change upon binding to the substrate binding pocket.

NorFES is a fluorogenic protease substrate which includes the relatively rigid, bent peptide sequence from α_1-antitrypsin's recognition/cleavage site and covalent linkage of the same fluorophore on each side of the cleavage site forming an intramolecular H-type excitonic homodimer (IHED). This probe which was synthesized to report the protease activity of neutrophil elastase (Packard *et al.*, 1996) illustrates how the substrate design elements described above (Packard *et al.*, 1997a) in addition to efficient cell permeability (*vide infra*) can be achieved.

The spectroscopic properties of NorFES as well as subsequently synthesized substrates for a wide variety of protease substrates all exhibit a blue shift in the absorption spectrum (an indication of ground state dimerization) and fluorescence quenching relative to the same fluorophore in dilute solution (Fig. 1.2); these data are consistent with formation of an IHED between the two covalently bound dyes (Packard *et al.*, 1996). Significantly, formation of this intramolecular dye dimer constrains the flexibility of the amino and carboxyl ends of the substrate and the end-to-end distance constraint imposes a stable loop-like conformation in the substrate.

Fortunately, fluorophores with spectral properties across the visible spectrum and into the near infrared exist so that a series of homodoubly labeled substrates with spectroscopic properties similar to those of NorFES can be synthesized (Packard *et al.*, 1997b,c, 1998a,b). By utilizing available databases and various sequence homology analyses amino acid sequences can be obtained where secondary structural elements similar to those found in the turns/loops of SERPINs can be incorporated into syntheses for substrates of other proteases as well. Thus, tying the spectroscopy together

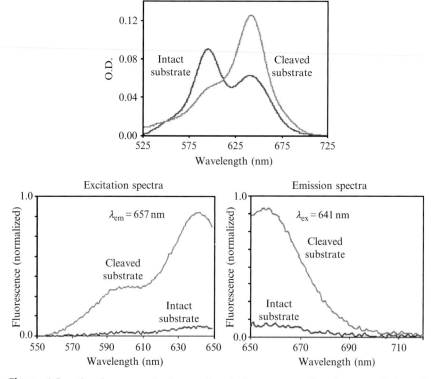

Figure 1.2 Absorbance, excitation, and emission spectra of an intact and cleaved IHED substrate.

with the available structural information enables synthesis of probes for simultaneous measurements of multiple activities inside live cells and tissues (*vide infra*).

2. PROTEASE SPECIFICITY

One of the long-standing issues in protease substrate design can be observed in the evolution of substrates which have been made to discriminate among members of the caspase family. The latter is a closely evolved group of proteases characterized by both a catalytic cysteinyl residue and a strong preference for an aspartyl residue in the P_1 position of their substrate recognition sequences. Structural and functional studies have shown that caspases also recognize the P_4 and sometimes the P_3 and P_2 amino acids. Even with the extensive database of putative target sequences available (Nicholson and Thornberry, 1997), most of the caspase literature, which is now quite vast, is based on substrates which do not clearly distinguish among the various caspase activities, largely due to limitations of the substrate design: tetrapeptides which contain only amino acids from one side (the amino side) of known caspase targets or amino side sequences derived from use of combinatorial protease substrate sequence determination/optimization methods (Harris *et al.*, 1998; Thornberry *et al.*, 1997). Thus, use of incomplete protease recognition sequences in these reporters and the resultant lack of native conformations has resulted in poor discrimination among caspase family members.

To illustrate that the activities of members of this close family of proteases can be clearly distinguished, five octadecapeptides were synthesized targeting five different caspases (caspases 1, 3, 6, 8, and 9) by incorporating the distinct protease recognition sequences for each caspase (Komoriya *et al.*, 2000). In cases where the physiologic target was known, the full amino acid sequence through the cleavage site was incorporated. In a series of criss-cross experiments each recombinant caspase was observed to cleave only the substrate containing the amino acid sequence from its physiologic target macromolecule.

In another example three different substrate sequences for interleukin converting enzyme (ICE) which is also known as caspase 1 were compared (Fig. 1.3) in live cells (for a discussion of live cell measurements, *vide infra*). First, the substrate sequence obtained from the combinatorial chemistry approach, WEHD, was compared with the tetrapeptide, YVHD; the latter is derived from the amino side of interleukin-1 type beta (IL-1$_\beta$), a naturally occurring target for caspase 1. In this first comparison, the substrate carboxyl side was GI, the common filler (*vide supra*), for both. Then, YVHD with the

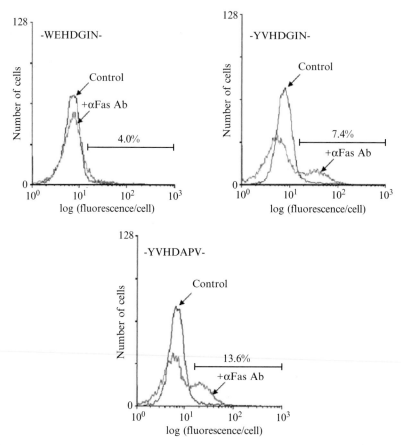

Figure 1.3 Comparison of cleavage among three caspase 1 substrates in live CTL. The percent of caspase 1 positive cells was measured by flow cytometry 3 h after addition of the apoptogen anti-Fas antibody and compared with control cells. (Measurements at later time points show proportionally greater percentage of positive cells.)

common filler on the carboxyl side was compared with YVHD followed by APV, the sequence found at the IL-1$_\beta$ processing site. A comparison of the three substrates shows caspase 1 activity detectable in only 4% of cells with the best combinatorial method-derived sequence, WEHDGI, whereas 7.4% of cells exhibit activity when measured with the substrate containing the amino side of a caspase 1 physiological macromolecular substrate, pro-IL-1$_\beta$. The third substrate with the full pro-IL-1$_\beta$ processing site sequence, YVHDAPV, yielded 13.8% cells with caspase 1 activity. These data clearly illustrate the superiority of using the amino acid sequence from a physiologic macromolecular target cleavage site, YVHDAPV, compared with the native but only tetrapeptide, YVHD, as well as the combinatorially derived WEHD.

3. Cell-Permeable Fluorogenic Probes

To make meaningful measurements of physiologic events, it is essential to perform each measurement in a physiologic environment. Since cells and tissues have evolved with certain structural protective elements, for example, plasma membranes, extracellular scaffolding, and membrane-bound intracellular compartments, permeability of probes into the microenvironment of interest is often a limiting factor. In each of the probes described above, the presence of an IHED is believed to induce an oval-shaped structure resulting in substrates with quenched fluorescence being able to cross all cell membranes in live cells by passive diffusion. This permeability trait is quite unusual for peptides of 18 amino acids. Once the probes reach the protease clefts in the cognate proteases, they are cleaved thereby abolishing the dye–dye dimer and fluorescence quenching. Thus, fluorescence is generated at the site of the active protease and, importantly, on the side of the membrane where the active protease resides. Speculation is that in the intact substrate the head-to-head bichromic complex forms a hydrophobic surface capable of interacting with lipid bilayers to allow passive diffusion across membranes; once the dimer is destroyed by cleavage of the peptide backbone, the two cleaved peptide fragments assume the properties of singly labeled peptides and, thus, their diffusion rates across lipid bilayers approach zero. These properties which extend to many fluorophores with spectral properties through the visible and into the near-infrared wavelength ranges have allowed the creation of probes in many colors. Most significantly, it has allowed simultaneous measurement of multiple protease activities inside live cells (Komoriya *et al.*, 2000; Packard *et al.*, 2001, 2007; Telford *et al.*, 2004).

In one such study (Komoriya *et al.*, 2000) pixel intensity of caspases 3 and 6 in intracellular areas of given cells were recorded as a function of time. These activity measurements were carried out using live single cells; in fact, it was as if intracellular compartments had been transferred into very small cell-free cuvettes. Thus, the simultaneous measurement of multiprotease activities in the same single cell provides the opportunity of turning a single cell into a nanoscale cuvette with all of the cellular organization, regulation and other necessary cofactors and machinery. This experiment illustrates that model solution studies may not be necessary as enzymatic kinetic measurements can be made in a truly physiologic environment.

3.1. Toxicity study of cell-permeable IHED substrates

Since the access of fluorogenic or fluorescent probes into intracellular compartments has traditionally necessitated hydrophobic protective groups which must be removed after entry into the cell or required some perturbation of the cell such as use of chemical fixatives, electroporation, or massive salt concentrations, for example, calcium phosphate, the question of intrinsic toxicity of the IHED probes deserves attention. To address this issue, the effect of two probes on the signal provided by a metabolic indicator, the mitochondrial membrane potential dye, 3,3′-dihexyloxacarbocyanine iodide ($DiOC_6$), in a viable cell culture was examined. The blue in the top panel of Fig. 1.4 shows strong mitochondrial polarization in each NT-2 cell; this image was acquired 18 h after loading cells with $DiOC_6$; substrates for caspase 9 and caspase 8 were present throughout the 18 h. This control image is in sharp contrast to the bottom panel which shows a parallel culture ($DiOC_6$ plus the same two substrates included) into which an apoptogen, staurosporine, had been present for the same 18 h time period. Thus, the loss of the blue signal from $DiOC_6$ in many cells with concomitant acquisition of green from caspase 9, followed by red from caspase 8, reveal the capability of the two caspase probes as intracellular protease activity reporter molecules. Most importantly, these two images demonstrate the lack of toxicity of the probes.

3.2. Cellular organization: Ordering of the activation of proteases in live cells

These cell-permeable fluorogenic protease substrates have been used in a multitude of cells to address a broad array of questions including localization of protease activities in granulosa cells of the mouse ovary (Robles *et al.*, 1999), embryonic pattern formation in plant cells (Bozhkov *et al.*, 2004), suppression of a matrix metalloprotease in human B cells (Guedez *et al.*, 1998), effect on survival of CrmA variants in yeast (Ekert *et al.*, 1999), proapoptotic activity in *Caenorhabditis elegans* (Kanuka *et al.*, 1999), branching of the caspase cascade in cells of the eye (Hirata *et al.*, 1998), analysis of villous cytotrophoblast and syncytial fragments during turnover of the human trophoblast (Huppertz *et al.*, 1999a,b), target cell death mediated by cytotoxic lymphocytes (CTLs) (Barber *et al.*, 2003; Liu *et al.*, 2002, 2004), enzymes associated with mitochondrial amplification (Metkar *et al.*, 2003) and depolarization (Finucane *et al.*, 1999; Johnson *et al.*, 2000), the role of Cdk activity (Harvey *et al.*, 2000; Lukovic *et al.*, 2003) and macromolecular synthesis (Chang *et al.*, 2002) in dying cells, the effect of the extract

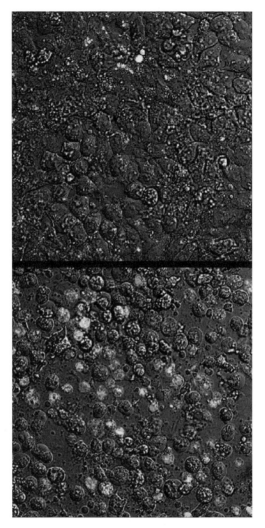

Figure 1.4 Comparison of the effect of two probes on the signal of a metabolic indicator, $DiOC_6$ in viable cells. Top panel: the blue shows strong mitochondrial polarization in each control NT-2 cell. Since two IHED caspase probes were present throughout the 18-h incubation period, the strong blue signal indicates a lack of toxicity of IHED probes. Bottom panel: in cells induced to undergo apoptosis the green due to caspase 9 activity followed by red due to caspase 8 and decreased level of blue illustrate the loss of mitochondrial membrane potential as cells undergo apoptosis. (See Color Insert.)

procedure on activation of proteases (Zapata *et al.*, 1998), and multiparametric analysis by laser scanning cytometry (Telford *et al.*, 2002). The studies described below illustrate how IHED probes have provided a means for eliciting unique information.

3.2.1. Ordering of activation of proteases involved in intracellular protease activation cascades

Since caspases are expressed in cells as proenzymes, they must be proteolytically processed to acquire activity. Various caspases have been found to activate other procaspases, and cascades of intracellular activating caspases have been described as cells begin to die following triggering of an internal program called apoptosis. While initiating events of caspase cascades are known with some certainty, for example, oligomerization of either procaspase 8 via a Fas-associated death domain protein (FADD) or procaspase 9 via mitochondrial release of the apoptotic protein-activating factor 1 (apaf-1), subsequent ordering in these cascades is less clear. Clarification of the latter has been attempted by ordering caspase processing events in cytoplasmic extracts of apoptotic cells, in conjunction with specific inhibitors and antibodies. However, this latter approach suffers from several problems. First, since cellular and molecular organization is lost during the extraction process, the search for order in a molecular cascade is virtually impossible as events controlled by the subcellular localization of regulatory components may not be accurately reproduced (Lukovic et al., 2003). An example is the autoactivation of long prodomain caspases which occurs in large complexes that are not well understood. Additionally, critical components often reside inside organelles, for example, mitochondrial intramembranous space, or move from the cytosol to a membrane during apoptosis, for example, Bcl-2. Second, the kinetic and thermodynamic parameters of enzymes and inhibitors may be significantly altered in extracts versus the structural integrity of whole cells. Third, use of antibodies to determine the level of activation of enzymes in lysates suffers from an unquantifiable level of antigenic site denaturation.

Using five octadecapeptide caspase substrates which are identical except for the seven central amino acids, quantitation by flow cytometry of intracellular activation of caspases was carried out in mouse thymocytes (Komoriya et al., 2000). When the five activities were assessed at various times after *in vitro* treatment with one apoptogen, dexamethasone, the following order of appearance was observed: caspases 9, 1, 6, 8, and 3. In contrast, when the same cells were examined after exposure to a different apoptogen, anti-Fas antibody, caspase 8 was the first measurable caspase and this was followed by 1, 3, 9, and 6. Thus, the order of caspase activation was found to depend on the apoptosis inducing agent. Confocal microscopy analysis further confirmed these different orders.

An additional piece of new information from this approach was the early transient increase in cell size during apoptosis. At first, this would appear to go against conventional wisdom as apoptosis is generally believed to result in cell shrinkage (Bortner and Cidlowski, 1999); however, the capability of looking at early caspase activation events as defined clearly by specific protease activities showed a transient increase followed by an ensuing decrease in cell size.

3.2.2. Subcellular localization of protease activities

Two different studies which have addressed this issue are described below.

In the first, the question of why following stimulation via the Fas death pathway caspase 8 activity was not the first caspase detectable by flow cytometry or in biochemical studies of some cell types (Packard et al., 2001; Scaffidi et al., 1998). This was surprising in view of the well-established model that Fas crosslinking leads to recruitment of procaspase 8 into a plasma membrane complex called the death inducing signaling complex or DISC, which promotes autoactivation, initiating the caspase cascade (Strasser et al., 2000). Time-lapse confocal imaging of lymphocytes in the presence of a fluorogenic caspase 8 substrate plus other caspase substrates provided critical insight into and clarification of this seeming inconsistency: the early appearance of caspase 8-containing plasma membrane vesicles and their release from the cell surface, followed by activation of other caspase activities into the cell interior showed the early, but low level of activation of caspase 8 at the cell surface from which the signal could be amplified into the cellular interior by initiating a caspase cascade (Stennicke et al., 1998). Once the cascade commenced, cells discarded the caspase 8 by blebbing off this enzyme in vesicles. This was a case in which the IHED probes provided clear physical evidence for a mechanism not previously understood.

In the second, the putative intracellular site of the aspartyl protease Cathepsin D as well as β-secretase was examined by comparing the subcellular localization of the cleaved fragments from a cell-permeable fluorogenic substrate bearing a sequence, SEVNLDAEF (cleavage site between L and D), found in the cells of a Swedish kindred showing a profound level of dementia (Felsenstein et al., 1994; Mullan et al., 1992); cleavage of this sequence was shown to be exquisitely sensitive to Cathepsin D. Thus, MCF-7 cells were coincubated with this probe and LysoTracker, a cell-permeable acidotrophic probe with high selectivity for acidic organelles. Figure 1.5 shows the fluorescence signals of the cleaved Cathepsin D substrate (red) and Lyso-Tracker (green) to only partially colocalize, suggesting the intracellular site of Cathepsin D to be of a higher pH than previously believed. These data including the absence of enzymatic activity in the cytosol illustrate not only the lack of complete agreement between the reported (from solution studies) acidic pH optimum for Cathepsin D but also suggest the apparently incorrect conclusion that Cathepsin D is only active in very acidic environments.

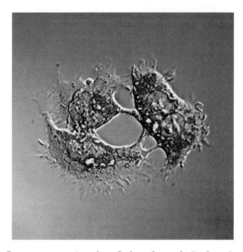

Figure 1.5 The fluorescence signals of the cleaved Cathepsin D substrate (red) and LysoTracker (green) in MCF-7 cells only partially colocalize (yellow) suggesting the intracellular site of Cathepsin D to be of a higher pH than previously believed. (See Color Insert.)

3.3. Quantification and visualization of cell-mediated cytotoxicity

Although cytotoxic T and NK cell killing of target, for example, tumor, cells is an essential component in the immunologist's analysis arsenal, the ^{51}Cr release assay, which is considered to be the gold standard assay, has been known to have numerous disadvantages. First, quantification is measured as a bulk property, in lytic units, compared with a more informative single cell level measurement. Second, the necessity of ^{51}Cr labeling of target cells limits the type of host targets that can be studied. Third, since target cell death is measured as an endpoint of the entire lytic process, kinetic information as well as the physiologic fate of effector and target cell populations at the molecular and cellular levels are quite limited. (This issue is present in the more recently published assays in which probes such as PI and 7-AAD fluoresce inside target cells after the latters' loss of plasma membrane integrity.) Fourth, the ^{51}Cr release assay is associated with a high radioactive background with the readout being small differences between larger numbers. Fifth, use of radioactivity requires special licensing and handling.

After considering the above, a nonradioactive assay to monitor and quantify target cell killing mediated by CTLs was developed by measuring an effector cell-induced biochemical event inside live target cells (Liu *et al.*, 2002, 2004). The assay was predicated on measurement of cytotoxicity in target cells through detection of the specific cleavage of fluorogenic caspase substrates. The data presented in Fig. 1.6 show induction of caspase activity

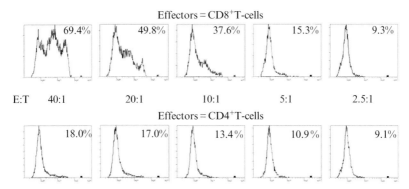

Figure 1.6 Cytotoxic (CD8$^+$) T-lymphocytes induce death via apoptosis in target cells more efficiently than helper (CD4$^+$) T-lymphocytes as indicated by the percent of caspase positive target cells. Data are from flow cytometric histograms of target cells only.

in target cells, L1210 cells, after labeling with a bifunctional antibody for recognition by a CTL (CD8$^+$) line. In contrast, a helper (CD4$^+$) T-lymphocyte line was ineffective in inducing caspase activity in the same targets. This assay now allows immunologists to examine killing with superior efficiency and sensitivity at lower effector:target ratios and shorter effector:target incubation times as well as ability to "see" T-cell memory (Barber *et al.*, 2003). Subsequently, an assay for measuring an even earlier event, that is, delivery of granzyme B activity from the granules of effector cells into the cytoplasm of target cells, was developed (Packard *et al.*, 2007).

3.4. Measurement of an early event in inflammation: Activation of caspase 1 in macrophages

A vast literature strongly supports pivotal roles for caspase 1 and IL-1$_\beta$ in inflammation and ischemia. However, due to the lack of a leader sequence in IL-1$_\beta$, its real-time detection in live systems, as well as the detection of caspase 1 activity in response to inflammatory stimuli in relevant cells, have been quite difficult. Using the caspase 1 IHED substrate containing the amino acid sequence found at the IL-1$_\beta$ processing site (*vide supra*) combined with a screening image analysis system, quantitative analysis of caspase 1 induction in live cells by a variety of inflammogens can be carried out. Figure 1.7A shows a comparison between images of murine peritoneal exudate cells (PEC) which were induced with lipopolysaccharide (LPS) and control cells, both in the presence of the fluorogenic caspase 1 substrate; Fig. 1.7B presents the dose response data for this system of study. Thus, the dose range of LPS utilized in this study induced caspase 1 activity without toxicity. Furthermore, the clear intracellular presence of caspase 1 activity after only a single inflammatory stimulus, that is, LPS, casts serious doubt on

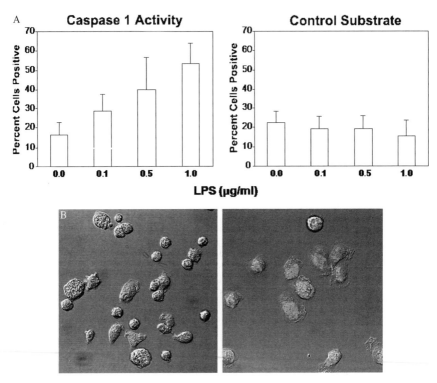

Figure 1.7 Caspase 1 activity was induced and measured in live peritoneal exudate cells (PEC) from mice by LPS in a dose-dependent manner. (A) High-throughput analysis of live cells. (B) Confocal imaging in the presence of a caspase 1 activity probe inside live cells induced by LPS (left) compared with a control population (right). (See Color Insert.)

the previously held belief that pro-IL-1_β remains unprocessed inside live cells in the absence of a second stimulus such as ATP, a bacterial toxin, or an encounter with a cytotoxic T-cell (Mariathasan *et al.*, 2004).

3.5. Measurement of migration and invasiveness in metastatic cancer cells

Cancer metastasis is the process whereby a cancer cell is enabled to migrate from its native tissue into other tissues and organs within the body. The biomedical literature strongly supports acquisition of metastatic capability to be coincident with expression of proteolytic activities in and around tumor sites. Much work has gone into identifying genes coding for proteins with proteolytic activities and many extant data implicate specific proteases in the metastatic process (Mook *et al.*, 2004; Nomura and Katunuma, 2005; Patten and Berger, 2005; Skrzydlewska *et al.*, 2005; Vihinen *et al.*, 2005; Wei and Shi, 2005). However, data defining the localization of protease activities with migration and invasiveness of live tumor cells are lacking. To achieve an accurate model for three-dimensional motility measurements fluorogenic substrates can be covalently coupled to three-dimensional macromolecular assemblies similar to extracellular matrices found in live tissues. As an example, it has recently been observed that as tumor cells invade and migrate through crosslinked collagen labeled with an IHED substrate containing the recognition/cleavage site found in interstitial collagen, cell-associated fluorescence is localized to the leading edge of invading cells (Packard et al., 2008).

3.6. Promise of new protease substrates: Measurement of viral protease activities in infected cells and tissues

In addition to the applications described above, the development of cell- and tissue-permeable fluorogenic protease substrates permits measurement of protease activities of specific viruses inside infected cells and tissues. Currently, much public health interest is focused on AIDS. The virus believed to cause this syndrome is a prime example of the need for new probes and methodologies as this virus is known to exist in locations inside the body beyond present detection capabilities (Chun et al., 2002). The IHED properties described above, for example, permeability, specificity, selectivity, and sensitivity, should allow significant improvements in AIDS as well as other virus detection.

3.7. Measurement of nuclease activity and hybridization with cell-permeable fluorogenic oligonucleotides inside live cells

Although the above discussion has focused entirely on fluorogenic protease substrates, the basic probe design is applicable to any class of hydrolytic enzyme including nucleases, glycosidases, and lipases. Moreover, any physical perturbation which can induce a separation of the two fluorophores which are covalently linked to the probe backbone will result in fluorescence. An example of the generic nature of this design is illustrated in Fig. 1.8 where a field of live K562 cells captured by confocal microscopy is shown. The turquoise inside each cell is derived from the dequenching of a homodoubly labeled piece of DNA composed of 24 nucleotides and having the complementary sequence to β-actin. The fluorescence results from dequenching of the IHED upon hybridization with its target, β-actin RNA. The robust morphology combined with the absence of fluorescence from propidium iodide (PI) present in the culture is consistent with a lack of toxicity of the probe. In addition to intracellular hybridization this approach can be used for the nontoxic introduction of siRNAs and other oligonucleotide probes for studies into live cells.

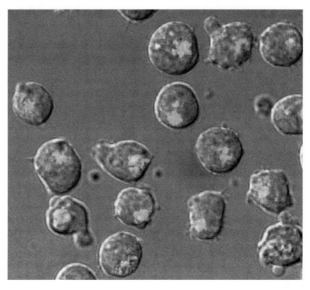

Figure 1.8 Confocal imaging of a field of live (PI-negative) cells following addition of a homodoubly labeled 24-mer oligonucleotide (turquoise) containing the complementary sequence to β-actin. (See Color Insert.)

4. Conclusions

In this chapter, the importance of and methodology to implement the direct measurement of biologically active molecules in physiologically relevant environments have been addressed. Achieving this objective has been illustrated by combining spectroscopic properties of IHED with knowledge gained from protease cleavage sites in macromolecular targets (structures of the resulting synthetic substrates are constrained to be in turn or loop conformations) and oligonucleotide databases. A method for synthesizing exceedingly specific probes with extremely high cellular permeability is thus provided. The potential for characterizing and defining physiologic processes at the molecular level while target enzymes are in physiological environments, for example, in live cells, is now possible.

The observation that the present IHED-based substrate design has no cellular toxicity is quite significant allowing: (1) measurement of intracellular enzyme activities as endpoints as well as by continuous kinetic readouts; duration of the latter range from seconds to hours and (2) introduction of oligonucleotides for RNA profiling, antisense and siRNA studies at the live, single cell level. Thus, observations of the entire course of physiological processes in live cells can be quantitated rather than merely envisioned as a snapshot of a model or reconstituted physiological system. Additionally, utilization of IHED probes in high-throughput analysis permits use in screening for drugs for many pathophysiologic conditions.

Any surprise observations derived from the present approach will undoubtedly deepen the understanding of cellular organization and regulation. To this end, the capability of this class of measurement represents a significant advance in methods in enzymology.

REFERENCES

Barber, D. L., Wherry, E. J., and Ahmed, R. (2003). Cutting edge: Rapid *in vivo* killing by memory CD8 T cells. *J. Immunol.* **171,** 27–31.

Bell, J. K., Goetz, D. H., Mhrus, S., Harris, J. L., Fletterick, R. J., and Craik, C. S. (2003). The oligomeric structure of human granzyme A is a determinant of its extended substrate specificity. *Nat. Struct. Biol.* **10,** 527–534.

Bortner, C. D., and Cidlowski, J. A. (1999). Caspase independent/dependent regulation of K(+), cell shrinkage, and mitochondrial membrane potential during lymphocyte apoptosis. *J. Biol. Chem.* **274,** 21953–21962.

Bozhkov, P. V., Filonova, L. H., Suarez, M. F., Helmersson, A., Smertenko, A. P., Zhivotovsky, B., and von Arnold, S. (2004). VEIDase is a principal caspase-like activity involved in plant programmed cell death and essential for embryonic pattern formation. *Cell Death Differ.* **11,** 175–182.

Chang, S. H., Harvey, K. J., Cvetanovic, M., Komoriya, A., Packard, B. Z., and Ucker, D. S. (2002). The effector phase of physiological cell death relies exclusively on the posttranslational activation of resident components. *Exp. Cell Res.* **277,** 15–30.

Chun, T. W., Justement, J. S., Pandya, P., Hallahan, C. W., McLaughlin, M., Liu, S., Ehler, L. A., Kovacs, C., and Fauci, A. S. (2002). Relationship between the size of the human immunodeficiency virus type 1 (HIV-1) reservoir in peripheral blood $CD4^+$ T cells and $CD4^+:CD8^+$ T cell ratios in aviremic HIV-1-infected individuals receiving long-term highly active antiretroviral therapy. *J. Infect. Dis.* **85,** 1672–1676.

Ekert, P. G., Silke, J., and Vaux, D. L. (1999). Inhibition of apoptosis and clonogenic survival of cells expressing crmA variants: Optimal caspase substrates are not necessarily optimal inhibitors. *EMBO J.* **18,** 330–338.

Felsenstein, K. M., Hunihan, L. W., and Roberts, S. B. (1994). Altered cleavage and secretion of a recombinant beta-APP bearing the Swedish familial Alzheimer's disease mutation. *Nat. Genet.* **6,** 251–255.

Finucane, D. M., Bossy-Wetzel, E., Waterhouse, N. J., Cotter, T. G., and Green, D. R. (1999). Bax-induced caspase activation and apoptosis *via* cytochrome c release from mitochondria is inhibitable by Bcl-xL. *J. Biol. Chem.* **274,** 2225–2233.

Guedez, L., Stetler-Stevenson, W. G., Wolff, L., Wang, J., Fukushima, P., Mansoor, A., and Stetler-Stevenson, M. (1998). *In vitro* suppression of programmed cell death of B cells by tissue inhibitor of metalloproteinases-1. *J. Clin. Invest.* **102,** 2002–2010.

Harris, J. L., Peterson, E. P., Hudig, D., Thornberry, N. A., and Craik, C. S. (1998). Definition and redesign of the extended substrate specificity of granzyme B. *J. Biol. Chem.* **273,** 27364–27373.

Harvey, K. J., Lukovic, D., and Ucker, D. S. (2000). Caspase-dependent Cdk activity is a requisite effector of apoptotic death events. *J. Cell Biol.* **148,** 59–72.

Hirata, H., Takahashi, A., Kobayashi, S., Yonehara, S., Sawai, H., Okasaki, T., Yamamoto, K., and Sasada, M. (1998). Caspases are activated in a branched protease cascade and control distinct downstream processes in Fas-induced apoptosis. *J. Exp. Med.* **187,** 587–600.

Huppertz, B., Frank, H.-G., Reister, F., Kingdom, J., Korr, H., and Kaufmann, P. P. (1999a). Apoptosis cascade progresses during turnover of human trophoblast: Analysis of villous cytotrophoblast and syncytial fragments *in vitro*. *Lab. Invest.* **79,** 1687–1702.

Huppertz, B., Frank, H.-G., and Kaufmann, P. (1999b). The apoptosis cascade—Morphological and immunohistochemical methods for its visualization. *Anat. Embryol.* **200,** 1–18.

Johnson, B. W., Cepero, E., and Boise, L. H. (2000). Bcl-xL inhibits cytochrome c release but not mitochondrial depolarization during the activation of multiple death pathways by tumor necrosis factor-alpha. *J. Biol. Chem.* **275,** 31546–31553.

Kanuka, H., Hisahara, S., Sawamoto, K., Shoji, S., Okano, H., and Miura, H. (1999). Proapoptotic activity of *Caenorhabditis elegans* CED-4 protein in Drosophila: Implicated mechanisms for caspase activation. *Proc. Natl Acad. Sci. USA* **96,** 145–150.

Komoriya, A., Packard, B. Z., Brown, M. J., Wu, M.-L., and Henkart, P. A. (2000). Assessment of caspase activities in intact apoptotic thymocytes using cell-permeable fluorogenic caspase substrates. *J. Exp. Med.* **191,** 1819–1828.

Liu, L., Chahroudi, A., Silvestri, G., Wernett, M. E., Kaiser, W. J., Safrit, J. T., Komoriya, A., Altman, J. D., Packard, B. Z., and Feinberg, M. B. (2002). Visualization and quantification of T cell-mediated cytotoxicity using cell-permeable fluorogenic caspase substrates. *Nat. Med.* **8,** 185–189.

Liu, L., Packard, B. Z., Brown, M. J., Komoriya, A., and Feinberg, M. B. (2004). Assessment of lymphocyte-mediated cytotoxicity using flow cytometry. *Methods Mol. Biol.* **263,** 125–140.

Lukovic, D., Komoriya, A., Packard, B. Z., and Ucker, D. S. (2003). Caspase activity is not sufficient to execute cell death. *Exp. Cell Res.* **289,** 384–395.

Mariathasan, S., Newton, K., Monack, D. M., Vucic, D., French, D. M., Lee, W. P., Roose-Flrma, M., Erickson, S., and Dixit, V. M. (2004). Differential activation of the inflammasome by caspase-1 adaptors ASC and Ipaf. *Nature* **430,** 213–218.

Metkar, S. S., Wang, B., Ebbs, M. L., Kim, J. H., Lee, Y. J., Raja, S. M., and Froelich, C. J. (2003). Granzyme B activates procaspase-3 which signals a mitochondrial amplification loop for maximal apoptosis. *J. Cell Biol.* **160,** 875–885.

Mook, O. R., Frederiks, W. M., and Van Noorden, C. J. (2004). The role of gelatinases in colorectal cancer progression and metastasis. *Biochim. Biophys. Acta* **1705,** 69–89.

Mullan, M., Crawford, F., Axelman, K., Houlden, H., Lilius, L., Winblad, B., and Lannfelt, L. (1992). A pathogenic mutation for probable Alzheimer's disease in the APP gene at the N-terminus of beta-amyloid. *Nat. Genet.* **1,** 345–347.

Nicholson, D. W., and Thornberry, N. A. (1997). Caspases: Killer proteases. *Trends Biochem. Sci.* **22,** 299–307.

Nomura, T., and Katunuma, N. (2005). Involvement of cathepsins in the invasion, metastasis and proliferation of cancer cells. *J. Med. Invest.* **52,** 1–9.

Packard, B. Z., Toptygin, D. D., Komoriya, A., and Brand, L. (1996). Profluorescent protease substrates: Intramolecular dimers described by the exciton model. *Proc. Natl Acad. Sci. USA* **93,** 11640–11645.

Packard, B. Z., Toptygin, D. D., Komoriya, A., and Brand, L. (1997a). Design of profluorescent protease substrates guided by exciton theory. *Methods Enzymol.* **278,** 15–28.

Packard, B. Z., Komoriya, A., Toptygin, D. D., and Brand, L. (1997b). Structural characteristics of fluorophores that form intramolecular H-type dimers in a protease substrate. *J. Phys. Chem. B* **101,** 5070–5074.

Packard, B. Z., Toptygin, D. D., Komoriya, A., and Brand, L. (1997c). Characterization of fluorescence quenching in bifluorophoric protease substrates. *Biophys. Chem.* **67,** 167–176.

Packard, B. Z., Toptygin, D. D., Komoriya, A., and Brand, L. (1998a). Intramolecular resonance dipole–dipole interactions in a profluorescent protease substrate. *J. Phys. Chem. B* **102,** 752–758.

Packard, B. Z., Komoriya, A., Nanda, V., and Brand, L. (1998b). Intramolecular excitonic dimers in protease substrates: Modification of the backbone moiety to probe the H-dimer structure. *J. Phys. Chem. B* **102,** 1820–1827.

Packard, B. Z., Komoriya, A., Brotz, T. M., and Henkart, P. A. (2001). Caspase 8 activity in membrane blebs after anti-Fas ligation. *J. Immunol.* **167,** 5061–5066.

Packard, B. Z., Telford, W. G., Komoriya, A., and Henkart, P. A. (2007). Granzyme B activity in target cells detects attack by cytotoxic lymphocytes. *J. Immunol.* **179,** 3812–3820.

Packard, B. Z., Artym, V. V., Komoriya, A., and Yamada, K. M. (2008). Direct visualization of protease activity on cells migrating in three dimensions. *Matrix Biol.* (in press).

Patten, L. C., and Berger, D. H. (2005). Role of proteases in pancreatic carcinoma. *World J. Surg.* **29,** 258–263.

Richardson, P. L. (2002). The determination and use of optimized protease substrates in drug discovery and development. *Curr. Pharm. Design* **8,** 2559–2581.

Robles, R., Tao, X.-J., Trbovich, A. M., Maravei, D. V., Nahum, R., Perez, G. L., Tilly, K. L., and Tilly, J. L. (1999). Localization, regulation and possible consequences of apoptotic protease-activating factor-1 (Apaf-1) expression in granulosa cells of the mouse ovary. *Endocrinology* **140,** 2641–2644.

Scaffidi, C. S., Fulda, S., Srinivasan, A., Friesen, C., Li, F., Tomaselli, K. J., Dabatin, K. M., Krammer, P. H., and Peter, M. E. (1998). Two CD95 (APO-1/Fas) signaling pathways. *EMBO J.* **17,** 1675–1687.

Silverman, G. A., Bird, P. I., Carrell, R. W., Church, F. C., Coughlin, P. B., Gettins, P. G., Irving, J. A., Lomas, D. A., Luke, C. J., Moyer, R. W., Pemberton, P. A., Remold-O'Donnell, E., *et al.* (2001). The serpins are an expanding superfamily of structurally similar but functionally diverse proteins. Evolution, mechanism of inhibition, novel functions, and a revised nomenclature. *J. Biol. Chem.* **276,** 33293–33296.

Skrzydlewska, E., Sulkowska, M., Koda, M., and Sulkowski, S. (2005). Proteolytic–antiproteolytic balance and its regulation in carcinogenesis. *World J. Gastroenterol.* **11,** 1251–1266.

Song, H. K., Lee, K. N., Kwon, K., Yu, M., and Suh, S. W. (1995). Crystal structure of an uncleaved alpha 1-antitrypsin reveals the conformation of its inhibitory reactive loop. *FEBS Lett.* **377,** 150.

Stennicke, H. R., Jurgensmeier, J. M., Shin, H., Deveraux, Q., Wolf, B. B., Yang, X., Zhou, Q., Ellerby, H. M., Ellerby, L. M., Bredesen, D., Green, D. R., Reed, J. C., *et al.* (1998). Procaspase-3 is a major physiologic target of caspase-8. *J. Biol. Chem.* **273,** 27084–27090.

Strasser, A., O'Connor, L., and Dixit, V. M. (2000). Apoptosis signaling. *Annu. Rev. Biochem.* **69,** 217–245.

Telford, W. G., Komoriya, A., and Packard, B. Z. (2002). Detection of localized caspase activity in early apoptotic cells by laser scanning cytometry. *Cytometry* **47,** 81–88.

Telford, W. G., Komoriya, A., and Packard, B. Z. (2004). Multiparametric analysis of apoptosis by flow and image cytometry. *Methods Mol. Biol.* **263,** 141–160.

Thornberry, N. A., Rano, T. A., Peterson, E. P., Rasper, D. M., Timkey, T., Garcia-Calvo, M., Houtzager, V. M., Nordstrom, P. A., Roy, S., Vaillancourt, J. P., Chapman, K. T., and Nicholson, D. W. (1997). A combinatorial approach defines specificities of members of the caspase family and granzyme B. Functional relationships established for key mediators of apoptosis. *J. Biol. Chem.* **272,** 17907–17911.

Vihinen, P., Ala-aho, R., and Kahari, V. M. (2005). Matrix metalloproteinases as therapeutic targets in cancer. *Curr. Cancer Drug Targets* **5,** 203–220.

Wei, L., and Shi, Y. B. (2005). Matrix metalloproteinase stromelysin-3 in development and pathogenesis. *Histo. Histopathol.* **20,** 177–185.

Zapata, J. M., Takahashi, R., Salvesen, G. S., and Reed, J. C. (1998). Granzyme release and caspase activation in activated human T-lymphocytes. *J. Biol. Chem.* **273,** 6916–6920.

CHAPTER TWO

HETEROGENEITY OF FLUORESCENCE DETERMINED BY THE METHOD OF AREA-NORMALIZED TIME-RESOLVED EMISSION SPECTROSCOPY

N. Periasamy

Contents

1. Introduction	22
2. TRES and TRANES Methods	24
3. TRES and TRANES Spectra of Simple Cases	25
4. Physical Significance of TRANES and Isoemissive Point	29
5. TRES and TRANES of Fluorophores in Microheterogeneous Media	30
6. Fluorescence in Microheterogeneous and Biological Media: Special Cases	32
Acknowledgment	33
References	33

Abstract

Heterogeneity of fluorescence due to multiple fluorophores or emitter species is a common problem in using fluorescent molecules to probe the structure, dynamics, and properties of microheterogeneous media (micelles, membranes, proteins, etc.) and biological media. Time-resolved emission spectra (TRES) and area-normalized TRES (TRANES) are useful to identify unambiguously emission from a single species (homogeneity) and emission from two species (dual emission). The features in TRANES spectra that are characteristic of solvation dynamics associated with single species and two species, and many other cases are also described.

Department of Chemical Sciences, Tata Institute of Fundamental Research, Homi Bhabha Road, Mumbai 400005, India

1. INTRODUCTION

Fluorescence has become a widely used method in biology, materials, and chemistry. Quantitative fluorescence methods involve measurement of one of the three properties of the photon emitted from the sample: frequency or wavelength (usually as spectrum), intensity (number of photons), and polarization. The spectrum is useful to identify the emitting species, intensity is useful to quantify the concentration of the species and polarization property is useful to identify the spatial orientation of the molecule. When a pulsed light source is used to excite the sample all the three fluorescence properties become time dependent, which gives rise to a host of new experiments to probe the molecular dynamics including chemical reactions of the excited molecule.

One popular fluorescence method is to use a small fluorescent molecule as a probe to investigate the physical properties (viscosity, solvent polarity, electric field, etc.) in the neighborhood of the probe molecule located in biological and microheterogeneous medium. This becomes possible when the probe molecule is homogeneously distributed and the fluorescence emission is identified with a single species. If the "single species" condition is not met in the experimental system, which is perhaps more likely in biological samples, then the interpretation of the fluorescence property is vastly more complex.

Present technology permits experimental measurements using a single molecule. The emission property of even a single molecule, as determined by several photons emitted by the same molecule, is time dependent and sensitive to the interaction with the molecules in the immediate neighborhood (Deniz et al., 2008; Kulzer and Omit, 2004; Wazawa et al., 2000). Therefore, it is necessary to define clearly the meaning of "a single species" in the context of experiments that are done on several thousands molecules in the excitation zone, as, for example, in a cuvette or microscopy sample.

In a dilute solution, instantaneous absorption and fluorescence properties of individual molecules fluctuate due to random translational and rotational motions, but the spatial average of the property over several molecules in a large volume is a time-invariant property. Similarly, the instantaneous property of a single molecule in a dilute solution is averaged out over a period of time that is sufficiently long compared to the fluctuation time. The net result is that the above two averages give identical results. Thus, a sample is said to be homogeneous and made up of a single species in the ground state if the spatial average of an ensemble of molecules and the time average of a single molecule randomly chosen in the same sample are identical. A very dilute solution of a fluorophore in a solvent of low viscosity is an example of single species and homogeneous in the ground state. The above definition is also valid, in principle, for dilute solutions in highly viscous, glassy or frozen (but not crystalline) solutions. The fluctuation time is proportional to the viscosity and the time average may take a longer time than the experiment duration, thus requiring special care in the interpretation of results from such samples.

With the above definition of homogeneity and single species in the ground state, heterogeneity gets readily defined; that is, a sample is heterogeneous if it is not homogeneous. Biological samples are expected to be heterogeneous by composition. Therefore, single species and homogeneity of a fluorophore has to be established by appropriate experimental evidence. Fluorescent molecules distributed in different locations of the membranes and micelles (e.g., core, inner and outer boundaries) are an example of site heterogeneity. A mixture of fluorophores is a trivial example of heterogeneity. Two tryptophans present in two sites (e.g., solvent exposed and buried) of a protein is heterogeneous. These and other examples give rise to fluorescence heterogeneity and dual or multiple emission.

Homogeneity in the ground state (S_0) does not guarantee fluorescence emission from a single species. This is simply because the chemical and physical properties of the excited state are different from those of ground state. Chemical reaction of the excited state produces new excited species, which are structurally distinct from the ground state and the fluorescence emission from the new species is characteristic of the new structure. Examples of the chemical reactions are protonation, deprotonation, excimer/exciplex formation, charge transfer, etc. In the simplest case, dual emission occurs from the parent and daughter species. Chemical reactions in the excited state is an important mechanism of dual and multiple emission even when the sample is homogeneous and single species in the ground state.

Intramolecular and intermolecular physical processes of the excited state lead to energy relaxation and thereby affect the emission property, especially the spectrum of the molecule. The time scale of nonradiative intramolecular relaxation is typically in picoseconds in liquids. Intermolecular relaxation arises due to a change in the physical property in the excited state (e.g., dipole moment) which affects the interaction with the solvent molecules in the neighborhood, leading to a solvent-relaxed structure. The time scale for solvent relaxation can vary from picoseconds to nanoseconds depending upon the viscosity and relative sizes of the molecules. In contrast to the chemical reaction where the concentration of the transition state is negligible throughout the reaction, the concentration of intermediate states in solvent relaxation is significant and emission from these states is observable during the relaxation process. Therefore, in cases of relaxation by a physical process, emission from a continuum of states is possible and observable in time-resolved emission spectroscopy. Thus, the above-described heterogeneity due to solvent relaxation is distinctly different from site heterogeneity and dual emission.

One of the experimental challenges is to distinguish heterogeneity due to solvent relaxation and dual emission due to chemical reaction or site heterogeneity. This is possible using time-resolved area-normalized emission spectroscopy (TRANES) method (Koti et al., 2001), which is a simple extension of the well-known time-resolved emission spectroscopy (TRES) method. TRANES is a model-free method and a model is tested only after a visual examination of the TRANES spectra. Numerous time-resolved fluorescence studies have validated the method by identifying unambiguously dual/multiple emission in homogeneous (Koti and Periasamy, 2002; Koti et al., 2001; Mukherjee and Datta, 2006; Rosspeintner et al., 2006) and microheterogeneous media (Choudhury et al., 2007; Dahl et al., 2005; Gasymov et al., 2007; Ira et al., 2003; Koti and Periasamy, 2001a,b; Maciejewski et al., 2003, 2005; Mukherjee et al., 2005a,b; Novaira et al., 2007; Sarkar et al., 2005) and solvent relaxation (Koti et al., 2001). The earliest reported use of area-normalized TRES spectra to explain dual emission and site heterogeneity in microheterogeneous media is to be found in (Pansu and Yoshihara, 1991), which is followed by (Laguitton-Pasquier et al., 1997).

2. TRES AND TRANES METHODS

Methods of recording time-resolved emission spectra (TRES) have been described (Lakowicz, 1999). The method using TCSPC data involves three steps (Easter et al., 1976). Area-normalized TRES spectra (called TRANES, which should have been acronymed as ANTRES) are obtained in an additional step (Koti et al., 2001):

Step 1. Fluorescence decays of the sample are obtained in the entire emission spectrum of the probe at 5nm or smaller interval. In TCSPC experiments, the peak count in the fluorescence decay could be $\sim 1 \times 10^4$ or higher for all wavelengths except in the wings of the spectrum for which the peak count can be less.

Step 2. Fluorescence decay at each wavelength is deconvoluted using the instrument response function and a multiexponential function, $I(t) = \sum \alpha_i \exp(-t/\tau_i)$, $i = 1-4$ or 5, where α_i can be negative (if excited-state kinetics or solvation dynamics is present) by the standard method of nonlinear least squares and iterative reconvolution (Grinvald and Steinberg, 1974; O'Connor and Phillips, 1984). Maximum entropy method with the proviso for negative amplitude can also be used (Livesey and Brochon, 1987). The sole aim of this step is to obtain a noise-free representation of the intensity decay function (for a hypothetical δ-function excitation), for each emission wavelength.

Step 3. TRES, plotted as intensity versus wavelength or wave number, are constructed using $\alpha_i(\lambda)$ and $\tau_i(\lambda)$, and steady-state emission spectrum. The equation used is

$$I(\lambda, t) = I_{ss}(\lambda) \frac{\sum_j \alpha_j(\lambda) e^{-t/\tau_j(\lambda)}}{\sum_j \alpha_j(\lambda) \tau_j(\lambda)}, \qquad (2.1)$$

where $I_{ss}(\lambda)$ is the steady-state fluorescence intensity at λ and $\alpha_j(\lambda)$ and $\tau_j(\lambda)$ are the values of the fit parameters obtained in Step 2.

Step 4. TRANES spectra are constructed by normalizing the area of each spectrum in TRES to a constant value; for example, the area of the spectrum at time t is made equal to the area of the spectrum at $t = 0$.

In practice, the following observations are pertinent while constructing the TRES and TRANES spectra:

1. Correction of steady-state spectrum for the sensitivity of the spectrometer (grating, detector, etc.) and plotting the spectra in frequency scale are desirable but not essential for the confirmation of single species or dual emission (isoemissive point in TRANES).
2. It is a common practice to use log-normal function to smooth the TRES spectra, which is usually done when there are fewer points to represent the spectrum. It may be noted that log-normal function was first proposed for smoothing a broad, featureless absorption spectrum as an empirical function (Siano and Metzler, 1969). There is no physical justification for fitting absorption or emission spectra using log-normal function. Smoothing TRES or TRANES spectra may distort spectral features; such as, loss of an isoemissive point, if present. Therefore, it is recommended that the raw TRES and TRANES spectra are constructed with larger number of data points to represent the spectrum.
3. TRES and TRANES spectra constructed from TCSPC decays may be noisy for the extrapolated time of $t = 0$ or at times which are less than the time per channel used. These extrapolated spectra may be omitted, or interpreted with due caution.
4. Area normalization of TRES spectra obtained by methods other than TCSPC, for example, using streak camera, is a straightforward procedure.

3. TRES AND TRANES SPECTRA OF SIMPLE CASES

The use of the TRES and TRANES spectra is to identify heterogeneity in samples, if present, and to understand the nature of heterogeneity, if possible. Dual emission has the easily identifiable feature of isoemissive point and solvation dynamics or extensive heterogeneity has the feature of continuous spectral shift with time. The distinguishing features of TRES and TRANES spectra for the simple cases are described below:

a. *Fluorophore in a dilute solution without excited-state reaction.* Nile red in methanol is an example of this case. Figure 2.1 shows the TRES and TRANES spectra for this sample. The TRES spectra decrease in intensity with time and the TRANES spectra overlap nearly identically. Identical TRANES spectra at all time is the identifying feature for emission from a single species.

b. *Fluorophore in dilute solution with excited-state reaction (an example of dual emission).* 2-Naphthol in water (pH 7) is an example of this case. The pKa of 2-naphthol is 9.5 in the ground state and 2.8 in the excited state (Laws and Brand, 1979). Thus, deprotonation occurs in the excited state giving rise to dual emission from naphthol and naphtholate anion. Figure 2.2 shows the TRES and TRANES spectra for this sample. TRES spectra decrease in intensity with time and the TRANES spectra show an isoemissive point. Observation of an isoemissive point is the unambiguous signature that emission occurs from two species. Examples of dual emission due to reversible and irreversible excited-state kinetics and mixture of two fluorophores have been discussed extensively (Koti et al., 2001). The meaning of the isoemissive point is discussed below.

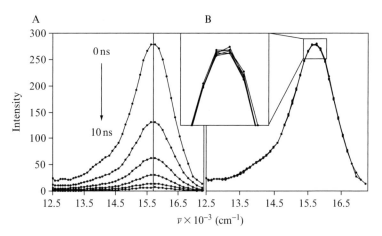

Figure 2.1 (A) Time-resolved emission spectra (TRES) and (B) time-resolved area-normalized emission spectra (TRANES) for nile red in methanol. Inset shows the peak region in greater detail. [nile red] = $2\mu M$. λ_{ex} = 571nm. The spectra are calculated for times: 0, 2, 4, 6, 8, and 10ns. The TRANES spectra overlap identically for all time. This is the signature for a single emissive species in the sample. Reprinted from Koti and Periasamy (2001a), with the permission of the Indian Academy of Sciences.

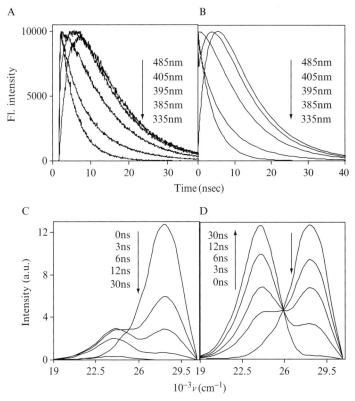

Figure 2.2 TRES and TRANES spectra for 2-naphthol in buffer (pH 6.6). [2-naphthol]≈5×10^{-6}M. (A) Fluorescence decays were taken at 5nm interval and the raw experimental data are shown only for a few wavelengths: 335, 385, 395, 405, and 485nm. (B) Noise-free intensity decays at 335, 385, 395, 405, and 485nm obtained by deconvolution of the experimental data which are shown in (A). (C) Time-resolved emission spectra (TRES) at 0, 3, 6, 12, and 30ns. (D) Time-resolved area-normalized emission spectra (TRANES) at 0, 3, 6, 12, and 30ns. TRANES spectra show an isoemissive point. This is the signature for emission from two species. Reprinted from Koti et al. (2001), with the permission of the American Chemical Society.

c. *Mixture of three (or more) fluorophores in dilute solution (no excited-state reaction).* A dilute solution of a mixture of fluorescent dyes, noninteracting in the ground and excited state is an example for this case. For a mixture of the three or more fluorophores, the TRES spectra appear similar to other cases namely, decreasing intensity with time. However, the TRANES spectra appear more complicated (Koti and Periasamy, 2001b). Upon closer analysis, the TRANES spectra reveal two isoemissive points in different time intervals. The analysis is described elsewhere (Koti and Periasamy, 2001b).

d. *Fluorophore in dilute, viscous solution, without excited-state reaction.* Nile red in 2-octanol (Koti and Periasamy, 2001a) or STQ dye in butanol (Koti et al., 2001) is an example of this case. The excited-state dipole moment of the dye is different from that of the ground state. As a result, solvent relaxation (also called solvation dynamics) occurs until the species is fully relaxed. This solvent relaxation process is ubiquitous in all cases. In high-viscosity solvents, solvation dynamics continues for a longer time. The TRES and TRANES spectra for STQ in butanol are shown in Fig. 2.3. There is a continuous red shift in TRES and TRANES spectra.

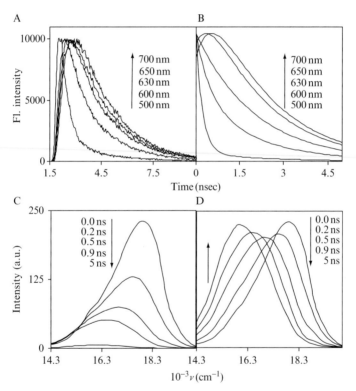

Figure 2.3 TRES and TRANES spectra for STQ dye in 2-octanol. [dye]≈5×10^{-6}M. (A) Raw experimental data of fluorescence decays for a few wavelengths at 500, 600, 630, 650, and 700nm. (B) Intensity decays at 500, 600, 630, 650, and 700nm obtained by deconvolution of the experimental decays shown in (A). (C) Time-resolved emission spectra (TRES) at 0, 0.2, 0.5, 0.9, and 5ns. (D) Time-resolved area-normalized emission spectra (TRANES) at 0, 0.2, 0.5, 0.9, and 5ns. TRANES spectra show red shift with time. This is the signature for solvation dynamics. Reprinted from Koti *et al.* (2001), with the permission of the American Chemical Society.

4. PHYSICAL SIGNIFICANCE OF TRANES AND ISOEMISSIVE POINT

The physical meaning of the TRES spectra is straightforward. Namely, they are the actual emission spectra from the sample at different times after excitation. TRANES spectra are obtained by normalizing the areas of TRES spectra to an arbitrarily constant value and therefore these are hypothetical spectra. The physical meaning of TRANES, namely, equal-area TRES spectra may be understood as follows. The total number of photons (or total energy emitted) is constant with time, which is equivalent to saying that the excited states do not decay and the number is conserved. However, the identity of the excited state changes due to kinetics (as in the case of 2-naphthol in Fig. 2.2). Change in the identity of the excited state implies change in spectrum. In the case of mixture of two molecules, the population of the species with a shorter lifetime decreases fast. In the equal area TRES spectra, decrease in the emission of one species (decrease in concentration due to kinetics and/or decay, multiplied by quantum yield) corresponds to an increase in the emission of the second species (hypothetical increase in concentration multiplied by its quantum yield). Thus, the equal area TRES spectra with a change in spectral shape with time indicate two or more emissive species. In case (a) (Fig. 2.1), there is only one emitting species and the equal area TRES spectra are identical at all time.

The proof for the existence of an isoemissive point in the TRANES spectra in dual emission cases, irrespective of the origin of the two species, has been given (Koti *et al.*, 2001). It was shown that the isoemissive point is observed at the frequency (wavelength) at which the ratio of the wavelength-dependent radiative rates ($k_r(\lambda)$) is equal to the ratio of the total radiative rates (k_r) of the two species. Use of the relation between quantum yield (φ) and radiative rate and fluorescence lifetime (τ), namely, $\varphi(\lambda) = k_r(\lambda)\tau$ and $\varphi = k_r\tau$, and recognizing that area of the fluorescence emission spectrum is proportional to the quantum yield, allows one to make the following simple statement: In dual emission samples, one or more isoemissive point in TRANES spectra are observed at those wavelengths where the ratio of the emission intensities of the two species is equal to the ratio of the areas under the emission spectra.

It is interesting to make a comparison of isoemissive point in TRANES spectra with the observation of isobestic point in absorption spectroscopy experiments involving two species. In the latter case, an isobestic point is observed at that wavelength where the extinction coefficients (i.e., fractional oscillator strengths) are equal. In the fluorescence case, the condition is based on the equality of fractional quantum yields of the two species. Thus, a molecular property forms the basis in both cases.

It is to be mentioned here that the observation of an isoemissive point in TRANES spectra relies upon the assumption that the spectrum and/or the fluorescence lifetime are different for the two species. Interesting situations arise in special cases which make the observation of isoemissive point impossible even though emission is from two species. These are (1) spectra and lifetimes of the two species are identical, (2) spectra are identical and lifetimes are different, and (3) spectra are different but lifetimes are identical. These cases of dual emission are difficult to resolve by TRES or TRANES methods. It is necessary to use other parameters such as anisotropy, quenching, solvent perturbation, etc., if TRANES fails.

5. TRES AND TRANES OF FLUOROPHORES IN MICROHETEROGENEOUS MEDIA

Surfactant micelles, lipid membrane vesicles, and microemulsions are submicrometer particles, easily dispersed in water. These are frequently used to mimic the biological systems (cell membrane, cytoplasm, protein milieu, etc., which are lot more complex), and fluorescent molecules are frequently used to probe the physical properties. The complexity of fluorescence in such samples could arise due to many reasons. Firstly, the fluorescent molecules are partitioned between the aqueous phase and nonaqueous phase, which depends upon the affinity of the dye. Secondly, there are different sites of solubilization for the molecule in the nonaqueous phase; that is, the molecule is partitioned in different sites based on affinity and density of sites. If the fluorescent molecule is sensitive to solvent polarity, viscosity, membrane potential, etc. (probes are chosen to be most sensitive to such parameters), then the molecules in different sites will have different fluorescence spectra and lifetimes. Thirdly, there is a dynamic equilibrium among the molecules in different sites and with the aqueous phase. If the exchange rate of molecules between sites is very slow compared to the fluorescence time scale then the molecules in different sites are equivalent to different emitting species. In general, use of fluorescent probes in microheterogeneous media gives rise to many possibilities, ranging from the simplest case of the molecule solubilized in one site (equivalent to a dilute solution in a viscous solvent) to the extreme case of extensive heterogeneity. Therefore, fluorescence results in these samples cannot be interpreted without ambiguity if the state of the sample is not correctly understood. Very often, a particular situation is assumed for the sake of simplicity (e.g., all molecules are in one type of site). We describe below cases where TRES and TRANES analysis are found to be useful to make correct choices about the status of the probe in microheterogeneous and biological media.

The following cases have been observed and results discussed for different dyes in surfactant micelles (Koti and Periasamy, 2001a) or lipid membranes (Ira et al., 2003). TRES and TRANES spectra are usually constructed for a single excitation wavelength. It will be desirable to confirm the features described below for other excitation wavelengths also:

a. TRANES spectra are identical. The emission is due to a single species. This has been observed for DPH in eggPC membrane (Ira et al., 2003). Interestingly, the fluorescence decay is multiexponential, which was attributed to difference in the radiative rate of DPH molecules oriented parallel and perpendicular to the membrane surface (Toptygin et al., 1992). It was possible to obtain the ratio populations in the two orientations (Krishna and Periasamy, 1998b). The spectrum of the molecule in the two orientations is the same and hence TRANES spectra are identical.
b. TRES spectra shift with time and TRANES spectra show an isoemissive point. This indicates that there is dual emission.
c. TRES and TRANES spectra shift with time for $t < t_x$ and thereafter TRANES spectra are identical. This indicates an energy relaxation (e.g., solvation dynamics) up to t_x which is followed by emission from the relaxed species.
d. TRES and TRANES spectra shift with time for $t < t_x$ and thereafter TRANES spectra show an isoemissive point. This indicates that solvation dynamics and dual emission are coupled. Many possibilities exist in such a case. For example, (a) two species in the ground state and solvation dynamics of one or both species being fast compared to fluorescence decay and (b) one species in the ground sate and fast solvation dynamics coupled with excited-state kinetics. Additional experiments (e.g., different excitation wavelength) and analysis (e.g., negative amplitude indicating excited kinetics) are required to differentiate them.
e. TRES and TRANES spectra shift continuously with time. This is possible when there is no heterogeneity, that emission is from one species, and further that solvation dynamics is a very slow process compared to the fluorescence lifetime. Another possibility is that the sample is extensively heterogeneous (many sites of solubilization) and the fluorescence lifetime increases with the red shifted species. Here, model-based analysis of fluorescence decays is required to resolve the problem. In the former case, negative amplitude for a multiexponential fit is expected.
f. Blue shift of TRES and isoemissive point in TRANES spectra. This unusual observation is demonstrated (Koti and Periasamy, 2001a). This will be seen in a sample in which there is heterogeneity in the ground state and longer fluorescence lifetime is associated with the species with blue shifted emission spectrum.

6. Fluorescence in Microheterogeneous and Biological Media: Special Cases

Site heterogeneity. Site heterogeneity is implicit in microheterogeneous (micelles, vesicle, proteins, etc.) and biological media. In the former, heterogeneity exists in nanometer scale structures made up of a single component (e.g., surfactant), which are larger than the size of the fluorescent probe. The site heterogeneity arises due to variation of physical property in different locations (e.g., surface, core, etc.). Fluorescent molecules have been identified in at least three distinct sites in lipid membranes (Krishna and Periasamy, 1998a). The interpretation of fluorescence signal in microheterogeneous and biological media requires special considerations. It is common to assume particular models of heterogeneity for the analysis. It is desirable that investigators choose models after a model-free TRANES analysis. Recent studies have validated the power of TRANES analysis in microheterogeneous media: Micelles (Koti and Periasamy, 2001a; Mukherjee *et al.*, 2005b; Shaw and Pal, 2007), reverse micelles (Mukherjee *et al.*, 2005a; Novaira *et al.*, 2007), vesicles (Choudhury *et al.*, 2007; Ira *et al.*, 2003), DNA/protein complexes (Gasymov *et al.*, 2007; Shaw and Pal, 2007), and probing confined water (Angulo *et al.*, 2006).

Fluorescence anisotropy. Fluorescence anisotropy decay of a molecule in neat solvent (absence of heterogeneity) is five exponential (Fleming, 1987), which may reduce to 1–3 exponential depending on the choice of the fluorescent molecule. It is natural to choose a molecule that exhibits single-exponential fluorescence anisotropy decay in neat solvents, so that the molecule can be used to probe site heterogeneity. The interpretation will be simpler if there is a prior knowledge about site heterogeneity. For example, if the molecule is present in one site only and wobbles freely (as if in a neat liquid) then the anisotropy decay is single exponential. The decay is double exponential if the wobbling is restricted (e.g., wobbling-in-cone model; Lakowicz, 1999). If the molecule is distributed in two sites then the molecular dynamics will be different in the two sites. Therefore, site-dependent anisotropy decay equation has to be associated with the site-dependent fluorescence intensity decay appropriate for that site. So far, this has not been tackled satisfactorily. The quality of anisotropy decay data by TCSPC or any other technique ought to be adequate to handle a complicated equation with several free parameters. Some approaches described in literature for separating fluorescence from different sites are useful (Davenport *et al.*, 1986; Krishna and Periasamy, 1997) and more are yet to be developed. However, analysis of fluorescence anisotropy decays assuming that only one site is populated in a potentially multisite, heterogeneous sample is highly speculative and the conclusions derived there from are questionable.

Solvation dynamics. Interpretation of time-dependent fluorescence emission spectral shift with solvation dynamics in a potentially multisite, heterogeneous sample presents the same challenge as the fluorescence anisotropy described above. That is, solvation dynamics and fluorescence intensity decay are site dependent. Multicomponent model (Vincent *et al.*, 2005) and a two-site model (Sachl *et al.*, 2008) have been used to separate fluorescence spectra and solvation dynamics in different sites. The studies which assume two sites would benefit if TRANES analysis is also included to confirm two sites by an isoemissive point. Quite often, a single site is assumed implicitly in complex media (Bhattacharya and Bagchi, 2000; Hof, 1999; Mitra *et al.*, 2008; and the references cited therein). TRANES analysis is required in these cases to confirm that site heterogeneity (i.e. isoemissive point) is absent.

ACKNOWLEDGMENT

The author wishes to acknowledge the contribution of Dr. A. S. R. Koti, a coauthor in the original publications.

REFERENCES

Angulo, G., Organero, J. A., Carranza, M. A., and Douhal, A. (2006). Probing the behavior of confined water by proton-transfer reactions. *J. Phys. Chem.* **110**, 24231–24237.

Bhattacharya, K., and Bagchi, B. (2000). Slow dynamics of constrained water in complex geometries. *J. Phys. Chem.* **104**, 10603–10613.

Choudhury, S. D., Kumbhakar, M., Nath, S., and Pal, H. (2007). Photoinduced bimolecular electron transfer kinetics in small unilamellar vesicles. *J. Chem. Phys.* **127**, 194901.

Dahl, K., Biswas, R., Ito, N., and Maroncelli, M. (2005). Solvent dependence of the spectra and kinetics of excited-state charge transfer in three (alkylamino)benzonitriles. *J. Phys. Chem. B* **109**, 1563–1585.

Davenport, L., Knutson, J. R., and Brand, L. (1986). Anisotropy decay associated fluorescence-spectra and analysis of rotational heterogeneity. 2. 1,6-Diphenyl-1,3,5-hexatriene in lipid bilayers. *Biochemistry* **25**, 1811–1816.

Deniz, A. A., Mukhopadhyay, S., and Lemke, E. A. (2008). Single-molecule biophysics: At the interface of biology, physics and chemistry. *J. R. Soc. Interface* **5**, 15–45.

Easter, J. H., DeToma, R. P., and Brand, L. (1976). Nanosecond time-resolved emission spectroscopy of a fluorescence probe adsorbed to l-alpha-egg lecithin vesicles. *Biophys. J.* **16**, 571–583.

Fleming, G. (1987). "Chemical Applications of Ultra-Fast Spectroscopy." Oxford University Press, London.

Gasymov, M. K., Abduragimov, A. R., and Glasgow, B. J. (2007). Characterization of fluorescence of ANS-tear lipocalin complex: Evidence for multiple-binding modes. *Photochem. Photobiol.* **83**, 1405–1414.

Grinvald, A., and Steinberg, I. Z. (1974). Analysis of fluorescence decay kinetics by method of least-squares. *Anal. Biochem.* **59**, 583–598.

Hof, M. (1999). *In* "Applied Fluorescence in Chemistry, Biology and Medicine," (W. Rettig, et al., eds.). Springer, Berlin (Chapter 18).

Ira, Koti, A. S. R., Krishnamoorthy, G., and Periasamy, N. (2003). TRANES spectra of fluorescence probes in lipid membranes: An assessment of population heterogeneity and dynamics. *J. Fluorescence* **13**, 95–103.

Koti, A. S. R., and Periasamy, N. (2001a). TRANES analysis of the fluorescence of nile red in organized molecular assemblies confirms emission from two species. *Proc. Indian Acad. Sci. (Chem. Sci.)* **113**, 157–163.

Koti, A. S. R., and Periasamy, N. (2001b). Application of time resolved area normalized emission spectroscopy to multicomponent systems. *J. Chem. Phys.* **115**, 7094–7099.

Koti, A. S. R., and Periasamy, N. (2002). Time resolved area normalized emission spectroscopy (TRANES) of DMABN confirms emission from two states. *Res. Chem. Intermediates* **28**, 831–836.

Koti, A. S. R., Krishna, M. M. G., and Periasamy, N. (2001). Time resolved area-normalized emission spectroscopy (TRANES): A novel method for confirming emission from two excited states. *J. Phys. Chem. A* **105**, 1767–1771.

Krishna, M. M. G., and Periasamy, N. (1997). Spectrally constrained global analysis of fluorescence decays in biomembrane systems. *Anal. Biochem.* **253**, 1–7.

Krishna, M. M. G., and Periasamy, N. (1998a). Fluorescence of organic dyes in lipid membranes: Site of solubilization and effects of viscosity and refractive index on lifetimes. *J. Fluorescence* **8**, 81–91.

Krishna, M. M. G., and Periasamy, N. (1998b). Orientational distribution of linear dye molecules in bilayer membranes. *Chem. Phys. Lett.* **298**, 359–367.

Kulzer, F., and Omit, M. (2004). Single-molecule optics. *Annu. Rev. Phys. Chem.* **55**, 585–611.

Laguitton-Pasquier, H., Pansu, R., Chauvet, J.-P., Pernot, P., Collet, A., and Faure, J. (1997). 10,10′-Bis(2-ethylhexyl)-9,9′-bianthryl (BOA) molecule: The first free aromatic probe for the core of micelles. *Langmuir* **13**, 1907–1917.

Lakowicz, J. R. (1999). "Principles of Fluorescence Spectroscopy." Plenum Press, New York(Chapter 7).

Laws, W. R., and Brand, L. (1979). Analysis of 2-state excited-state reactions—Fluorescence decay of 2-naphthol. *J. Phys. Chem.* **83**, 795–802.

Livesey, A. K., and Brochon, J. C. (1987). Analyzing the distribution of decay constants in pulse-fluorometry using the maximum entropy method. *Biophys. J.* **52**, 693–706.

Maciejewski, A., Kubicki, J., and Dobek, K. (2003). The origin of time-resolved emission spectra (TRES) changes of 4-aminophthalimide (4-AP) in SDS micelles. The role of the hydrogen bond between 4-AP and water present in micelles. *J. Phys. Chem. B* **107**, 13986–13999.

Maciejewski, A., Kubicki, J., and Dobek, K. (2005). Shape and position of 4-aminophthalimide (4-AP) time-resolved emission spectra (TRES) versus sodium dodecyl sulfate SDS concentration in micellar solutions: The partitioning of 4-AP in the micellar phase and in water surrounding the micelles. *J. Phys. Chem. B* **109**, 9422–9431.

Mitra, R. K., Sinha, S. S., and Pal, S. K. (2008). Temperature-dependent solvation dynamics of water in sodium bis(2-ethylhexyl)sulfosuccinate/isooctane reverse micelles. *Langmuir* **24**, 49–56.

Mukherjee, T. K., and Datta, A. (2006). Regulation of the extent and dynamics of excited-state proton transfer in 2-(2′-pyridyl)benzimidazole in Nafion membranes by cation exchange. *J. Phys. Chem.* **110**, 2611–2617.

Mukherjee, T. K., Panda, D., and Datta, A. (2005a). Excited-state proton transfer of 2-(2′-pyridyl)benzimidazole in microemulsions: Selective enhancement and slow dynamics in aerosol OT reverse micelles with an aqueous core. *J. Phys. Chem.* **109**, 18895–18901.

Mukherjee, T. K., Ahuja, P., Koner, A. L., and Datta, A. (2005b). ESPT of 2-(2′-pyridyl) benzimidazole at the micelle–water interface: Selective enhancement and slow dynamics with sodium dodecyl sulfate. *J. Phys. Chem.* **109**, 12567–12573.

Novaira, M., Biasutti, M. A., Silber, J. J., and Correa, N. M. (2007). New insights on the photophysical behavior of PRODAN in anionic and cationic reverse micelles: From which state or states does it emit? *J. Phys. Chem.* **111**, 748–759.

O'Connor, D. V., and Phillips, D. (1984). "Time Correlated Single Photon Counting". Academic Press, London.

Pansu, R. B., and Yoshihara, K. (1991). Diffusion of excited bianthryl in microheterogeneous media. *J. Phys. Chem.* **95**, 10123–10133.

Rosspeintner, A., Angulo, G., Weiglhofer, M., Landgraf, S., and Grampp, G. (2006). Photophysical properties of 2,6-dicyano-N,N,N',N'-tetramethyl-p-phenylenediamine. *J. Photochem. Photobiol. A: Chem.* **183**, 225–235.

Sachl, R., Stepanek, M., Procheka, K., Humpolickova, J., and Hof, M. (2008). Fluorescence study of the solvation of fluorescent probes prodan and laurdan in poly(ε-caprolactone)-*block*-poly(ethylene oxide) vesicles in aqueous solutions with tetrahydrofurane. *Langmuir* **24**, 288–295.

Sarkar, R., Ghosh, M., and Pal, S. K. (2005). Ultrafast relaxation dynamics of a biologically relevant probe dansyl at the micellar surface. *J. Photochem. Photobiol. B Biology* **78**, 93–98.

Shaw, A. K., and Pal, S. K. (2007). Fluorescence relaxation dynamics of acridine orange in nanosized micellar systems and DNA. *J. Phys. Chem.* **111**, 4189–4199.

Siano, D. B., and Metzler, D. E. (1969). Band shapes of electronic spectra of complex molecules. *J. Chem. Phys.* **51**, 1856–1861.

Toptygin, D., Svobodova, J., Konopasek, I., and Brand, L. (1992). Fluorescence decay and depolarization in membranes. *J. Chem. Phys.* **96**, 7919–7930.

Vincent, M., de Foresta, B., and Gallay, J. (2005). Nanosecond dynamics of a mimicked membrane–water interface observed by time-resolved stokes shift of LAURDAN. *Biophys. J.* **88**, 4337–4350.

Wazawa, T., Ishii, Y., Funatsu, T., and Yanagida, T. (2000). Spectral fluctuation of a single fluorophore conjugated to a protein molecule. *Biophys. J.* **78**, 1561–1569.

CHAPTER THREE

Multiparametric Probing of Microenvironment with Solvatochromic Fluorescent Dyes

Andrey S. Klymchenko[*,†] *and* Alexander P. Demchenko[†]

Contents

1. "Universal" and "Specific" Noncovalent Interactions	39
2. The Methodology of Multiparametric Approach with Application of 3-Hydroxyflavone Dyes	40
3. Correlations of Spectroscopic Data with Solvatochromic Variables	43
3.1. Position of absorption band	43
3.2. Positions of N* and T* fluorescence bands	44
3.3. Stokes shift of the N* and T* bands	44
3.4. Ratio of intensities of the N* and T* bands (I_{N*}/I_{T*})	45
4. Algorithm for Multiparametric Probing Based on Parameters of Absorption and Dual Emission	47
5. Application of Multiparametric Probing	49
5.1. Probing binary solvent mixtures	49
5.2. Probing protein-binding sites	50
6. Limitations of the Multiparametric Approach	51
6.1. H-bonding heterogeneity within the sites of the same polarity	51
6.2. Site heterogeneity: Lipid bilayers and protein-labeling sites	52
6.3. Applicability of multiparametric approach to other solvatochromic dyes	53
7. Concluding Remarks	55
References	56

Abstract

We describe new methodology for multiparametric probing of weak non-covalent interactions in the medium based on response of environment-sensitive fluorescent dyes. The commonly used approach is based on correlation of one spectroscopic parameter (e.g. wavelength shift) with environment polarity, which

---

[*] Lab. de Pharmacologie et Physicochimie, UMR 7034 du CNRS, Faculté de Pharmacie, Université Louis Pasteur, 67401 Illkirch, France
[†] A.V. Palladin Institute of Biochemistry, Kiev 252030, Ukraine

describes a superposition of universal and specific (such as hydrogen bonding) interactions. In our approach, by using several independent spectroscopic parameters of a dye, we monitor simultaneously each individual type of the interactions. For deriving these extra parameters the selected dye should exist in several excited and/or ground states. In the present work, we applied 4′-(diethylamino)-3-hydroxyflavone, which undergoes the excited-state intramolecular proton transfer (ESIPT) and thus exhibits an additional emission band belonging to an ESIPT product (tautomer) form of the dye. The spectroscopic characteristics of the excited normal and the tautomer states as well as of the ESIPT reaction of the dye are differently sensitive to the different types of interactions with microenvironment and therefore can be used for its multiparametric description. The new methodology allowed us to monitor simultaneously three fundamental physicochemical parameters of probe microenvironment: polarity, electronic polarizability and H-bond donor ability. The applications of this approach to binary solvent mixtures, reverse micelles, lipid bilayers and binding sites of proteins are presented and the limitations of this approach are discussed. We believe that the methodology of multiparametric probing will extend the capabilities of fluorescent probes as the tools in biomolecular and cellular research.

The present fluorescence probe methodology in molecular and cellular research that originated about 50 years ago is focused on single-parametric probing. The probes are so designed or selected that the sensitivity to one parameter thought to sense the most important property is reinforced, whereas the sensitivity toward other properties is intentionally diminished (or even ignored). This is true for "site-sensitive" or "polarity-sensitive" fluorescence probes that characterize intermolecular interactions with one parameter, the position of fluorescence band maximum, λ_F^{max} or, more rarely, fluorescence excitation band maximum λ_{Ex}^{max}. The λ_F^{max} data are calibrated against an empirical "polarity scale" that is constructed based on the spectra obtained in series of model solvents (Reichardt, 1994). It should not be forgotten however that the binding sites in macromolecular systems present a real complexity, and a variety of noncovalent interactions can influence the spectra. The empirical "polarity" has to incorporate a number of physically defined variables, such as electronic and nuclear polarizabilities, as well as hydrogen (H–) bond donor and acceptor potentials. A variety of other interactions can be found in specific cases—the charge-transfer complexes and the complexes formed in the excited states—excimers and exciplexes (Mataga and Kubata, 1970; Suppan and Ghoneim, 1997). The scaling in some of these parameters, in principle, can be provided by selection of proper series of model solvents (Catalan, 1997), but characterization of unknown media or binding sites of biological macromolecules with this approach is ambiguous. Additional problems with this approach are caused by absent or incomplete structural equilibrium of excited fluorophores with their binding sites, their structural anisotropy and a variety of excited-state reactions. Therefore the primary spectroscopic data that can be obtained from fluorescence probes in an ordinary way by observing the shift of one emission band is not sufficient to resolve these interactions and their dynamics. To obtain several spectroscopic parameters and to provide their direct connection with several types of intermolecular interactions, the probe molecule has to be present in several ground and excited states. Then the transitions between these states have to be represented by separate absorption/emission bands that can be differently sensitive to different types of perturbations.

Common fluorescence probes cannot operate in this manner. Their application is not sufficient and often misleading in the cases when the probe spectral response is analyzed in complex systems, such as solvent mixtures, surfactant micelles, biomembranes, proteins, etc., where a number of parameters of the probe microenvironment, especially the H-bonding ability, can be involved in interactions producing spectroscopic effect. Thus, to provide simultaneous description of several different physicochemical parameters, more advanced fluorescence probes exhibiting a more extended multiparametric response should be found and applied.

1. "Universal" and "Specific" Noncovalent Interactions

In finding the connection between spectroscopic data and parameters characterizing intermolecular interactions we can start from the use of the theory of liquid dielectrics (Bakhshiev, 1972; Lippert, 1975; Mataga and Kubata, 1970; Suppan and Ghoneim, 1997). In this theory the universal solvent–solute interactions are composed of those that are determined by electronic and nuclear solvent polarizabilities. In the simplest approximation (Bakhshiev, 1972; Lippert, 1975) they can be described by the functions of refractive index n, $f(n) = (n^2 - 1)/(2n^2 + 1)$, and low-frequency dielectric constant ε, $f(\varepsilon) = (\varepsilon - 1)/(2\varepsilon + 1)$. This theory predicts that the position of emission maximum on a wavenumber scale, ν_F^{max}, should be a linear function of $f(\varepsilon)$, and the Stokes shift (the shift between absorption and emission band maxima, $\nu_{abs}^{max} - \nu_F^{max}$) should show linear dependence on the so-called Lippert parameter, $L = f(\varepsilon) - f(n)$. This allows incorporating into analysis already two variables, ν_{abs}^{max} and ν_F^{max} that allow evaluation of two fundamental parameters of the environments, $f(\varepsilon)$ and $f(n)$.

However, specific interactions, which are not accounted in this model, can provide systematic deviations from these linear relationships. The H-bonding is the most important specific interaction influencing dramatically absorption and emission spectra, and its effects are different dependent on whether the solvent molecules can serve the donors or acceptors in H-bonding with the fluorescent probe. The problem is that the common solvatochromic fluorescent dyes that exhibit broad site-dependent variations of v_{abs}^{max} and v_F^{max} contain electron-donor and electron-acceptor groups that could provide the strongly increased excited-state dipole moment. These groups can participate in intermolecular H-bonding with strong effects on the absorption and emission spectra. Thus, for a popular probe PRODAN (6-propionyl-2-dimethylaminonaphthalene) nearly half of the total solvent-dependent shift in emission is due to intermolecular H-bonding of its carbonyl group with an H-bond donor solvent, and this effect is indistinguishable from nonspecific effects of solvent polarity (Catalan et al., 1991). Thus, distinguishing specific (H-bonding) from nonspecific (universal) interactions remains a crucial problem in the fluorescence probing of the microenvironment.

A quantitative description of these effects can be provided based on such solvent parameters as H-bond acceptor ability (basicity), α, and H-bond donor ability (acidity), β introduced by Taft, Kamlet, and Abraham (Abraham, 1993; Kamlet et al., 1993; Abraham et al., 1994). Since all specific interactions provide the influence on electronic spectra that differ in sign and magnitude, their involvement cannot be evaluated with certainty without sufficient amount of spectroscopic information. In order to be sensitive to different H-bonding properties of solvent, a fluorescent probe has to contain functional groups that are able to form the H-bonds with the solvent molecules and provide modulation of spectra by the binding event. In 3-hydroxychromone (3HC) and 3-hydroxyflavone (3HF) derivatives (Scheme 3.1) there are two such groups. The 3-hydroxy group can form intermolecular bonds with H-bond proton acceptors and report on proton-acceptor basicity of environment, while the 4-carbonyl can form H-bonds with proton-donor groups and report on proton donor acidity. In these derivatives the donor and acceptor groups are coupled by intramolecular H-bond. The ability of formation of intermolecular bond by 4-carbonyl can be eliminated by proper chemical substitution, which does not influence the intramolecular bond (Klymchenko et al., 2003b).

2. THE METHODOLOGY OF MULTIPARAMETRIC APPROACH WITH APPLICATION OF 3-HYDROXYFLAVONE DYES

The exploration of shifts in absorption (or excitation) spectra and construction of Lippert plots is only a partial solution of the problem of search for "specific" interactions, since from deviations on these plots one can deduce the presence of specific interactions but cannot characterize them. Therefore, we need additional spectroscopic parameters that could be derived from additional electronic transitions in the probe between the states that interact differently with the environment. These new transitions

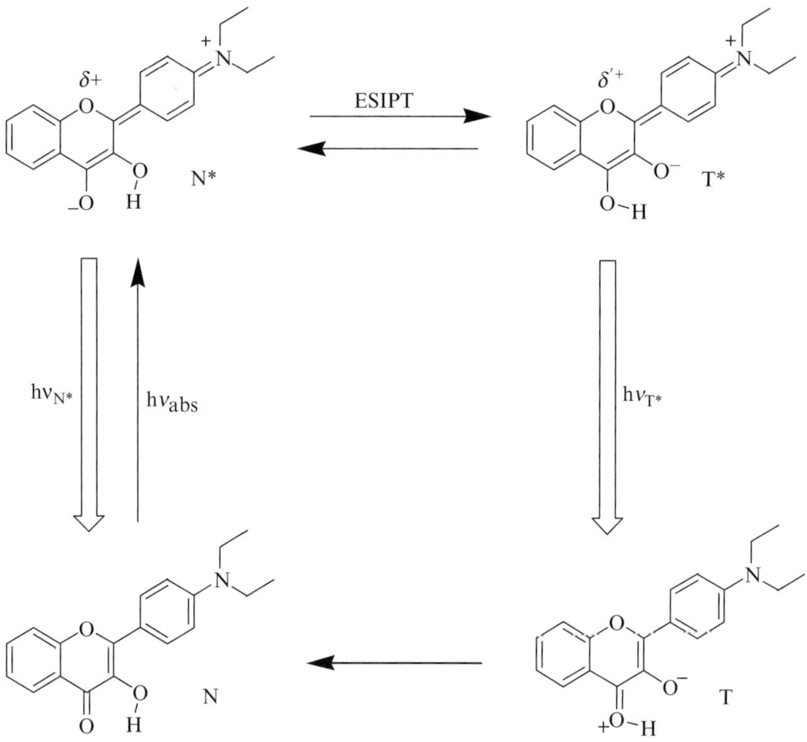

Scheme 3.1 Four-level model of the ESIPT transformation of 3-hydroxyflavone dye FE. Inequilibrium Franck–Condon states are not shown. Interplay between normal (N*) and tautomer (T*) forms is observed in a variety of conditions and can be expressed as a ratio of relative intensities of correspondent bands.

can be obtained due to excited-state reaction that produces a new fluorophore with the appearance of additional band in emission. The other possibility is to use a ground-state reaction, the light-absorbing and fluorescent product of which can provide new excitation and new fluorescence bands. Among intramolecular reactions in excited states are isomerizations, deprotonation, intramolecular charge transfer (ICT) and excited-state intramolecular proton transfer (ESIPT). The latter is probably the most interesting because it features a four-level diagram of electronic states, and the electronic transitions to and from these states can be highly different by energy (Formosinho and Arnaut, 1993). The presence of two emissive states connected by excited-state reaction allows obtaining additional parameters: the position of the second emission band and the ratio of intensities between the two bands. Together with the positions of excitation spectrum and of the first (initially excited) fluorescence band they can form a set of parameters that can characterize simultaneously different types of intermolecular interactions with the environment. This will happen if all these parameters are orthogonal, that is, respond differently to variations of different types of these interactions.

In this respect, 3-hydroxychromone (3-HC) dyes, such as 4′-(dialkylamino)-3-hydroxyflavones (3-HFs) and also 2-furyl-, 2-benzofuryl-, and 2-thiophenyl-3-hydroxychromone analogues, are prospective candidates for multiparametric fluorescence probes. The unique property of 3-HCs is the ability to exhibit two intensive well separated bands in the fluorescence spectra that respond differently to variation of solvent properties (Chou et al., 1993; Kasha, 1986; Swiney and Kelley, 1993). One of the emission bands belongs to the normal excited state, N*, and the other to the ESIPT reaction product tautomer, T* (Scheme 3.1). The sequence of events that results in two emission bands is the following. Absorption of light quantum $h\nu_{abs}$ generates the normal Franck–Condon state N^{F-C}, which then relaxes to N* state, possessing ICT character (Chou et al., 2005; Yesylevskyy et al., 2005), followed by relaxation of the solvent shell. Then the N* state can emit the light quantum $h\nu_{N^\star}$ or exhibit ESIPT reaction—the transition to the T* state. The emission of the light quantum $h\nu_{T^\star}$ results in populating the solvent-unrelaxed ground T state with subsequent relaxation and, occurring in the dark, back proton transfer to the ground N state, which closes the four-state cycle.

It is important that the four states, N, T, N*, and T*, possess different distribution of charges and interact differently with the environment. The strongest dipole moment is in the N* state, the formation of which is connected with electronic charge transfer from 4′-dialkylamino group to the chromone moiety (Chou et al., 2005; Yesylevskyy et al., 2005). Therefore, it shows significant solvatochromy (Chou et al., 1993; Swiney and Kelley, 1993). Meanwhile, the separation of charges in the T* state due to the proton transfer is significantly smaller, which is in line with the reported poor solvatochromy of the T* band (Chou et al., 1993). Due to this significant difference, their band positions can be considered as independent variables.

An important parameter is the ratio of intensities of the N* and T* bands in emission, $I_{N^\star}/I_{T^\star}$, which is a very sensitive indicator of solvent polarity (Ercelen et al., 2002; Klymchenko et al., 2003c). Thus, this system allows simultaneous determination of a set of four parameters, ν_{abs}, $\nu_{N^\star}$, $\nu_{T^\star}$, and $I_{N^\star}/I_{T^\star}$, which can provide different characteristics of the physical properties of microenvironment.

4′-(Dialkylamino)-3-hydroxyflavone derivatives are becoming very popular as fluorescence probes. They were already applied as fluorescence indicators of polar impurities in solvents (Liu et al., 1999) and water content in reverse micelles (Klymchenko and Demchenko, 2002a), as ion sensors (Roshal et al., 1998), electrochromic dyes (Klymchenko and Demchenko, 2002b) and membrane probes, which being fixed at particular locations of the bilayer respond to different properties of model and cellular membranes

(Klymchenko et al., 2002, 2003a, 2006; Shynkar et al., 2005, 2007). These applications are mainly due to extreme sensitivity to the environment of I_{N^*}/I_{T^*} ratio, which enables simple two-band ratiometric analysis. Positions of the spectral bands play an additional important role in specifying intermolecular interactions.

Recently we made an attempt to develop a more general algorithm for providing multiparametric correlations between spectroscopic parameters and variables that characterize the environment (Klymchenko and Demchenko, 2003). It is based on the found linear correlations of spectroscopic properties of dye 4'-(diethylamino)-3-hydroxyflavone (FE) with physicochemical parameters characterizing the properties of a selected set of model solvents with different abilities to universal and specific interactions (Klymchenko and Demchenko, 2003). This approach can be applied to different dyes with similar properties.

3. Correlations of Spectroscopic Data with Solvatochromic Variables

Choosing FE as a model multiparametric probe we studied the correlation of experimental spectroscopic parameters with four solvatochromic variables: solvent polarity function $f(\varepsilon)$, electronic polarizability $f(n)$, H-bond donor ability (acidity), α, and H-bond acceptor ability (basicity), β (Klymchenko and Demchenko, 2003). For providing maximal variability in chemical structure and physical properties, 21 organic solvents were chosen with broad variation of polarity, electronic polarizability and the ability for H-bonding. For convenience of data presentation they were classified into the following groups: protic, H-bond acceptors, and neutral. Chloroform and dichloromethane, the neutral solvents with significant H-bond donor ability, were considered separately from those groups. The following correlations between spectroscopic and solvent properties were found.

3.1. Position of absorption band

Within the studied range of solvents v_{abs} exhibits a broad variation, from 24,900 to 23,600 cm^{-1} (400–423 nm). For neutral solvents an excellent correlation for v_{abs} with $f(n)$ is observed. The found correlation equation is: $v_{abs} = 29,166 - 20,901 f(n)$ (correlation coefficient $r = -0.999$, standard deviation SD $= 28$). The presence of this correlation indicates the key role of solvent electronic polarizability in determining the solvent-dependent shifts of absorption spectra. Such correlation is common for chromophores, which are low polar in their ground states and become polar in the Franck–Condon excited states (Mataga and Kubata, 1970; Suppan and Ghoneim, 1997). For the solvents with H-bond acceptor properties we observe a systematic deviation of v_{abs} to the red, which signifies the reduction of energy of electronic transition, although the dependence on $f(n)$ is still present. A much stronger deviation from the linear function to the red is provided by protic solvents. According to our studies it is connected with the formation of the ground state complex of a protic solvent with 4-carbonyl group of the dye (Klymchenko et al., 2003b). This deviation is also observed in the neutral solvents dichloromethane and chloroform, which are known to show a considerable H-bond acidity. These results allow concluding that the positions of absorption maxima can serve as useful parameters for evaluating $f(n)$ function, but only for the case of neutral solvents.

3.2. Positions of N* and T* fluorescence bands

In line with the results of previous studies (Chou et al., 1993; Swiney and Kelley, 1993) the N★ band position exhibits strong solvent-dependent variation in the range 22,000–19,000 cm^{-1} (455–526 nm) in the studied set of solvents. In contrast, the variations in positions of T★ band are not significant and not systematic. They are observed between 18,000 and 17,000 cm^{-1} (556–588 nm). The solvent dependence of the N★ band position exhibits a strong correlation with dielectric constant ε and fits the linear function of $f(\varepsilon)$ (the correlation equation for aprotic solvents: $v_{N\star} = 24{,}890 - 10{,}850 f(\varepsilon)$, $r = -0.956$, $SD = 240$; $v_{T\star} = 18{,}120 - 1716 f(\varepsilon)$, $r = -0.717$, $SD = 175$). Small systematic deviations to the red are observed only in protic solvents. The position of T★ band maximum does not show a strong correlation with either $f(\varepsilon)$ or $f(n)$.

3.3. Stokes shift of the N* and T* bands

The Stokes shift of the N★ band maximum demonstrates a strong linear correlation with the Lippert solvent parameter L, $L = f(\varepsilon) - f(n)$ (Fig. 3.1A). Surprisingly, this correlation holds for all studied solvents, including protic solvents. In our case the parameter L being derived on the basis of simplified continuous dielectric solvent model (Bakhshiev, 1972; Lippert, 1975) holds better than the empirical solvent polarity scales (Reichardt, 1994). This means that independently of the type of microenvironment, the parameter L can be evaluated from the values of Stokes shifts. It is interesting to note that the Stokes shift for the T★ band does not show any correlation with that of the N★ band (Fig. 3.3B), and therefore with the Lippert parameter L. Meantime, it appears to be the smallest for protic solvents and the largest for the highly polar solvents with H-bond accepting ability. A more systematic study is needed to consider if the quantitative measure of solvent basicity can be derived from this effect.

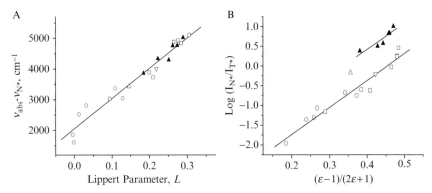

Figure 3.1 Correlations between spectroscopic parameters of probe FE and physicochemical variables in a series of organic solvents. (A) The Stokes shift of the N★ band versus Lippert parameter L. Correlation equation: Stokes shift $(cm^{-1}) = 2047 + 9864L$, $r = 0.976$, $SD = 234$. (B) $\log(I_{N*}/I_{T*})$ versus polarity $f(\varepsilon)$. The correlation equation for aprotic solvents: $\log(I_{N*}/I_{T*}) = -3.111 + 6.849f(\varepsilon)$, $r = 0.976$, $SD = 0.159$; for protic solvents: $\log(I_{N*}/I_{T*}) = -2.245 + 6.653f(\varepsilon)$, $r = 0.909$, $SD = 0.111$. Solvents were neutral (○—hexane, toluene, carbon disulfide, thiophene, anisole, and bromobenzene), protic (▲—*tert*-pentanol, *n*-octanol, *tert*-butanol, *n*-butanol, isopropanol, and ethanol), and H-bond acceptor (□—di-*n*-butyl ether, ethyl acetate, tetrahydrofuran, acetone, *N*,*N*-dimethylformamide, acetonitrile, and dimethyl sulfoxide), chloroform (△) and dichloromethane (▽).

3.4. Ratio of intensities of the N* and T* bands (I_{N*}/I_{T*})

The most dramatic solvent-dependent variations are observed in the I_{N*}/I_{T*} ratios, up to almost complete disappearance of one of the bands. We observe strong correlation of I_{N*}/I_{T*} with $f(\varepsilon)$: the increase of solvent polarity increases dramatically the relative contribution of the N★ band. The function $\log(I_{N*}/I_{T*})$ versus $f(\varepsilon)$ consists of two linear segments, one of which is formed by all aprotic including H-bond acceptor solvents, and the other—by protic solvents (Fig. 3.1B). Interestingly, chloroform, the solvent with considerable H-bond donor ability fits closely to the function of protic solvents. The obtained strong correlations demonstrate that to a good approximation there are two linear functions that can be applied for estimation of $f(\varepsilon)$, separately for aprotic and protic solvents. This shows the ability of these functions based on linear fits to separate the polarity and H-bonding effects. The separation between these functions on the scale of relative band intensities is very significant, ca. 0.8 on the $\log(I_{N*}/I_{T*})$ scale (which is the factor of 6.3 on the I_{N*}/I_{T*} scale), while the standard deviations for these fits are 0.159 and 0.110 for aprotic and protic solvents, respectively (see Fig. 3.1B). The behavior of H-bond acceptor solvents does not differ from that of neutral solvents. Thus the measurement of $\log(I_{N*}/I_{T*})$ allows estimation of $f(\varepsilon)$ independently from H-bond acceptor ability (basicity β) of the analyzed microenvironment.

It is well established that in the N★ state 3-HF possesses much higher dipole moment than the N ground state (Kasha, 1986). Stabilization of the N★ state is achieved both by electronic polarization and by the dipole–dipole relaxations in solute–solvent interactions. Based on the Onsager model of liquid dielectrics, the position of the normal N★ emission band is predicted to change linearly with $f(\varepsilon)$, which is in line with our data. In contrast to the N★ band, the correlation of T★ band position with solvent polarity function $f(\varepsilon)$ is very poor, which is in accordance with the low dipole moment in the T★ state.

Recently we demonstrated that the dynamic equilibrium between N★ and T★ states in the studied flavone dyes is established at the earliest steps of fluorescence emission (Shynkar et al., 2003). Therefore, the relative integral emission intensities from these states should follow a linear function of their relative populations at equilibrium. According to the Boltzmann distribution the latter is an exponential function of the free energy difference between the N★ and T★ states. This energy difference can be approximated by the difference in energies of emitted quanta from these states expressed as the difference in positions of band maxima $(\nu_{N\star} - \nu_{T\star})$ + const. If it is so, then $\log(I_{N\star}/I_{T\star})$ should be a linear function of $\nu_{N\star} - \nu_{T\star}$. Our spectroscopic data show that the linear correlation between these two parameters is observed and holds for all types of solvents (Fig. 3.2A) (Klymchenko and Demchenko, 2003). The observed linear correlation between $\log(I_{N\star}/I_{T\star})$ versus $\nu_{N\star} - \nu_{T\star}$ provides an intrinsic criterion of achievement of equilibrium in ESIPT reaction between N★ and T★ species: in the cases when the data point is out of the correlation function, we can conclude that the system did not reach the ESIPT equilibrium.

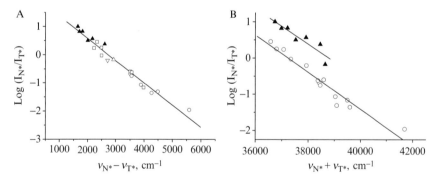

Figure 3.2 Correlations between $\log(I_{N\star}/I_{T\star})$ and functions of spectral positions of N★ and T★ bands for probe FE in a series of organic solvents. (A) $\log(I_{N\star}/I_{T\star})$ versus band separation $\nu_{N\star} - \nu_{T\star}$. Correlation equation $\log(I_{N\star}/I_{T\star}) = 2.149 - 7.879 \times 10^{-4}(\nu_{N\star} - \nu_{T\star})$, $r = -0.983$, SD $= 0.156$. (B) $\log(I_{N\star}/I_{T\star})$ versus $\nu_{N\star} + \nu_{T\star}$ for protic (σ) and all the other ($\bigcirc$) solvents. Correlation equation for aprotic solvents: $\log(I_{N\star}/I_{T\star}) = 19.06 - 5.12 \times 10^{-4}(\nu_{N\star} + \nu_{T\star})$, $r = -0.971$, SD $= 0.17$; for protic solvents: $\log(I_{N\star}/I_{T\star}) = 18.94 - 4.88 \times 10^{-4}(\nu_{N\star} + \nu_{T\star})$, $r = -0.916$, SD $= 0.17$. Symbols of the points correspond to that in Fig. 3.1.

4. ALGORITHM FOR MULTIPARAMETRIC PROBING BASED ON PARAMETERS OF ABSORPTION AND DUAL EMISSION

Our results show that there are five valuable spectral parameters, v_{abs}, $v_{N^\star}$, $v_{T^\star}$, $v_{abs} - v_{N^\star}$, and $\log(I_{N^\star}/I_{T^\star})$, that can describe different properties of microenvironment expressed by three solvatochromic variables: polarity $f(\varepsilon)$, electronic polarizability $f(n)$ and hydrogen-bonding acidity α. Since four of these parameters are obtained from the independent measurements and three of them are mechanistically independent, this may provide sufficient matrix of data to evaluate the latter three variables. The most efficient algorithm for their determination is the following:

1. Primarily we determine the Stokes shift, $v_{abs} - v_{N^\star}$, of the N* band, from which using the linear function of Fig. 3.1A we derive the Lippert parameter L. In our case the latter is independent or low sensitive to specific solvent–solute interactions. This may be due to the fact that intermolecular H-bonding is not formed or disrupted in the excited state and it contributes equally to the shifts of absorption and emission bands (Shynkar et al., 2004b).
2. Parameter $\log(I_{N^\star}/I_{T^\star})$ is evidently the most sensitive and convenient for estimation of polarity $f(\varepsilon)$. Since it can operate as polarity measure only in the case of equilibrium in ESIPT reaction, the correlation equation $\log(I_{N^\star}/I_{T^\star})$ versus $v_{N^\star} - v_{T^\star}$ can be considered as a general criterion determining the validity of this approach. Therefore, on the next step we specify if the measured parameters fit the dependence $\log(I_{N^\star}/I_{T^\star})$ versus $v_{N^\star} - v_{T^\star}$ presented in Fig. 3.2A. Thus, if the fluorescence spectrum is such that the corresponding point fits to this function, it is then possible to estimate $f(\varepsilon)$ using $\log(I_{N^\star}/I_{T^\star})$ values. At present the cases of deviation form this fit cannot be analyzed quantitatively, however some qualitative conclusions can be made. Thus, if the deviation of particular point exceeds significantly the standard deviation of the fit (SD = 0.156) and locates above the fitting straight line, this may be an indication of a very specific interaction (beyond common H-bonding). It hampers ESIPT reaction probably by breaking intramolecular H-bond, which is the pathway of ESIPT reaction. If the corresponding point deviates from this fit downward, this can be considered as the case of "abnormally high" ESIPT. Practically, this condition can be observed if the N* and T* emission bands originate from the dye molecules located at two or more different environments, as it will be illustrated below.
3. Before the estimation of $f(\varepsilon)$ from the $\log(I_{N^\star}/I_{T^\star})$ value, it has to be determined if the probe microenvironment is H-bond donor or not. The H-bond donor ability, α, in this algorithm is characterized by all-or-none

criteria: this bonding with protic environment either exists or is totally absent. We can use several of these criteria:
   a. The cases when $\log(I_{N\star}/I_{T\star})$ exceeds certain limits. For aprotic environments in the high polarity limit, $f(\varepsilon) \to 0.5$, so that $\log(I_{N\star}/I_{T\star})$ cannot be larger than 0.313, that is, $I_{N\star}/I_{T\star}$ cannot be larger than 2.1. Therefore, if $I_{N\star}/I_{T\star} > 2.1$, then the environment is definitely protic. Similarly, the N★ and T★ band separation for aprotic solvents, $v_{N\star} - v_{T\star}$, cannot be smaller than 2300 cm^{-1}. Therefore, if $v_{N\star} - v_{T\star} < 2300$ cm^{-1}, then it is the case of intermolecular H-bonding.
   b. The Stokes shift of the T★ band in protic solvents is the lowest (Fig. 3.3B), namely in the range of 6670–6330 cm^{-1}. Therefore, if $v_{abs} - v_{T\star} < 6670$ cm^{-1}, then there is a high probability for the existence of intermolecular H-bonding.
   c. The N★ band position, $v_{N\star}$, is a linear function of $f(\varepsilon)$, which in protic environment shows a negative deviation (to the red), while the position of the T★ band, $v_{T\star}$, almost does not change with $f(\varepsilon)$ but increases in protic environment (shift to the blue). Therefore, we suggest using the sum $v_{N\star} + v_{T\star}$ as a convenient linear function of $f(\varepsilon)$, which is practically independent on H-bonding. Since $\log(I_{N\star}/I_{T\star})$ is also a linear function of $f(\varepsilon)$, but with the strong deviation for the protic environment, then the dependence $\log(I_{N\star}/I_{T\star})$ versus $v_{N\star} + v_{T\star}$ should also be a linear function with the strong deviation for the case of H-bonding. Indeed, this is in line with our data (Fig. 3.2B) (Klymchenko and Demchenko, 2003): aprotic and protic solvents demonstrate parallel linear segments separated from each other on the $\log(I_{N\star}/I_{T\star})$ scale by up to 0.8, while the standard deviation for both of the fits is much lower, ca. 0.17. Therefore, the conclusion whether the probe environment is protic or aprotic can be made according to the closeness of the point obtained for unknown environment to the corresponding fitting straight line. If this point does not fit to any of the functions and locates between the linear segments in Fig. 3.2B, this should be considered as the case of partial H-bonding, the analysis of which is not provided by the present algorithm.
4. On the next step, after finding out whether the microenvironment is protic or not, the value of $f(\varepsilon)$ can be estimated from $\log(I_{N\star}/I_{T\star})$ using the corresponding linear function of Fig. 3.2B.
5. After obtaining $f(\varepsilon)$ the value of $f(n)$ is evaluated by $f(n) = f(\varepsilon) - L$.

Thus at the output of this algorithm we obtain information on three independent properties of microenvironment: polarity expressed as $f(\varepsilon)$, electronic polarizability expressed as $f(n)$ and the presence/absence of H-bond donor groups in the probe environment. To demonstrate the power of the proposed multiparametric approach the following examples of applications are presented.

5. APPLICATION OF MULTIPARAMETRIC PROBING

5.1. Probing binary solvent mixtures

To test the applicability of our approach we can first analyze the spectra in solvent mixtures with different types of preferential solute–solvent interactions. Toluene is an aprotic solvent with high electronic polarizability ($f(n) = 0.2261$) and low polarity ($f(\varepsilon) = 0.2387$), while acetonitrile is an aprotic solvent with low polarizability ($f(n) = 0.1747$) and high polarity ($f(\varepsilon) = 0.4802$). The study of their mixtures with flavone FE used as multiparametric probe led to interesting results. Since $\log(I_{N^\star}/I_{T^\star})$ shows perfect linear correlation with the band separation for the mixtures of different content and its linear fit is close to that obtained for extended number of solvents, this system meets our general criterion (Figs. 3.2A and 3.3A). This is a signature of the absence of any specific (beyond H-bonding) solvent–solute interactions and, in particular, of those interactions that break intramolecular H-bond. Since this system is aprotic, using the linear fit for aprotic solvents presented in Fig. 3.1B it is possible to estimate the polarity function $f(\varepsilon)$ from the obtained $I_{N^\star}/I_{T^\star}$ ratios. Then the solvent polarizability function $f(n)$ can be found as described above. The obtained values of $f(\varepsilon)$ and $f(n)$ as a function of acetonitrile concentration (v/v) are presented in Fig. 3.3B. According to our results, the variation of $f(\varepsilon)$ is not a linear function of polar solvent concentration, it exhibits a steep initial rising part and saturation. This is typical for the effect of preferential solvation (Suppan and Ghoneim, 1997). Meantime, the polarizability function $f(n)$ does not decrease in the same manner. It demonstrates a more complex dependence on concentration of polar component. At low acetonitrile concentrations $f(n)$ changes steeply but slower than $f(\varepsilon)$. This can be explained by the fact that the probe FE structure is composed of highly polar and low polar interaction sites. Therefore it can be assumed that the solvation shell is not homogeneous, the polar sites are saturated first by the polar component with stronger effect on $f(\varepsilon)$ than on $f(n)$. In the middle of titration curve both functions follow the same trend. At the final step of titration acetonitrile molecules substitute the remaining toluene molecules in low polar sites with stronger influence on $f(n)$ than on $f(\varepsilon)$. Thus, toluene can contribute to its solvation even when the concentration of this solvent in acetonitrile is low. This explanation is reasonable in view of the fact that the flavone dye is a conjugated π-electronic system, which should show increased ability to interact with aromatic toluene molecules. Thus, deconvolution of spectroscopic information into polarity and electronic polarizability functions allows characterizing intermolecular interactions in more detail, it allows revealing the effects that are commonly hidden when the analysis is based on common "polarity scaling" approach.

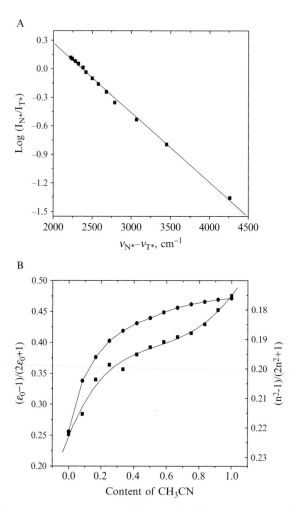

Figure 3.3 Results of spectroscopic studies of probe FE in the mixtures of toluene with acetonitrile. (A) $\log(I_{N\star}/I_{T\star})$ versus band separation $v_{N\star} - v_{T\star}$. Correlation equation $\log(I_{N\star}/I_{T\star}) = 1.735 - 7.319 \times 10^{-4}(v_{N\star} - v_{T\star})$, $r = -0.999$, SD = 0.022. (B) Estimated values of polarity $f(\varepsilon)$ (●) and polarizability $f(n)$ (■) as a function of volume content of acetonitrile in toluene.

5.2. Probing protein-binding sites

On binding to bovine serum albumin (BSA) the probe FE demonstrates the presence of both N★ and T★ bands in fluorescence emission with predominance of the N★ band (Demchenko, 1994). We analyzed the binding site properties of BSA using the proposed algorithm. It was found, that $v_{abs} = 23{,}860$ cm^{-1}, $v_{N\star} = 19{,}980$ cm^{-1}, $v_{T\star} = 17{,}680$ cm^{-1}, and $I_{N\star}/I_{T\star} = 4.33$ ($\log(I_{N\star}/I_{T\star}) = 0.636$). Sequential steps of analysis of these data were the following. First we evaluated $v_{abs} - v_{N^*} = 3880$ cm^{-1}. Then from the linear function presented in Fig. 3.1A we obtained $L = 0.186 \pm 0.023$. Since $I_{N^*}/I_{T^*} > 2.1$ and the recorded $\log(I_{N^*}/I_{T^*})$ value corresponds to that obtained from the correlation equation $\log(I_{N^*}/I_{T^*}) = 18.94 - 4.88 \times 10^{-4}(v_{N^*} + v_{T^*})$ for protic solvents (0.56 ± 0.17; Fig. 3.2B), we concluded that the microenvironment of the probe is saturated by interactions with H-bond donor groups. Then according to the linear fit for protic solvents on Fig. 3.1B we obtained $f(\varepsilon) = 0.433 \pm 0.017$, which corresponds to this function for alcohols of medium-low polarity. On the final step we evaluated the polarizability function. $f(n) = f(\varepsilon) - L$; $f(n) = 0.247 \pm 0.029$. The obtained results demonstrate that the binding site of the flavone probe FE to BSA is protic. The probe interacts either with OH and NH$_2$ groups of the protein or with hydration water. The polarity of the binding site is close to the polarity of relatively low polar protic solvent 1-octanol. It is interesting that electronic polarizability function $f(n)$ is rather high, corresponding to that of aromatic compounds. This suggests that the aromatic residues Trp, Tyr, or Phe participate in formation of the probe-binding site, which finds confirmation in this analysis of three-dimensional structures of serum albumins and their ligand-binding sites (Bhattacharya et al., 2000; Kragh-Hansen et al., 2001).

The same methodology was applied to the binding of 3-hydroxychromone analog of FE, 2-(6-diethylaminobenzo[b]furan-2-yl)-3-hydroxychromone (FA), the probe which was also well characterized in organic solvents (Ercelen et al., 2002). According to titration data, this dye shows extremely high affinity to a sole binding site of BSA ($K_d = 5 \times 10^{-7}$ M) (Ercelen et al., 2003). We found that the local refractive index of this site is $f(n) = 0.24 \pm 0.01$, which demonstrates that the binding site of this dye also contains aromatic amino acid residues.

Thus, the binding site of probe FA can be considered as low polar, it shows similar features with that of FE. The principal difference between them is that in the case of FA this site is low-hydrated whereas it is fully hydrated in the case of FE. Since probe FA is highly specific to the binding site of BSA ($K_d \sim 10^{-7}$ M), the probe molecules fits well to this site so that no water molecules are capable to interact with the probe. Meantime, probe FE is smaller and fits looser to this site, so that there might be a space for locating water molecules forming H-bonds with the probe.

6. LIMITATIONS OF THE MULTIPARAMETRIC APPROACH

6.1. H-bonding heterogeneity within the sites of the same polarity

As it was already mentioned, in the present multiparametric algorithm we can analyze only the cases when the fluorophore does not form H-bonds with the solvent or all its molecules are H-bonded with protic solvent

(Klymchenko and Demchenko, 2003). The intermediate cases when only a part of the dye molecules is involved in hydrogen-bonding cannot be evaluated because this H-bonding heterogeneity would result in a superposition of bands corresponding to the H-bonded and H-bond free forms in absorption and emission spectra in unknown proportions. To resolve this problem of heterogeneity one could apply correlations shown on Fig. 3.2B for direct determination of the part of the fluorophore molecules bound with the protic solvent. Thus, position of the data point on this plot with respect to the two curves corresponding to protic and aprotic environments could determine semi-quantitatively the extent of the H-bonding of the dye with the system. The other approach is based on deconvolution of the experimental spectrum into corresponding H-bonded and H-bond free components.

As an example, we consider a mixture of two solvents of very close dielectric constant: ethyl acetate ($\varepsilon = 6.02$) and 2-methyl-2-butanol ($\varepsilon = 5.8$). As it is shown in Fig. 3.4, fluorescence spectrum of the mixture can be presented as a superposition of the spectra in neat ethyl acetate and 2-methyl-2-butanol. By using this principle any experimental spectrum obtained from a homogeneous mixture of protic and aprotic solvents of close polarity could be presented as a superposition of the two spectra corresponding to neat solvents.

6.2. Site heterogeneity: Lipid bilayers and protein-labeling sites

Recently, we have shown that dimethyl analog of dye FE, probe F, exhibits bimodal distribution in the lipid bilayers, so that it locates at two different membranes sites (Klymchenko *et al.*, 2004b). The first site

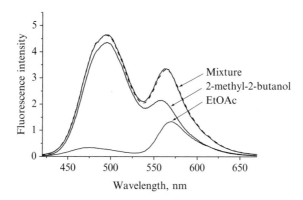

Figure 3.4 Fluorescence spectrum of dye FE in a mixture of 25% of 2-methyl-2-butanol and 75% of ethyl acetate can be presented as a sum of the spectra of the dye in corresponding neat solvents. The experimental curves are solid; the sum of the spectra is a dashed curve.

corresponds to the deep location in the bilayers, where the probe does not form H-bonds with surrounding (aprotic environment) and the second site corresponds to the location at the bilayer interface where it is H-bonded with water molecules. This ground-state heterogeneity represents the case when the dye is redistributed between two cites characterized by different polarity and H-bond donor ability. This complicated case cannot be directly analyzed with the present multiparametric method, because the experimental absorption and emission spectra are the superpositions of the profiles of the deep and surface located species of probe F. We succeeded to obtain individual emission profiles for these forms that allowed us to evaluate polarity of the deep location of the dye (the ratio of N* and T* emission bands) in the bilayer as well as probe hydration in the bilayer (as a ratio of the emission intensities of surface located hydrated form of the dye with respect to the deeply located nonhydrated form) (Klymchenko et al., 2004b). The parameter of electronic polarizability could not be easily evaluated in this case because individual absorption spectra of hydrated and nonhydrated forms of the dye cannot be directly obtained from the data.

The other example is the heterogeneity in the labeling sites of proteins. Recently, we introduced a new SH-reactive derivative of 3-hydroxyflavone, BMFE (Klymchenko et al., 2004a). However, labeling of α-crystallin (the protein containing a single SH group per monomer) with this dye was heterogeneous. We identified two-label species that were attached to SH and to NH_2 groups of the protein (Klymchenko et al., 2004a). Due to the presence of local electric field effect in the case of the NH_2-label species (the conjugation site retains a positive charge), their absorption and emission spectra were significantly shifted to the red. Thus, the experimental spectrum was heterogeneous and required deconvolution of the individual spectroscopic profiles for SH- and NH_2-label species. We succeeded to resolve these profiles in emission by using a specially developed method based on recording of excitation-dependent emission (Klymchenko et al., 2004a). The obtained emission profiles allowed us to estimate polarity and presence of H-bonding within the sites of labeling. Some other examples of the ground-state heterogeneity and their analysis are described in Chapter 4 of this book (Demchenko, 2008).

6.3. Applicability of multiparametric approach to other solvatochromic dyes

Primarily, we have to stress several key requirements to a solvatochromic dye to work as a multiparametric probe of the microenvironment. Most importantly, the recorded spectroscopic parameters from this dye should be *independent* (uncorrelated or low-correlated), namely, the absorption and fluorescence spectra of a dye should depend differently on such properties of microenvironment as polarity, electronic polarizability and H-bonding. The difference in absorption and emission behavior originates from the value and directionality of the ground and excited-state dipoles of the dye (Bakhshiev, 1972; Mataga and Kubata, 1970). Thus, when the dipole moment of the dye in the excited state is much larger than that in the ground state and the direction of the dipole change significantly on excitation then the Stokes shift of the dye and the position of the emission maximum will be highly sensitive to $f(\varepsilon) - f(n)$ and $f(\varepsilon)$, respectively. The common solvatochromic fluorescent dyes, such as PRODAN, Nile Red, ketocyanines, stylbene, and dansyl derivatives, due to excited-state charge transfer (ICT) exhibit significant increase in the dipole moment on electronic excitation, so that, in principle they could be used to evaluate $f(\varepsilon) - f(n)$ and $f(\varepsilon)$ parameters of their environment. However, the problem appears in the protic environments. Thus, the typical solvatochromic dye PRODAN shows strong increase of the Stokes shift on H-bonding with protic component (Catalan et al., 1991), which is the result of formation of new H-bonds between dye and solvent in the excited state.

This behavior is commonly observed for different solvatochromic dyes, because ICT results in a strong increase in the basicity of the electron-acceptor group of the dye. Meantime, 3-hydroxyflavone dye FE does not show deviation of the Stokes shift in the presence of protic solvents (Fig. 3.1A), thus allowing simultaneous measurements of $f(\varepsilon) - f(n)$ and $f(\varepsilon)$ even in protic environments. This very special behavior we attribute to already pre-existing *intramolecular* H-bond that probably strengthen in the excited state and prevent the molecule from the formation of new H-bonds in the excited state (Shynkar et al., 2004b).

The second problem is a number of spectroscopic parameters that can be obtained from a single solvatochromic dye. For common dyes emitting one-band fluorescence, one obtains only two spectroscopic parameters: the positions of the absorption and emission bands. We do not consider here other band characteristics like bandwidth and band asymmetry because these parameters are poorly sensitive to the properties of the environment and usually represent intrinsic properties of the fluorophore and, often, heterogeneity of its location. Theoretically, these two spectroscopic parameters allow calculation of only two parameters of the environment. Thus, the one-band dyes can be used only to evaluate polarity, $f(\varepsilon)$, and electronic polarizability, $f(n)$, in aprotic environments, or polarity and H-bonding effects keeping constant an electronic polarizability value. Therefore, evaluation of more than two parameters simultaneously should always require an application of a dye with more than one emission bands. This was successfully realized in the case of dyes of 3-HC–3-HF family, as shown above for dye FE. We believe that the present multiparametric approach can be adapted to some other advanced dyes showing multiple bands due to excited-state reactions.

In discussion on informative fluorescence parameters we did not mention the quantum yield Q. Operating with Q is difficult in view of a variety of quenching effects on which it depends. The same can be said for the ratios of two bands in emission if these bands are formed by inequilibrium (kinetically controlled) excited-state reaction. In these cases the band intensity ratio is determined by the rate of this reaction and also by quenching rates of both these states, which, in a general case, are different. In contrast, if the excited-state equilibrium is achieved and the reaction becomes thermodynamically controlled, then this ratio becomes the measure of relative energies of the two excited states. In 3-HFs this is observed in all studied cases when Q is sufficiently high and the lifetime is sufficiently long (Oncul and Demchenko, 2006; Shynkar et al., 2003; Tomin et al., 2007). Then any quenching effect does not change the intensity distribution between the bands, and band intensity ratio becomes an informative parameter. This rare case is realized in those 3-HCs and 3-HFs, in which the N* state possesses the ICT character.

3-Hydroxychromones are probably not the only type of "multiparametric" fluorescent environment-sensitive probes. The search for advanced dyes with these properties should continue. They should provide sufficient number of independent spectroscopic parameters highly sensitive to fundamental physicochemical properties of microenvironment: polarity, electronic polarizability, and H-bonding.

7. Concluding Remarks

Fluorescence probes feature broad and rapidly increasing range of applications in the studies of conformational changes of proteins and nucleic acids, intramolecular dynamics and intermolecular interactions, folding of macromolecules and phase transitions in molecular ensembles such as biomembranes. The new range of applications involves nucleic acid hybridization, drug discovery, and fluorescence sensing. In many of these applications the "site-sensitive" fluorescence dyes are used and the common way of obtaining the information about their interactions in the binding sites and their changes in different molecular events is the registration of fluorescence emission spectra. The transitions between electronic states that are in the background of light absorption and fluorescence offer the possibility to observe these events with high resolution in energy as the shifts of recorded spectra. Presently the use of this information is very limited. The positions of fluorescence band maxima are calibrated in different physical variables, commonly, in empirical units of polarity. We suggest a different approach to analyze intermolecular interactions in the binding sites of fluorescence probes. It is based on multiparametric correlation of spectroscopic data with a set of independent physically justifiable parameters of microenvironment. To obtain sufficient amount of primary spectroscopic data, we apply the probe molecules exhibiting the excited-state reaction intramolecular proton transfer (ESIPT) that produce additional band in fluorescence spectra belonging to a tautomer form of the probe. Spectroscopic characteristics of the reactant and product states as well as the reaction itself provide information on different interactions of unknown microenvironment with the probe and allow realizing efficient multiparametric sensing of this microenvironment. The new methodology allowed us to obtain simultaneously several fundamental physicochemical parameters of probe microenvironment: polarity, electronic polarizability and H-bond donor ability. In aqueous systems, in which biological macromolecules, their complexes and biomembranes are studied, the latter may be connected with hydration. In addition, local electric fields can be detected and characterized by their spectroscopic effects (Klymchenko and Demchenko, 2002b). Moreover, in environments with strong H-bond proton acceptor properties the ground-state and excited-state anionic forms can be observed and characterized (Demchenko et al., 2003; Mandal and Samanta, 2003; Shynkar et al., 2004a). It is remarkable that all these forms behave as distinct species that interact differently with the environment and are differently sensitive to a variety of interactions. We believe that multiparametric approach can become a valuable tool for analysis of spectroscopic information provided by fluorescence probes exhibiting transitions between several ground and/or excited states and stimulate rapid development of novel probes based on this principle. They will extend possibilities of fluorescence probes as the tools in biomolecular and cellular research.

REFERENCES

Abraham, M. H. (1993). Hydrogen bonding. 31. Construction of a scale of solute effective or summation hydrogen-bond basicity. *J. Phys. Org. Chem.* **6,** 660.

Abraham, M. H., Chadha, H. S., Whiting, G. S., and Mitchell, R. C. (1994). Hydrogen bonding. 32. An analysis of water-octanol and water-alkane partitioning and the Δlog P parameter of Seiler. *J. Pharm. Sci.* **83,** 1085.

Bakhshiev, N. G. (1972). "Spectroscopy of Intermolecular Interactions." Nauka, Leningrad.

Bhattacharya, A. A., Curry, S., and Franks, N. P. (2000). Binding of the general anesthetics propofol and halothane to human serum albumin: High resolution crystal structures. *J. Biol. Chem.* **275,** 38731.

Catalan, J. (1997). On the E_T (30), $\pi^{\star}$, P_y, S', and SPP empirical scales as descriptors of nonspecific solvent effects. *J. Org. Chem.* **62,** 8231.

Catalan, J., Perez, P., and Blanco, F. G. (1991). Analysis of the solvent effect on the photophysics properties of 6-propionyl-2-(dimethylamino)naphthalene (PRODAN). *J. Fluorescence* **1,** 215.

Chou, P.-T., Martinez, M. L., and Clements, J. H. (1993). Reversal of excitation behavior of proton-transfer vs. charge-transfer by dielectric perturbation of electronic manifolds. *J. Phys. Chem.* **97,** 2618.

Chou, P.-T., Pu, S.-C., Cheng, Y.-M., Yu, W.-S., Yu, Y.-C., Hung, F.-T., and Hu, W.-P. (2005). Femtosecond dynamics on excited-state proton/ charge-transfer reaction in 4'-N,N-diethylamino-3-hydroxyflavone. The role of dipolar vectors in constructing a rational mechanism. *J. Phys. Chem. A* **109**, 3777.

Demchenko, A. P. (1994). Protein fluorescence, dynamics and function: Exploration of analogy between electronically excited and biocatalytic transition states. *Biochim. Biophys. Acta* **1209**, 149.

Demchenko, A. P., Klymchenko, A. S., Pivovarenko, V. G., Ercelen, S., Duportail, G., and Mely, Y. (2003). Multiparametric color-changing fluorescence probes. *J. Fluorescence* **13**, 291.

Demchenko, A. P. (2008). Site-selective red-edge effects. *Methods in Enzymology* **450R21**: 59 (This volume).

Ercelen, S., Klymchenko, A. S., and Demchenko, A. P. (2002). Ultrasensitive fluorescent probe for the hydrophobic range of solvent polarities. *Anal. Chim. Acta* **464**, 273.

Ercelen, S., Klymchenko, A. S., and Demchenko, A. P. (2003). Novel two-color fluorescence probe with extreme specificity to Bovine Serum Albumin. *FEBS Lett.* **538**, 25.

Formosinho, S. J., and Arnaut, L. G. (1993). Excited-state proton transfer reactions II. Intramolecular reactions. *J. Photochem. Photobiol. A: Chem.* **75**, 21.

Kamlet, M. J., Abboud, J. L. M., Abraham, M. H., and Taft, R. W. (1983). Linear solvation energy relationships. 23. A comprehensive collection of the solvatochromic parameters, .pi.*, .alpha., and .beta., and some methods for simplifying the generalized solvatochromic equation. *J. Org. Chem.* **48**, 2877.

Kasha, M. (1986). Proton-transfer spectroscopy. Perturbation of the tautomerization potential. *J. Chem. Soc. Faraday Trans. 2* **82**, 2379.

Klymchenko, A. S., and Demchenko, A. P. (2002a). Probing AOT reverse micelles with two-color fluorescence dyes based on 3-hydroxychromone. *Langmuir* **18**, 5637.

Klymchenko, A. S., and Demchenko, A. P. (2002b). Electrochromic modulation of excited-state intramolecular proton transfer: The new principle in design of fluorescence sensors. *J. Am. Chem. Soc.* **124**, 12372.

Klymchenko, A. S., and Demchenko, A. P. (2003). Multiparametric probing of intermolecular interactions with fluorescent dye exhibiting excited state intramolecular proton transfer. *Phys. Chem. Chem. Phys.* **5**, 461.

Klymchenko, A. S., Duportail, G., Ozturk, T., Pivovarenko, V. G., Mely, Y., and Demchenko, A. P. (2002). Novel two-color ratiometric fluorescence probes with different location and orientation in phospholipid membranes. *Chem. Biol.* **9**, 1199.

Klymchenko, A. S., Duportail, G., Mely, Y., and Demchenko, A. P. (2003a). Ultrasensitive two-color fluorescence probes for dipole potential in phospholipid membranes. *Proc. Natl. Acad. Sci. USA* **100**, 11219.

Klymchenko, A. S., Pivovarenko, V. G., and Demchenko, A. P. (2003b). Elimination of hydrogen bonding effect on solvatochromism of 3-hydroxyflavones. *J. Phys. Chem. A* **107**, 4211–4216.

Klymchenko, A. S., Pivovarenko, V. G., Ozturk, T., and Demchenko, A. P. (2003c). Modulation of solvent-dependent dual emission in 3-hydroxychromones by chemical substituents. *New J. Chem.* **27**, 1336.

Klymchenko, A. S., Avilov, S. V., and Demchenko, A. P. (2004a). Resolution of Cys and Lys labeling of alpha-crystallin with site-sensitive fluorescent 3-hydroxyflavone dye. *Anal. Biochem.* **329**, 43.

Klymchenko, A. S., Duportail, G., Demchenko, A. P., and Mely, Y. (2004b). Bimodal distribution and fluorescence response of environment-sensitive probes in lipid bilayers. *Biophys. J.* **86**, 2929.

Klymchenko, A. S., Stoeckel, H., Takeda, K., and Mely, Y. (2006). Fluorescent probe based on intramolecular proton transfer for fast ratiometric measurement of cellular transmembrane potential. *J. Phys. Chem. B* **110**, 13624.

Kragh-Hansen, U., Hellec, F., de Foresta, B., le Maire, M., and Moller, J. V. (2001). Detergents as probes of hydrophobic binding cavities in serum albumin and other water-soluble proteins. *Biophys. J.* **80,** 2898.

Lippert, E. L. (1975). Laser-spectroscopic studies of reorientation and other relaxation processes in solution. *In* "Organic Molecular Photophysics," (J. B. Birks, ed.), Vol. 2, pp. 1–31. Wiley, New York.

Liu, W., Wang, Y., Jin, W., Shen, G., and Yu, R. (1999). Solvatochromogenic flavone dyes for the detection of water in acetone. *Anal. Chim. Acta* **383,** 299.

Mandal, P. K., and Samanta, A. (2003). Evidence of ground-state proton-transfer reaction of 3-hydroxyflavone in neutral alcoholic solvents. *J. Phys. Chem. A* **107,** 6334.

Mataga, N., and Kubata, T. (1970). "Molecular interactions and electronic spectra." Marcel Dekker, New York.

Oncul, S., and Demchenko, A. P. (2006). The effects of thermal quenching on the excited-state intramolecular proton transfer reaction in 3-hydroxyflavones. *Spectrochim. Acta A* **65,** 179.

Reichardt, C. (1994). Solvatochromic dyes as solvent polarity indicators. *Chem. Rev.* **94,** 2319.

Roshal, A. D., Grigorovich, A. V., Doroshenko, A. O., Pivovarenko, V. G., and Demchenko, A. P. (1998). Flavonols and crown-flavonols as metal cation chelators. The different nature of Ba^{2+} and Mg^{2+} complexes. *J. Phys. Chem.* **102,** 5907.

Shynkar, V. V., Mely, Y., Duportail, G., Piemont, E., Klymchenko, A. S., and Demchenko, A. P. (2003). Picosecond time-resolved fluorescence studies are consistent with reversible excited-state intramolecular proton transfer in 4′-(dialkylamino)-3-hydroxyflavones. *J. Phys. Chem. A* **107,** 9522.

Shynkar, V. V., Klymchenko, A. S., Mely, Y., Duportail, G., and Pivovarenko, V. G. (2004a). Anion formation of 4′-(dimethylamino)-3-hydroxyflavone in phosphatidylglycerol vesicles induced by hepes buffer: A steady-state and time-resolved fluorescence investigation. *J. Phys. Chem. B* **108,** 18750.

Shynkar, V., Klymchenko, A. S., Piemont, E., Demchenko, A. P., and Mely, Y. (2004b). Dynamics of intermolecular hydrogen bonds in the excited states of 4′-dialkylamino-3-hydroxyflavones. on the pathway to an ideal fluorescent hydrogen bonding sensor. *J. Phys. Chem. A* **108,** 8151.

Shynkar, V. V., Klymchenko, A. S., Duportail, G., Demchenko, A. P., and Mely, Y. (2005). Two-color fluorescent probes for imaging the dipole potential of cell plasma membranes. *Biochim. Biophys. Acta* **1712,** 128.

Shynkar, V. V., Klymchenko, A. S., Kunzelmann, C., Duportail, G., Muller, C. D., Demchenko, A. P., Freyssinet, J.-M., and Mely, Y. (2007). Fluorescent biomembrane probe for ratiometric detection of apoptosis. *J. Am. Chem. Soc.* **129,** 2187.

Suppan, P., and Ghoneim, N. (1997). "Solvatochromism." The Royal Society of Chemistry, Cambridge, UK.

Swinney, T. C., and Kelley, F. D. (1993). Proton transfer dynamics in substituted 3-hydroxyflavones: Solvent polarization effects. *J. Chem. Phys.* **99,** 211.

Tomin, V. I., Oncul, S., Smolarczyk, G., and Demchenko, A. P. (2007). Dynamic quenching as a simple test for the mechanism of excited-state reaction. *Chem. Phys.* **342,** 126.

Yesylevskyy, S. O., Klymchenko, A. S., and Demchenko, A. P. (2005). Semi-empirical study of two-color fluorescent dyes based on 3-hydroxychromone. *J. Mol. Struct.: Theochem.* **755,** 229.

CHAPTER FOUR

Site-Selective Red-Edge Effects

Alexander P. Demchenko

Contents

1. Introduction	60
2. Molecular Disorder and the Origin of Red-Edge Effects	61
3. The Principle of Photoselection	62
4. Ground-State Heterogeneity	65
5. The Magnitude of Red-Edge Excitation Fluorescence Shift and Its Connection with Dielectric Relaxations	67
6. Red-Edge Effect with High Resolution in Time	71
7. Red-Edge Effects in Excited-State Reactions	72
8. Fluorescent Probes for Optimal Observation of Red-Edge Effects	74
9. Peculiarities of the Red-Edge Effects of Indole and Tryptophan	75
10. Conclusions	76
References	77

Abstract

Observation of Red Edge effects is the basis of unique methodology that allows combination of site-photoselection with dynamics of molecular relaxations. The important dynamic information on molecular level can be obtained even by simple recording of steady-state fluorescence using the lifetime as the time marker. The extension to time domain allows distinguishing these relaxations from other dynamic processes that influence the excited-state energies. In this Chapter I briefly discuss the background of this technique and concentrate on quantitative measure of these effects and on importance of their distinction from ground-state heterogeneity. The peculiarity of Trp emission in proteins and the optimal selection of fluorescence probes are discussed. The Red Edge excitations influence dramatically the excited-state reactions that are coupled with dielectric relaxations and this opens a new fascinating prospect for protein and biomembrane studies.

A.V. Palladin Institute of Biochemistry, Kiev 252030, Ukraine

Methods in Enzymology, Volume 450   © 2008 Elsevier Inc.
ISSN 0076-6879, DOI: 10.1016/S0076-6879(08)03404-6   All rights reserved.

1. INTRODUCTION

When a measured experimental parameter starts to depend on some factor in an unforeseen and unpredictable manner, it is first considered as a curiosity and a trouble for researchers. And only after this phenomenon is studied in more detail and there appears an understanding of its mechanism and conditions of observation, then an efficient application may start with deriving new valuable information. This evolution is observed for a group of phenomena that refer to fluorescence spectroscopy and got a common name "Red-Edge effects" (Demchenko, 2002). It was found that in some special conditions the fluorescence spectroscopic properties do not conform to classical rules. Thus, when fluorescence is excited at the red (long-wavelength) edge of the absorption spectrum, the emission spectra start to depend on excitation wavelength, and excited-state energy transfer, if present, fails at the "red" excitation edge. The Red-Edge excitation can influence profoundly the excited-state reactions, such as electron and proton transfers and fluorescence resonance energy transfer (FRET). Presently the conditions for observation of these effects are well established. The chromophore should be solvatofluorochromic, that is, it should respond to the changes in interaction energy with its environment by the shifts in fluorescence spectra. And also this environment should be relatively polar but rigid or highly viscous, so that the relaxation times of its dipoles, τ_F, should be comparable or longer than the fluorescence lifetime τ_F (in the case of recording the steady-state spectra) or on the time scale of observation (in time-resolved spectroscopy). This allowed coupling these effects with molecular dynamics in condensed media and obtaining valuable tools for study of these dynamics in unknown, often microheterogeneous media. It is therefore not surprising that these phenomena developed into new methods of analysis and their major fields of application are biological macromolecules and biomembranes.

Basic physical background of Red-Edge effects was described, together with its manifestations in different model systems (Demchenko, 1982, 1986a, 1988, 1989; Demchenko and Ladohin, 1988; Nemkovich et al., 1991). Overviews of their application appear frequently (Raghuraman et al., 2005; Demchenko, 1991, 2002; Mulkherjee and Chattopadhyay, 1995) some of them with particular focus of studying proteins (Demchenko, 1986b; Lakowicz, 2000) and biomembranes (Chattopadhyay, 2003; Lakowicz, 2000). The present work presents a condensed modern view on the Red-Edge phenomena with particular focus on methodology in research.

2. MOLECULAR DISORDER AND THE ORIGIN OF RED-EDGE EFFECTS

Structural disorder is characteristic of many condensed matter systems (Richert and Blumen, 1994). Unlike ideal crystals, where only vibrational motions are allowed, real systems usually display structural, energetic, and dynamic microscopic inhomogeneity, which results in a variety of new properties: broadening or even disappearance of first-order phase transitions, nonexponential reaction kinetics, non-Arrhenius-type activated processes and others. Polymers and solvent glasses commonly display these properties. In many cases the disorder is combined with microscopic structural heterogeneity, formation of microscopic structures and microphases. These properties are characteristic of colloid systems and also of biological macromolecules, their assemblies, and biological membranes. In all these cases, the electronic absorption and fluorescence emission spectra are usually broad with vibrational structure smoothened or even entirely lost. Cooling to cryogenic temperatures usually does not result in improvement in structural resolution. This means that in addition to common strongly temperature-dependent homogeneous broadening (which is mainly due to electron–lattice and electron–vibrational interactions), in the systems with molecular disorder there exists the so-called inhomogeneous broadening of the spectra (Nemkovich et al., 1991). The latter originates from nonequivalence of chromophore environments in an ensemble that results in the distribution in solute–solvent interaction energies. As a result, the electronic transition energies for every species become distributed on the scale of energy and their superposition forms inhomogeneously broadened contour. Thus this contour contains valuable information on the extent of molecular disorder.

In a condensed medium this distribution should always exist at the time of excitation. But its display in a variety of spectroscopic phenomena depends on how fast are the transitions between the species forming this ensemble of states. Dependent on these conditions, the broadening of spectra can be static or dynamic (Nemkovich et al., 1991). The signatures of static broadening are observed in rigid environments, when the dynamics is slower than the rate of emission. The broadening is dynamic if the motions in the chromophore environment occur simultaneously or faster than the emission. Thus, the inhomogeneous broadening effects contain also information about the dynamic properties of condensed systems, and the rate of fluorescence emission provides the necessary time scale for these observations.

The scheme presented below shows the correlation between different mechanisms of spectral broadening. The strongly temperature-dependent homogeneous broadening always exists, even in ensembles of identical molecules in identical environments. Inhomogeneous broadening (static and dynamic) appears in an ensemble of identical molecules when there are variations of their interactions with the environment so that the average properties of the environment are the same. For static inhomogeneous broadening, the correlation times for dipolar relaxations, and for dynamic broadening, $\tau_C \leq \tau_F$. If the molecules are not identical or if they are distributed between different localized environments, we will call this case "the ground-state heterogeneity."

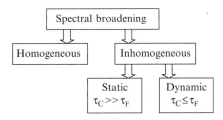

3. THE PRINCIPLE OF PHOTOSELECTION

Light has a selective power to excite exclusively those chromophores, the properties of which match the energy and polarization of its electronic transition. Thus, if a chromophore is excited by polarized light, its emission will be also highly polarized. Depolarization occurs only when the time correlation in the excited state is lost due to chromophore rotation or its participation in some photophysical process, such as FRET. Similarly, photoselection can be provided by variation of excitation energy. The chromophore molecule can absorb only the light quanta that correspond to its electronic transition energy. Then it will emit light at a particular wavelength, which is determined by the intrinsic properties of the fluorophore and its interaction with the environment, unless this interaction changes during its fluorescence lifetime or its intrinsic properties change due to participation in some photophysical or photochemical reaction.

Presently fluorescence spectroscopy has reached the absolute limit of sensitivity, which is the limit of detection of individual molecules. This allows studying the chromophore interactions on a level of very high structural resolution. But in more common cases the researcher deals with ensembles composed of a great number of chromophore molecules. The broad distribution on the energy of noncovalent interactions with the environment is the characteristic feature of these ensembles. These distributions and the observed effects of photoselection within them depend strongly on the properties of chromophore. Let us consider these properties.

In every case the absorption of light quanta results in increase of energy of electronic states, which causes the redistribution of electronic density within molecule. Its dipole moment may increase, decrease, or change the orientation. Correspondingly, the dipole–dipole and dipole–induced dipole interactions with surrounding molecules may change substantially. The stronger interactions always result in a broader distribution on the energy of these interactions. Consider first an ideal case when these interactions are negligibly small in the ground state but become significant due to increased dipole moment in the excited state (Fig. 4.1A). If we neglect homogeneous broadening and vibronic structure of the bands, we can represent the ground state by a single energy level and the excited state as the state exhibiting a broad distribution on chromophore–environment interaction

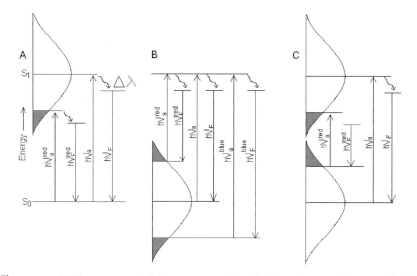

Figure 4.1 Different cases in inhomogeneous broadening of spectra that provide the means for wavelength photoselection. (A) Chromophore dipole moment increases dramatically on excitation. The ground state S_0 does not interact strongly with the environment and can be represented by a single energy level, whereas the excited state S_1 exhibits a broad distribution on chromophore–environment interaction energies. The light quantum produces photoselection within the excited-state distribution. (B) Chromophore dipole moment decreases dramatically on excitation. In this case the ground state exhibits a distribution on chromophore–environment interaction energy and the excited state can be represented by a single energy level. The light quantum produces photoselection within the ground-state distribution. (C) Chromophore dipole moment is high both in the ground and in the excited state, and on electronic transition it changes its orientation. Thus both ground-state and excited-state distributions have to be considered. The dipole reorientation produces reversal of the distribution in such a way that the highest in energy part of the ground-state distribution becomes the lowest in the excited state. The low-energy quanta can selectively excite these species. The shaded areas show photoselected parts of the distributions. The upward vertical arrows stand for light absorption, and downward arrows for light emission. The energy gap $\Delta\lambda$ indicates the Stokes shift.

energy. The most probable case is the excitation of chromophores residing in the center of excited-state distribution and emission from the center of this distribution. According to well-known Kasha's rule, excitation by quanta of higher energies will result in relaxation to this lowest energy level prior to emission, and no site selection by high-energy quanta can be achieved. But if we decrease substantially the energy of excitation quanta (shift the excitation to red edge on the wavelength scale), then this energy will become so small that it will not be able to excite all the chromophores in the ensemble but only those which can interact with the environment in the excited state much stronger than the average and which possess individual excited-state levels shifted down along the energy scale. Thus we achieve photoselection in excitation energy. When excited these selected species will emit fluorescence differently than that of the mean of the distribution, their emission spectra will be shifted in the direction of low energies, to longer wavelengths. This is the site selective in excitation Red-Edge effect that is most popular in many applications.

Consider now the other limiting case, when the chromophore dipole moment is high in the ground state but decreases substantially in the excited state (Fig. 4.1B). In this case the ground state should exhibit a broad distribution on chromophore–environment interaction energy and the excited state can be represented by a single energy level. By variation of the energy of the light quanta we can produce photoselection within the ground-state distribution. Again, with the quanta of energies higher than the mean, photoselection cannot be achieved because of the presence of higher-energy vibronic levels in the excited state (not shown). These levels are almost not populated in the ground state. So if we decrease the energy of excited light quanta, these quanta will be absorbed mostly by species, the ground-state energy of which is higher than the mean (they interact weaker with the environment than the average species in the distribution), so that the separation of energy between ground and excited states for them is smaller than the mean. The energy of emitted quanta will also be small, and we will have the same Red-Edge effect—the shift of fluorescence spectra to longer wavelengths.

And now we can resolve a more difficult case when the chromophore dipole moments are relatively high both in the ground and excited states and the distributions on interaction energies are broad but due to redistribution of electronic density in the excited state the chromophore dipole moment changes its orientation (Fig. 4.1C). Since the dipole–dipole interactions with environment molecules are directional, the weakest ground-state interactions may become the strongest in the excited state, and vice versa. In this case, again, no photoselection is achievable by excitation at wavelengths shorter than the band maximum. Meantime, the shift to the red-excitation edge will select out of the whole ensemble the chromophore molecules that interact weaker in the ground state (upper part of the distribution) but stronger in the excited state (the lower part of its distribution). This selective excitation by low-energy quanta will result in the long-wavelength shifts of fluorescence spectra compared to that excited at the band maximum, which is the same common Red-Edge effect.

The difference between these three cases appears when we consider another site-selective effect—the dependence of excitation spectra on emission wavelength. For observing it, the ground-state distribution becomes important. Emission from the excited state can occur to different vibronic high-energy ground-state levels, which makes impossible the photoselection by collecting the low-energy emission quanta. Therefore, excitation spectra measured by setting the emission wavelength longer than the band maximum will be emission wavelength independent. The situation becomes different only at the blue (short-wavelength) edge of emission band. In this case we can collect the emitted quanta that are emitted from vibrationally relaxed excited state to also vibrationally relaxed ground state, which possesses the distribution on interaction energy. We can then collect the emitted quanta that possess higher energies than the mean, since they correspond to lower part of the distribution in the ground state (see Fig. 4.1B). As a result, the excitation spectrum will shift to the blue.

These simplified schemes allow illustrating the essence of wavelength-photoselection phenomena in the case of static inhomogeneous broadening of spectra. More detailed description of physical background of these phenomena can be found elsewhere (Demchenko, 2002; Nemkovich *et al.*, 1991).

4. GROUND-STATE HETEROGENEITY

If two or several chromophore forms are not identical (for instance, they are conformers of the same organic dye) or if they represent a mixture of species both participating and not participating in specific bonding (e.g., charge-transfer complexes or hydrogen bonds) and if these forms absorb light at different wavelengths they can be in principle distinguished by individual absorption bands. But very often the separation between these bands is small. And if their broadening is significant, then instead of individual bands one can observe a single band with additional broadening. These individual forms can generate individual spectra of fluorescence emission that can also remain unresolved. In these cases the simple recording of spectra is not sufficient and additional parameters are needed to distinguish these individual ground-state forms from the effects of inhomogeneous broadening. Fluorescence lifetime or anisotropy can be used for this purpose if these parameters exhibit variation between these forms. In a more general case the experiments with variation of both excitation and emission wavelengths become useful. The ground-state heterogeneity can be analyzed and distinguished from inhomogeneous broadening by studying the site-selective effects in excitation and in emission.

The display of heterogeneity in the absorption and fluorescence spectra cannot be easily compared quantitatively. This is because band separation in the ground state may not correspond to band separation in the excited state due to different Stokes shifts. Relative intensities can also not correspond due to differences in quantum yields. The widths of individual bands can also differ substantially. Meantime some qualitative correlations can be derived that could allow distinguishing the ground-state heterogeneity and the effects of inhomogeneous broadening. Consider two typical cases, when two species are represented by wavelength-shifted absorption and emission bands (Fig. 4.2). For simplicity we assume that their Stokes shifts are similar, so that the same wavelength separation exists between their absorption and emission spectra. The first case (Fig. 4.2A) represents the situation when the two species possess wavelength-shifted positions of absorption and emission spectra of comparable intensities. In this case by scanning the excitation wavelength we have to observe gradual shift from domination of one form to their equal contribution and further to domination of the second form in emission. As a result, fluorescence spectrum will gradually shift to longer wavelengths. This shift will be observed not only at the red edge, but through the whole range of excitation wavelengths, starting from the blue edge. Recording of excitation spectra (not shown) allows observing their gradual shift to the red with the increase of emission

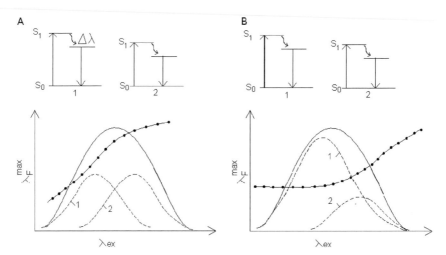

Figure 4.2 The idealized cases of two-species ground-state heterogeneity. The ground state is represented by two single ground-state levels for species 1 and 2 that differ in the energies of electronic transition and the positions of absorption and fluorescence bands. (A) The absorption bands for 1 and 2 are of comparable intensities, and variation of position of fluorescence band maximum, λ_F^{max}, as a function of excitation wavelength, λ_{ex}, is a monotonous function extending over the whole spectrum. (B) Specie 2 is a minor component, and its presence is revealed at the long-wavelength slope of the spectrum.

wavelength, over the whole emission spectrum but not only at the blue edge. These features are very different from that observed in the case of Red-Edge effects.

Consider next the case depicted in Fig. 4.2B, in which the major component is the species with short-wavelength location of absorption and emission spectra, and the species with long-wavelength absorption and emission is a minor less intensive component. This component can only be detected at the red-excitation edge as the shift of fluorescence spectrum to longer wavelengths (which is very similar to site-selective Red-Edge effect). But variation of excitation wavelength will show the motion of excitation spectra at much longer wavelengths than that observed in the case of blue-edge photoselective emission that was discussed above.

In real situations the effects of ground-state heterogeneity can overlap on the Red-Edge effects, which may produce additional difficulties in analysis. Therefore the knowledge of chromophore properties as well as the ability of the medium to provide differences in its environment is very important.

Such heterogeneity can appear in the case of presence of two forms of the same dye that differ by formation of weak intermolecular hydrogen bond. As discussed in Chapter 3 (Klymchenko and Demchenko, 2008), such bonding observed in protic solvents shifts absorption spectra of 3-hydroxychromone dyes to longer wavelengths. Photoselection of this form at long-wavelength slope of the spectrum has no relation to Red Edge effects.

5. THE MAGNITUDE OF RED-EDGE EXCITATION FLUORESCENCE SHIFT AND ITS CONNECTION WITH DIELECTRIC RELAXATIONS

The first and widely explored Red-Edge effect is the long-wavelength shift of fluorescence spectra at the red-excitation edge. Shining narrow bandwidth light onto the sample and shifting the wavelength from that corresponding to band maximum further and further to the red edge, a smaller and smaller number of chromophores are excited. Fluorescence intensity will be reduced dramatically. But if we account carefully for background fluorescence and Raman scattering and normalize band intensities we can observe that in certain cases the spectra move to longer wavelengths with the retention of their shapes and with sometimes clearly observed decrease of the bandwidths (but not the increase as in the case of ground-state heterogeneity). This is the result of photoselection. Out of the total population of chromophores those subpopulations are photoselected, which happen to have their light absorption energies shifted to smaller values, and fit to the energy of illuminating light (Fig. 4.1). Upon absorption of light quanta, this fraction of molecules becomes emissive, and its emissive properties differ from the mean values more and more significantly. The experiment on shifting the excitation wavelength stops when the emission intensity becomes very low and the spectrum becomes indistinguishable from the background. Sometimes one can even reach the anti-Stokes region, where the excitation wavelength, λ_{ex}, becomes longer than the position of the maximum of fluorescence spectrum, λ_F^{max}, excited at the band maximum, λ_{ex}^{max}. In this far-edge region the shift of emission spectrum approaches in value the shift of applied excitation wavelength. Thus, a typical dependence of λ_F^{max} on λ_{ex} is observed only at the red edge, and no such dependence is detected at the excitation band maximum and shorter wavelengths. This dependence becomes steep at the far-red edge (Fig. 4.3).

Depicted in Fig. 4.3 is the behavior of fluorescence excitation spectra as a function of emission wavelength, which is observed in the systems with broad site distributions in the ground state. The shift of excitation spectra to the blue is observed when fluorescence emission is collected at wavelengths shorter than the fluorescence band maximum. This dependence increases with the further shift to the blue edge, and no saturation point is observed.

Such dependence is typical for every system with static ($\tau_R \gg \tau_F$) or slow dynamic ($\tau_R \geq \tau_F$) inhomogeneous broadening. With an increase of relaxation rate the fluorescence spectra at the main-band excitation shift to longer wavelengths (the common relaxational shift of spectra). At the same time the excitation wavelength dependence decreases (Fig. 4.4) because the relaxation produces new distribution of chromophore sites that is not correlated with original distribution. In this dependence on λ_{ex} there is one characteristic point, λ_{ex}^*, in which the energy of electronic transition corresponds to that of the relaxed state. At $\lambda_{ex} < \lambda_{ex}^*$ the relaxation occurs

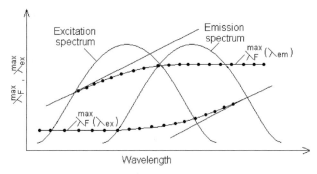

Figure 4.3 Typical dependences of positions of fluorescence band maxima on excitation wavelength, $\lambda_F^{max}(\lambda_{ex})$, and positions of excitation band maxima, $\lambda_{ex}^{max}(\lambda_{em})$, on emission wavelength for the case of inhomogeneous broadening of spectra.

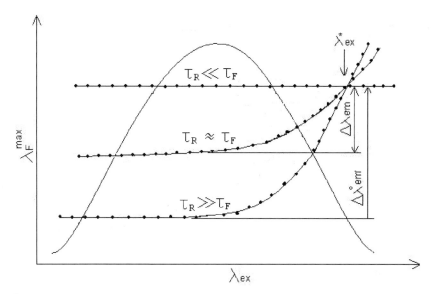

Figure 4.4 Dependences of positions of fluorescence band maxima on excitation wavelength, $\lambda_F^{max}(\lambda_{ex})$, for different correlations between the fluorescence lifetime, and the dipole relaxation time. λ_{ex}^* is the isorelaxation point. $\Delta\lambda_{em}^0$ and $\Delta\lambda_{em}$ are the magnitudes of Red-Edge effect.

with the decrease of the energy, and the spectra have to move in time to longer wavelengths, while at $\lambda_{ex} > \lambda_{ex}^*$ the relaxation results in increase in energy and to achieve the equilibrium the spectra move to shorter wavelengths (uprelaxation; Nemkovich et al., 1991). We call λ_{ex}^* an isorelaxation point. Its presence allows introducing the quantitative characteristics of Red-Edge effect as the shift of fluorescence spectrum $\lambda_F^{max}(\text{mean}) - \lambda_F^{max}(\text{edge})$ on variation of λ_{ex} from that at excitation band maximum λ_{ex}^{mean} to λ_{ex}^*. Thus, the relaxational shift of emission spectra can be observed at any λ_{ex} beside λ_{ex}^*.

The optical electronic excitation results in rapid redistribution of electronic density in the chromophore and the change of its dipole moment. The interaction with surrounding molecules becomes out of equilibrium. The process of attaining a new equilibrium (relaxation) may be faster, slower, or occur simultaneously with the emission decay. Since the relaxation is the decrease of excited-state energy as a function of time, in the case when it occurs simultaneously with the decay, a complex emission wavelength dependence should be observed for the decay, and the spectra should move as a function of time in the direction of lower energies.

If N_0 molecules are excited at an instant $t = 0$, the number dN of emitted quanta within a time interval dt and a frequency range $d\nu$ can be determined from the formula:

$$dN = N_0 Q_F I(\nu, t) d\nu dt, \quad (4.1)$$

where Q_F is the quantum yield of emission and

$$I(\nu, t) = (1/\tau_F) I(\nu - \xi) \exp(-t/\tau_F). \quad (4.2)$$

The latter function determines a number of quanta emitted per unit time within a unit frequency interval. At a fixed time it can be regarded as an "instantaneous" emission spectrum, while at a fixed frequency ν it represents a law of emission decay. ξ is the maximum or, more precisely, the center of gravity of the spectrum (in cm^{-1}). It was shown (Mazurenko and Bakhshiev, 1970) that within the Debye model of relaxation (single relaxation time τ_R) the position of the spectrum ξ has to change exponentially with time:

$$\xi(t) = \xi_{t \to \infty} + (\xi_{t=0} - \xi_{t \to \infty}) \exp(-t/\tau_R). \quad (4.3)$$

Since the relaxation is the time-dependent loss of correlation between initial site distribution and the distribution at time t, a time-dependent correlation function $C(t)$ for $\xi = \xi(t)$ can be used for describing the process of relaxation:

$$C(t) = [\xi(t) - \xi_{t \to \infty}] / (\xi_{t=0} - \xi_{t \to \infty}). \quad (4.4)$$

It normalizes the spectral shifts to unity and allows comparing the effects produced by different chromophores. The experiments show that in the relaxation range, the spectra move to longer wavelengths as a function of time (Brand and Gohlike, 1971). With its aid we can describe the behavior of the steady-state spectra. By assuming τ_F to be unchanged in relaxation process the following expression can be obtained (Mazurenko and Bakhshiev, 1970):

$$(\xi_{St} - \xi_{t \to \infty}) / (\xi_{t=0} - \xi_{t \to \infty}) = \tau_R / (\tau_R + \tau_F) \quad (4.5)$$

Here ξ_{St} is the position of steady-state spectrum. To achieve $\xi_{t \to \infty}$ and $\xi_{t=0}$ as the limiting cases by analyzing only the positions of steady-state spectra at common excitations at the band maxima the variations of temperature are usually applied, in such broad ranges that the limit of slow relaxations $\xi_{t=0}$ should be achieved at low temperatures (when $\tau_R \gg \tau_F$) and the limit of fast relaxations $\xi_{t \to \infty}$—at high temperatures (when $\tau_R \ll \tau_F$).

In real systems and especially in biophysical applications it is hard to achieve these conditions. Whereas reducing the temperature with maintenance of observed structure is frequently possible, subjecting to high temperatures often cannot be tolerated. Therefore in our earlier works

(Demchenko, 1986a) we suggested an extension of this approach by incorporating the information obtained in the study of Red-Edge effect. When the relaxation is complete, the dynamic distribution of sites becomes uncorrelated with initial distribution, so that the spectra become λ_{ex}-independent and the Red-Edge effect lost. If we assume that τ_R and τ_F at the Red-Edge excitation do not differ from their mean values, we obtain very simple means of obtaining the dynamic information from the steady-state spectra. By rewriting Eq. (4.5) for the mean and for the edge excitations and taking into account that when the relaxation is complete, $\xi^{\text{mean}}_{t\to\infty} = \xi^{\text{edge}}_{t\to\infty}$, and we obtain

$$\xi - \xi^{\text{edge}} = \left(\xi_{t=0} - \xi^{\text{edge}}_{t=0}\right)\tau_R/(\tau_R + \tau_F). \quad (4.6)$$

Equation (4.6) uses only the data on steady-state spectra and allows to analyze the Red-Edge excitation shifts of fluorescence maxima, λ^{max}_F. ξ-values on wave number scale can be easily transformed into λ-values, $\xi(\text{cm}^{-1}) = 10^7/\lambda(\text{nm})$. Thus, dynamic information about molecular relaxations can be obtained in simple steady-state measurements using τ_F as a time marker and provides the analysis of the Red-Edge effects.

6. RED-EDGE EFFECT WITH HIGH RESOLUTION IN TIME

The spectroscopic observations of dielectric relaxation are well known. Relaxation results in structural rearrangements in the surrounding of a fluorophore, relative reorientation of their dipoles to achieve their mutual equilibrium configuration (Brand and Gohlike, 1971). When excited at the band maximum the relaxation is observed as the shifts of emission spectra to longer wavelengths (Suppan and Ghoneim, 1997; Vincent et al., 1995). As a result, when observed at the blue edge, the emission contains short-decaying positive component(s) due to fast temporal decrease of a number of emitters possessing higher energies. When observed at the red emission edge the decay contains a negative component due to increase with time of the number of excited-state species emitting at low energies. The major features of this process can be adequately described based on the Bakhshiev–Mazurenko model of dipolar relaxations (Brand and Gohlike, 1971) that uses Eqs. (4.1)–(4.3). The Red-Edge effects introduce new dimension into this picture and allows probing the redistribution on emission energy as a function of time between the sites in an ensemble (Nemkovich et al., 1991). These studies introduce new concept of relaxation, which is the reorganization in ensemble of distributed states on the achievement of excited-state equilibrium.

In time-resolved emission decays we observe that at the Red-Edge excitation the short components of emission decay at the blue and red slopes of emission spectrum that reflect the relaxation, disappear (Vincent et al., 1995). The emission kinetics excited at the red edge became unimodal and almost single exponential. In time-resolved spectra the evolution of fluorescence bandwidth is much more pronounced at the red edge than at the main-band excitation—the spectra are initially more narrow (since a part of the distribution is selected) and are broadened in the course of relaxation due to redistribution between different sites.

These experiments allow better understanding of the temporal formation of the contour of fluorescence band that is observed in the steady state. The spectroscopic behavior of the system is determined by both the properties of initial distribution (inhomogeneous broadening) and the changes, which occur with this distribution in time (relaxation) (Rubinov et al., 1982). A variety of experimental data demonstrate clearly that inhomogeneous broadening is a universal property. If it is static and the environment is immobile at the time of emission, then the behavior of photoselected species will not change as a function of time. When it is dynamic and the relaxation occurs on the time scale of emission, then the initial (existing at the time of excitation) distribution exhibits rearrangements during the emission decay. As a result, the photoselected species with time "forget about that," they loose their specific properties and rearrange over the whole distribution. The static effect that is integrated over the time of emission depends upon the time window. In viscous media (when $\tau_R \approx \tau_F$) not only the freezing (reducing τ_R) but also the fluorescence quenching (reduction of τ_F) may cause the appearance of Red-Edge effects (Nemkovich et al., 1991).

7. Red-Edge Effects in Excited-State Reactions

Solvent-reorganizational coordinate is important for many excited-state reactions (Demchenko, 1994; Maroncelli et al., 1989), but not for all of them. There are reactions that are uncoupled with dipolar relaxations and that can occur on a short time scale even at extremely low temperatures [for instance, intramolecular proton tunneling (Bader et al., 2004) and nonadiabatic electron transfer (Marcus and Sutin, 1985)]. On the other extreme are the slow reactions that occur in solution after attaining dielectric equilibrium, for instance, the reactions of diffusional bimolecular quenching. But there are many examples of those reactions that are coupled with these relaxations, and the study of this coupling may be used as an important clue for elucidating their mechanisms. In unrelaxed states the distribution of excited-state species on their interaction energies with the environment may result in distributed reaction kinetics. It was shown that the part of this distribution that interacts stronger with the environment may exhibit an extreme increase in reactivity in intramolecular electron transfer reaction and a decreased reactivity in proton transfer and energy transfer reactions (Demchenko and Sytnik, 1991a,b). Site-selective spectroscopy allows not only to characterize the site-selective reactivity but also to provide the means to model the reactions occurring in the ground states (Demchenko, 1994), especially those of them which possess low intrinsic activational barriers and depend on dynamics in the environment. Those are many enzyme reactions (Demchenko, 1992).

Site-selective effects in FRET between identical molecules (homo-FRET) got a lot of attention. Because of the presence of inhomogeneous broadening, the homo-FRET reaction is not random, it is directed from those members of the ensemble which emit at shorter wavelengths to those which absorb at longer wavelengths. As a result of this directed transfer the fluorescence spectra in concentrated dye solutions in rigid and highly viscous environments are shifted to longer wavelengths. This shift can be observed as a function of time (Nemkovich *et al.*, 1981), even if the environment is completely immobile, at very low temperatures. The discovery by Weber and Shinitzky (1970) of failure of energy transfer in rigid and highly viscous solutions at red-excitation edge (the Weber Red-Edge effect) can be easily explained by site selectivity. Since both donors and acceptors are chemically identical molecules in the same medium, their excited-state site distributions (see Fig. 4.1A) are the same. The species from the upper part of the distribution can transfer their energy to another species on the same or lower-energy level. In contrast, the species from the lower part of the distribution (their effective concentration is low) can transfer their energy only to other species of the lower part of the distribution, and this is an event of low probability. In the studies of tryptophan dimers, it was shown that failure of directed transfer correlates with temperature-dependent activation of relaxational dynamics in the solvent (Demchenko and Sytnik, 1991a; Nemkovich *et al.*, 1981). In time-resolved experiment it was shown that transition to the red-excitation edge suppresses homo-FRET and modifies dramatically the emission decay by decreasing its emission wavelength dependence. Producing the shift at the main-band excitation, homo-FRET decreases the Red-Edge shift in steady-state spectra (Demchenko, 1986a).

The increased interest in molecular and cellular biology of fluorescent dyes producing the excited-state reactions is not only because they model the natural reactions (electron transfer in photosynthetic reaction centers or homo-FRET between antenna pigments) but because as fluorescence probes they offer new possibilities for amplification of their sensing signal and making it more specific. In particular, one can design fluorescence dyes that can sense the distribution of intermolecular interactions and their dynamics by using two-band ratiometric response and its variation on the application of Red-Edge effects. This possibility was demonstrated for adiabatic intramolecular electron transfer (Demchenko and Sytnik, 1991a, b), homo-FRET (Demchenko and Sytnik, 1991a), and intramolecular proton transfer (Demchenko, 1994).

8. FLUORESCENT PROBES FOR OPTIMAL OBSERVATION OF RED-EDGE EFFECTS

Usually in the studies of site-selective Red-Edge effects we are dealing with chromophores that display substantial homogeneous broadening. It can be so significant that the vibronic structure becomes hidden under a smooth featureless contour of absorption band. The typical inhomogeneous contribution to spectral bandwidth is small, commonly about 300 cm^{-1}, which is only 4.7 nm at 400 nm and 2.4 nm at 280 nm, while the total width of the absorption band is sometimes 50–60 nm and broader. Thus the site-selective Red-Edge effects should not be large in terms of the magnitude of spectral shifts, and the maximal shifts that could be observed under optimal conditions vary between several and several tens of nanometers. These effects can be sufficiently strong and easily detectable with optimal choice of fluorescent dyes as molecular probes.

The general rule for the optimal probe selection is the following. The dyes that are sensitive to some interactions should also be sensitive to distribution of these interactions between the sites. So, the chromophores that produce the strongest spectral shifts as a function of solvent polarity should be of primary concern. A number of these "polarity probes" have been described, and they are commercially available (Haugland, 1996). They are characterized by the presence of substitutions in aromatic heterocycles by groups that can serve as excited-state electron donors (e.g., dialkylamino groups) and electron acceptors (e.g., carbonyls). The dipole moment of these molecules being small in the ground state increases dramatically in the excited state, which allows achieving a broad distribution in excited-state energies (see Fig. 4.1A). In some of the charge-transfer probes the redistribution of electronic density in the excited state is dramatic, and they provide unusually large Red-Edge effects (Chou et al., 1995).

In contrast, if the researcher is interested to probe the distribution in the ground state, the probes with a strong dipole moment decrease should be preferred. They are less common. Usually they exhibit strong polarity dependence of absorption spectra with so-called negative solvatochromy— a shift of absorption spectra to shorter wavelengths with the increase of solvent polarity (Chou et al., 1995). Good examples of this behavior are merocyanine dyes (Al-Hassan and El-Byoumi, 1980). In rigid media in addition to red-excitation shifts of emission spectra they exhibit blue-emission shifts of the spectra of excitation (A. P. Demchenko, unpublished results).

9. Peculiarities of the Red-Edge Effects of Indole and Tryptophan

Tryptophan as the derivative of indole is not only an important protein light emitter, it has very interesting spectroscopic properties. A small indole chromophore ring exhibits a dramatic increase of the dipole moment on its major ^1L_a transition, which makes it strongly polarity sensitive. Meantime the solvent-dependent variations of absorption spectra are small. Due to these features the Red-Edge effects for Trp residues in proteins probe mainly the site distributions in the excited but not in the ground state. The imino nitrogen of indole can form intermolecular hydrogen bonds, but their impact on spectra is not significant. Due to these features the effects of structural heterogeneity of location of Trp residues in multi-Trp proteins and peptides are not as significant as it could be expected. The positions of Trp absorption spectra between different model environments and between different locations in proteins rarely exceed 2–3 nm (Demchenko, 1986a), which does not allow to ignore completely the factor of ground-state heterogeneity but still allows attributing the major part of observed Red-Edge effects to inhomogeneous broadening (static or dynamic), as it was discussed above.

Overlap of two transitions ^1L_a and ^1L_b with very different properties produces a dramatic impact on all Red-Edge effects. The narrow band of the low-polar ^1L_b transition overlaps with the low-energy slope of 0–0 origin of the highly polar ^1L_a band (Callis, 1997; Demchenko, 1986a). As a result, when λ_{ex} is shifted gradually from the absorption maximum at 280 nm to λ_{ex}^* located at 307–308 nm, the start of the Red-Edge selectivity for ^1L_a is overlapped by the ^1L_b maximum that is observed at 289–290 nm, so that no static Red-Edge effect is detected in this range. Then starting at about 292 nm the Red-Edge effect appears as a very steep function of λ_{ex}. This is due to both display of the Red-Edge effect of ^1L_a and to a decrease of the relative contribution of the ^1L_b transition to the fluorescence spectrum (Demchenko, 1986a; Demchenko and Ladohin, 1988). As a result, we have a good opportunity not only to observe a very pronounced Red-Edge fluorescence shift, but also to avoid the contribution of Tyr emission, the red edge of which terminates at about 290 nm. The latter is the second important fluorophore of proteins with the fluorescence spectrum located at about 310 nm (Demchenko, 1986a). The point at 292 nm is the characteristic wavelength where the tyrosine contribution already disappears from the protein spectrum, and the Red-Edge effect for tryptophan does not yet start (Demchenko, 1981). Therefore, it is recommended that measurements be obtained by variation of the excitation wavelength in the 15 nm range of 292–307 nm and to take the difference of λ_F^{max}-values recorded at these excitation wavelengths as a quantitative measure of the effect.

10. Conclusions

The observation and analysis of Red-Edge effects is a powerful methodology for studying molecular disorder and its coupling with molecular dynamics. This methodology is simple in application, it can easily complement other fluorescence spectroscopic methods in studying macroscopically heterogeneous systems of different kinds: biomacromolecules in solutions and more complex systems such as biomembranes. Its application demonstrates a significant extent of molecular disorder within the well-organized structures of protein molecules and lipid bilayers of biomembranes.

The attractive feature of this technique is that it can provide the dynamic information about dielectric relaxations in the system using simple steady-state measurements and using τ_F as the time marker of dynamic events. It allowed demonstrating the slow rates of molecular relaxations on the nanosecond time scale in proteins and membranes and specifying the conditions in which these dynamics change as a result of conformational transitions that activate fast molecular motions.

The time-resolved fluorescence methods can be easily extended for experiments with site-selective excitation. These results allow one to achieve a better understanding of the molecular relaxation phenomena. Relaxation can be viewed as not only the decrease in time of the energy of the average (or most probable) species integrated into the molecular ensemble but also reorganization in this ensemble, the loss of time correlation of site-selected species as a function of time.

Although the application of this technique is straightforward and simple, the research work and the analysis of data require the knowledge of the processes behind the observed results and the mechanisms that lead to disorder and to site photoselection. The most important in this respect is the separation of site-selective Red-Edge effects and ground-state heterogeneity, which often needs additional experiments, such as the measurements of spectral shifts, anisotropy, and lifetime as a function of emission wavelength.

The Red-Edge effects are important tools for studying the mechanisms of different excited-state reactions, such as isomerizations, electron and proton transfers, and fluorescence resonance energy transfers. They allow studying the coupling of these reactions with dielectric relaxations in the environment and the involvement of inhomogeneous kinetics in these reactions. The dyes exhibiting these reactions are expected to provide prospective means for constructing fluorescence probes with strongly amplified two-color ratiometric response (Demchenko, 2006).

Whereas many common fluorescence methods exhibit difficulties in application to the tryptophan chromophore in peptides and proteins due to its low molar extinction, nonexponential fluorescence decay, etc., the studies of Red-Edge effects appear to be not only applicable but offers some advantages. This is mainly because the small site-dependent shifts of absorption spectra reduce the significance of the effects of ground-state heterogeneity and also because the transition to red-excitation edge results in interplay between exciting the ^1L_a and ^1L_b transitions producing an unusually strong effect in rigid environments.

REFERENCES

Al-Hassan, K. A., and El-Byoumi, M. A. (1980). Excited-state phenomena associated with solvation site heterogeneity: A large edge-excitation red-shift in a merocyanine dye-ethanol solution. *Chem. Phys. Lett.* **76,** 121.

Bader, A. N., Pivovarenko, V. G., Demchenko, A. P., Ariese, F, and Gooijer, C. (2004). Influence of redistribution of electron density on the excited state and ground state proton transfer rates of 3-hydroxyflavone and its derivatives studied by Spol'skii spectroscopy. *J. Phys. Chem. A.* **108,** 10589.

Brand, L., and Gohlike, J. R. (1971). Nanosecond time-resolved fluorescence of a protein-dye complex BSA+TNS. *J. Biol. Chem.* **246,** 2317.

Callis, P. R. (1997). ^{1}La and ^{1}Lb transitions of tryptophan: Application of theory and experimental observations to fluorescence of proteins. *Methods Enzymol.* **278,** 113.

Chattopadhyay, A. (2003). Exploring membrane organization and dynamics by the wavelength-selective fluorescence approach. *Chem. Phys. Lipids* **122,** 3.

Chou, P.-T., Chang, C.-P., Clements, J. H., and Kuo, M.-S. (1995). Synthesis and spectroscopic studies of 4-Formyl-4′-N, N-dimethylamino-1, 1′-biphenyl: The unusual Red Edge Effect and efficient laser generation. *J. Fluorescence* **5,** 369.

Demchenko, A. P. (1981). Dependence of human serum albumin fluorescence spectrum on excitation wavelength. *Ukr. Biochim. Zh.* **53,** 22.

Demchenko, A. P. (1982). On the nanosecond mobility in proteins. Edge excitation fluorescence red shift of protein-bound 2-(p-toluidinylnaphthalene)-6-sulfonate. *Biophys. Chem.* **15,** 101.

Demchenko, A. P. (1986a). "Ultraviolet spectroscopy of proteins," Heidelberg, Springer Verlag.

Demchenko, A. P. (1986b). Fluorescence analysis of protein dynamics. *Essays Biochem.* **22,** 120.

Demchenko, A. P. (1988). Red-edge-excitation fluorescence spectroscopy of single-tryptophan proteins. *Eur. Biophys. J.* **16,** 121.

Demchenko, A. P. (1989). Site-selective excitation: A new dimension in protein and membrane spectroscopy. *Trends Biochem. Sci.* **10,** 374–377.

Demchenko, A. P. (1991). Fluorescence and dynamics in proteins. *In* "Topics in Fluorescence Spectroscopy" (J. R. Lakowicz ed.), Vol. **3,** pp. 61–111. Plenum Press, New York.

Demchenko, A. P. (1992). Does biocatalysis involve inhomogeneous kinetics? *FEBS Lett.* **310,** 211.

Demchenko, A. P. (1994). Protein fluorescence, dynamics and function: Exploration of analogy between electronically excited and biocatalytic transition states. *Biochim. Biophys. Acta* **1209,** 149.

Demchenko, A. P. (2002). The red-edge effects: 30 years of exploration. *Luminescence* **17,** 19–42.

Demchenko, A. P. (2006). Visualization and sensing of intermolecular interactions with two-color fluorescent probes. *FEBS Letters* **580,** 2951.

Demchenko, A. P., and Ladohin, A. S. (1988). Red-edge fluorescence spectroscopy of indole and tryptophan. *Eur. Biophys. J.* **15,** 369.

Demchenko, A. P., and Sytnik, A. I. (1991a). Site-selectivity in excited-state reactions in solutions. *J. Phys. Chem.* **95,** 10518–10524.

Demchenko, A. P., and Sytnik, A. I. (1991b). Solvent reorganizational red-edge effect in intramolecular electron transfer. *Proc. Natl. Acad. Sci. USA* **88,** 9311.

Haugland, R. P. (1996). Handbook of fluorescent probes and research chemicals. Eugene, OR, Molecular Probes.

Klymchenko, A. S., and Demchenko, A. P. (2008). Multiparametric probing of microenvironment with solvatochromic fluorescent dyes. *Methods in Enzymology* **450**R21: 37 (This volume).

Lakowicz, J. R. (2000). On spectral relaxation in proteins. *Photochem. Photobiol.* **72,** 421.

Marcus, R., and Sutin, N. (1985). Electron transfers in chemistry and biology. *Biochim. Biophys. Acta* **811,** 265.

Maroncelli, M., McInnis, J., and Fleming, G. R. (1989). Polar solvent dynamics and electron transfer reactions. *Science* **243,** 1674.

Mazurenko, Y. T., and Bakhshiev, N. G. (1970). The influence of orientational dipolar relaxation on spectral, temporal and polarization properties of luminescence in solutions. *Opt. Spektr.* **28,** 905.

Mulkherjee, S., and Chattopadhyay, A. (1995). Wavelength-selective fluorescence as a novel tool to study organization and dynamics in complex biological systems. *J. Fluorescence* **5,** 237.

Nemkovich, N. A., Rubinov, A. N., and Tomin, V. I. (1981). Kinetics of luminescence spectra of rigid dye solutions due to directed electronic energy transfer. *J. Lumin.* **23,** 349.

Nemkovich, N. A., Rubinov, A. N., and Tomin, V. I. (1991). Inhomogeneous broadening of electronic spectra of dye molecules in solutions. *In* "Topics in fluorescence spectroscopy," (J. R. Lakowicz, ed.), **2,** p. 367. Plenum Press, New York.

Raghuraman, H., Kelkar, D. A., and Chattopadhyay, A. (2005). Novel insights into protein structure and dynamics utilizing the Red Edge Excitation Shift approach. *In* "Reviews in Fluorescence," (C. D. Geddes and J. R. Lakowicz, eds.), pp. 199–222.

Richert R., Blumen A., Eds. (1994). Disorder effects on relaxational processes: Glasses, polymers, proteins. Springer-Verlag, Berlin.

Rubinov, A. N., Tomin, V. I., and Bushuk, B. A. (1982). Kinetic spectroscopy of orientational states of solvated dye molecules in polar solution. *J. Lumin.* **26,** 377.

Suppan, P., and Ghoneim, N. (1997). Solvatochromism. Cambridge, UK, Royal Society of Chemistry.

Vincent, M., Gallay, J., and Demchenko, A. P. (1995). Solvent relaxation around the excited state of indole: Analysis of fluorescence lifetime distributions and time-dependent spectral shifts. *J. Phys. Chem.* **99,** 34931.

Weber, G., and Shinitzky, M. (1970). Failure of energy transfer between identical aromatic molecules on excitation at the long wave edge of the absorption spectrum. *Proc. Natl. Acad. Sci. USA* **65,** 823.

CHAPTER FIVE

Fluorescence Approaches to Quantifying Biomolecular Interactions

Catherine A. Royer* *and* Suzanne F. Scarlata[†]

Contents

1. Introduction	80
2. Fluorescence Observables	80
2.1. Intensity decay measurements	81
2.2. Förster resonance energy transfer	81
2.3. Fluorescence correlation spectroscopy	82
2.4. Emission energy	84
2.5. Fluorescence anisotropy	84
3. Designing a Fluorescence Experiment	85
3.1. Choosing fluorescence probes	85
3.2. Protein–ligand interactions	86
3.3. Protein–nucleic acid interactions	91
3.4. Aqueous phase protein–protein interactions	96
3.5. Protein–protein interactions in or at the cell membrane	98
4. Conclusions	103
References	103

Abstract

This review is conceived as an introductory text to aid in the understanding and conception of fluorescence-based measurements of biomolecular interactions. The major fluorescence observables are introduced briefly. Next, the criteria that are involved in the choice of the fluorescent probe are discussed in terms of their advantages and disadvantages for different types of experiments. The last sections deal with the experimental design for fluorescence-based assays aimed at detecting different types of biomolecular interactions. Included in our examples are protein-ligand interactions, protein-nucleic acid interactions, aqueous phase protein-protein interactions and protein interactions in or at the

* Centre de Biochimie Structurale, 29 rue de Navacelles, 34090 Montpellier Cedex, France
[†] Department of Physiology and Biophysics, School of Medicine, State University of New York at Stony Brook, Stony Brook, New York

Methods in Enzymology, Volume 450   © 2008 Elsevier Inc.
ISSN 0076-6879, DOI: 10.1016/S0076-6879(08)03405-8   All rights reserved.

cell membrane. We hope that this introduction will be of use to students and researchers considering the use of fluorescence in their work.

1. Introduction

Progress in functional genomics will require the characterization and quantification of the interactions between the multiple players in cellular signaling networks. This is particularly true for functional genomics programs with the goal of developing therapeutic approaches to modulate these signaling networks for the targeted treatment of specific diseases. Such detailed molecular medicine cannot be conceived without a detailed understanding of the structure–function relationships governing these molecular interactions. For this reason, methodologies for measuring and analyzing these interactions both *in vitro* and *in vivo* are experiencing rapid development. Among the approaches most widely used are those based on fluorescence techniques. Since the late 1950s and the seminal work of Gregorio Weber in the application of fluorescence techniques to study biomolecular interactions (Weber, 1952a,b), the field has grown significantly, and fluorescence is routinely used in both high-throughput screening applications and in fundamental research programs.

The widespread application of fluorescence in the study of biological interactions can be explained by a combination of its sensitivity, ease of use, rapidity, reproducibility, and adaptability. With appropriate visible dyes and sensitive instrumentation, detection of very low concentrations (picomolar), and indeed even single molecules, has become fairly routine. This high sensitivity allows for binding studies to be carried out under true equilibrium conditions (i.e., the binding profile is determined by the mass action of the added partner, while the target fluorescent molecule is present at concentrations below the dissociation constant). Moreover, there is no need to separate bound from free species in a fluorescence-based titration. Thus, there are no kinetic artifacts associated with the measurement. Moreover, because the timescale of fluorescence is in the nanosecond range, binding and dissociation kinetics can be studied on the microsecond timescale. Since fluorescence measurements can be made quickly on very small volumes of sample, they are well adapted to high-throughput applications as well. This rapidity and small sample requirements also render fluorescence quite useful in the global analysis of binding reactions for fundamental studies of structure–function relationships because a large number of experiments can be carried out with limited sample.

2. Fluorescence Observables

There are three basic fluorescence observables that can be used in various ways to detect and characterize an interaction: intensity, color, and anisotropy. These three observables can be detected in steady-state or in time-resolved mode depending upon the degree of detail necessary for determining the parameters of the interaction. The experimental design can be more or less complicated, allowing for the very specific detection of interactions in inhomogeneous media.

2.1. Intensity decay measurements

In addition to simple detection of changes in the intensity of fluorescence, much more complex, and hence informative intensity measurements can be made. For the simplest fluorophores, the intensity decay is exponential and the fluorescence lifetime measurements as the lifetime is defined as the inverse of the rate constant for emission and is the time required for the intensity to decay to $1/e$ of its initial value. In biological systems, the intensity decay is often much more complex and can reflect the multiple conformational states of the biomolecule. For example, fluorescence intensity decay measurements on the pico- to nanosecond timescale, especially if they are carried out in conjunction with an investigation of the wavelength dependence of their values, correspond to absolute values that do not depend upon the instrumentation, and can provide direct information on the fractional population of bound and unbound species (Knutson et al., 1982). Lifetime measurements can be coupled with quenching experiments or anisotropy, or other variations of physical parameters for an in-depth characterization of the interaction parameters (Beechem and Brand, 1986).

2.2. Förster resonance energy transfer

Intensity measurements of donor and acceptor emission (steady-state or time-resolved) give access to measurements that monitor Förster resonance energy transfer (FRET) between donor and acceptor fluorophores on two interacting molecules of interest in the presence of multiple interacting or noninteracting molecules (Clegg, 1995). FRET occurs when an acceptor fluorophore (i.e., a probe whose absorption spectrum overlaps with the emission spectrum of the donor) is close enough (see below) to the donor that the excited-state energy of the donor can be transferred directly to the acceptor before emission from the donor. Emitted light will, then, come from the acceptor. This process is distinct from trivial reabsorption in which the emission from a donor is subsequently absorbed by the acceptor. Note that FRET is critically dependent on the distance between the donor and acceptor, which makes it valuable to detect associations between two proteins labeled with a donor/acceptor pair.

FRET is one of several events that can compete for the emission of a photon from an excited fluorophore during the time that a probe is in the excited state; others include quenching, thermal relaxation and conversion to a triplet state. The rate of energy transfer (k_t) can be written as

$$k_t = 1/\tau_D (R_o/R_{DA})^6, \qquad (5.1)$$

where τ_D is the fluorescence lifetime of the donor, R_{DA} is the distance between the donor and the acceptor, and R_o is the critical transfer distance, or the distance at which 50% of the donor fluorescence is lost to transfer. R_o can be calculated from the quantum yield of the donor (Q_D, which refers to the number of photons emitted over number of photons absorbed), the index of refraction of the solvent between the donor and acceptor (n), the overlap integral between the donor emission and the acceptor absorption spectra, and the orientation between the donor and acceptor dipoles (κ^2) where

$$R_o = (8.8 \times 10^{25} \kappa^2 Q_D J/n^4)^{1/6} \text{cm}, \tag{5.2}$$

$$J = \mu_A(\lambda) f_D(\lambda) \lambda^4 d\lambda, \tag{5.3}$$

where μ_A is the extinction coefficient of the acceptor at wavelength λ, and $f_D(\lambda)$ is the emission intensity of the donor normalized so that the integral of $f_D(\lambda)$ over its emission band equals Q_D:

$$\kappa^2 = (\sin\zeta_D \sin\zeta_A \cos\theta - 2\cos\zeta_D \cos\zeta_A)^2, \tag{5.4}$$

where ζ_D and ζ_A are the angles between the vector of R_{DA} and the donor and acceptor dipole directions, and θ is the azimuthal angle between the donor and acceptor dipoles.

The above expressions show that the extent of FRET depends on the ability of the donor to absorb the light to be transferred to the acceptor, and so a donor with a high extinction coefficient and high quantum yield is desired. Similarly, the acceptor should have a high extinction coefficient, as well, for efficient resonance with the excited-state dipole of the donor. A very difficult parameter to assess is the orientation factor, or κ^2. The simple reason for its importance is that transfer can only occur when some population of the dipoles are similarly aligned. If the dipoles are perpendicular, transfer will not occur. Fortunately, in solution, the donor and acceptor molecules typically can rotate and sample a large number of orientations during the excited state of the donor. Under conditions of free rotation, κ^2 is 2/3. In most circumstances, free rotation is assumed, and the extent of rotational motion can always be checked by fluorescence anisotropy (see below). Unless one is using energy transfer to determine exact intermolecular distances, then the 2/3 assumption is usually reasonable.

2.3. Fluorescence correlation spectroscopy

The measurement of fluorescence intensity fluctuations and their analysis using various histogram or correlation methods has become a major approach for studying interactions between biomolecules (Elson, 2004;

Haustein and Schwille, 2004). We will limit the present discussion of fluorescence correlation spectroscopy (FCS) to measurements of changes in the diffusion time by autocorrelation analysis as a means of characterizing biomolecular interactions. Fluctuation spectroscopy is based on the measurement of the time course of fluorescence intensity emanating from a very small observation volume (1 fl or less). Typically, this is achieved by either using a confocal or a two-photon microscope setup. In the absence of chemical or photophysical events giving rise to fluctuations, these will occur arise uniquely from the diffusion of fluorescent species into and out of the observation volume. These intensity fluctuations based on local concentration fluctuations will show a characteristic time dependence that is linked to the time the molecule spends in the volume. Since larger molecules diffuse more slowly that small ones, FCS measurements can be used to study diffusion times, and hence, binding. The intensity fluctuations profile is autocorrelated, which amounts to probing how similar the profile is at a time delay, τ, after time, t:

$$G(\tau) = \langle \delta F(t) \delta F(t+\tau) \rangle / \langle F(t) \rangle^2. \tag{5.5}$$

If no other photophysical or physical process gives rise to intensity changes, then this autocorrelation function can be analyzed in terms of the diffusion time, τ_D, and the shape and size of the volume defined by the constants, r_o and z_o:

$$G(\tau) = 1/N(1 + \tau/\tau d)^{-1}(1 + r_o^2 \tau / z_o^2 \tau_d)^{-1/2}. \tag{5.6}$$

At time delay 0, the amplitude of the function allows to calculate the concentration, C, of the fluorescent species:

$$G(0) = 1/N = 1/(\langle C \rangle V). \tag{5.7}$$

The diffusion coefficient, D, is related to the diffusion time differently for one- and two-photon FCS, respectively:

$$\tau d = r_o^2 / 4Dot, \quad \tau d = r_o^2 / 8D. \tag{5.8}$$

From it, one can calculate the radius of an approximately spherical diffusing particle:

$$D = kT/6\pi\eta R. \tag{5.9}$$

It should be noted that the radius is a cube root function of the molecular weight

$$R = (3MWV_h/4\pi N_A)^{1/3} \qquad (5.10)$$

such that one needs an eightfold increase in the molecular weight to obtain a doubling of the diffusion time.

2.4. Emission energy

Fluorescence emission energy is highly sensitive to the fluorophore environment, and changes in environment often occur upon complex formation. Complex formation between two proteins for instance, buries considerable surface area. A fluorophore (such as an intrinsic tryptophan residue or even an extrinsic probe) placed in or near the interaction interface will likely become less exposed to the solvent, water, when the complex is formed. In such cases, it is typical that the intensity will increase and the energy of emission will shift to shorter wavelengths due to the concomitant decrease in solvent quenching and in solvent relaxation. To construct binding profiles based on the change in emission energy, it is best to calculate the center of spectral mass, $\bar{v}_g$, rather than using the emission wavelength of maximum intensity since the former is more sensitive to changes in shape or breadth of the spectrum, and less sensitive to noise:

$$\bar{v}_g = \frac{\sum_i F_i v_i}{\sum F_i}. \qquad (5.11)$$

2.5. Fluorescence anisotropy

Polarized fluorescence intensity measurements can provide information about the size of the fluorescent species through the determination of the rotational correlation time, which is particularly useful in characterizing interactions. Fluorescence anisotropy, A, is defined as the difference between vertically and horizontally polarized emission, normalized to the total fluorescence intensity, under conditions of vertically polarized exciting light, and is related to the rotational correlation time of the fluorophore, τ_c, through the limiting anisotropy, A_o, the fluorescence lifetime, τ, the solution viscosity, η, the hydrated molecular volume, V_h, and temperature, T:

$$A = (I_\| - I_\perp)/(I_\| + 2I_\perp) \qquad (5.12)$$

and

$$A_o/A - 1 = \tau/\tau_c = RT\tau/\eta V_h. \qquad (5.13)$$

Thus, anisotropy depends on molecular volume or size and therefore, is sensitive to changes in size that occur upon complexation. Because steady-state anisotropy measurements of labeled biomolecules combine the depolarization due to macromolecular tumbling and that due to local probe dynamics, time-resolved anisotropy measurements are often required to differentiate between the two.

3. Designing a Fluorescence Experiment

The exceptional versatility in the design of fluorescence experiments carries many advantages, but the experimenter must be aware of the possibilities and limitations of these different approaches and design the experiment based on the possibilities afforded by the technique and on the requirements for the particular system under study. Rather than to attempt to review the enormous literature on the subject of the use of fluorescence to study interactions between biomolecules, this review will attempt to explain how to design a fluorescence experiment for particular purposes and discuss the advantages and limitations of the possible approaches. Thus, we will focus on three types of biological complexes: protein–ligand interactions, protein–nucleic acid interactions, and protein–protein interactions, with special attention given to protein complexes that are formed in or at biological membranes.

3.1. Choosing fluorescence probes

In addition to the different fluorescence observables and the combinations of ways in which they can be exploited, fluorescence experiments can be based on the observation of either intrinsic fluorescence or extrinsic probes or some combination of the two. Intrinsic probes, most particularly tryptophan residues present the enormous advantage of not introducing a perturbation into the system. On the other had, there are often more than one tryptophan residues in the protein of interest and even if there is only one, it is not necessarily located at an ideal position in the structure. Moreover, for protein–protein interactions it is almost impossible to distinguish between the fluorescence of one partner and that of the other. Finally, the quantum yield of tryptophan is typically about 10%, which severely limits the sensitivity of tryptophan-based assays. Tyrosine and phenylalanine have even lower extinction coefficients and quantum yields, and so their usefulness is even more limited. For these reasons, it is often advantageous to use extrinsic fluorescent dyes coupled (usually) covalently to the biomolecules of interest. In the case of nucleic acids, the extremely low quantum yields of the nucleotides, precludes their use in most applications. There are highly fluorescent nucleotide analogues, such as 2-amino purine, however, that can be used without introducing an enormous perturbation. For protein research, some fluorescent dyes, such as the naphthalene derivatives, ANS,

PRODAN, Acrylodan (Weber, 1976; Weber and Farris, 1979) that are excited in the near UV and fluoresce in the visible, exhibit very large excited-state dipoles, and thus, their emission energy is very sensitive to their environment. Thus, placed correctly, they report quite well on binding events. One note of caution is that one must be very careful to correct the color-based binding curves obtained with these probes for changes in intensity, since these are also rather large for this family of probes. Alternately, a large number of dyes, among them the Cyanine Cy3 and Cy5 (Amersham Pharmacia), and the Alexa derivatives (Molecular Probes, Eugene Oregon), that are excited and emit in the visible region show very little variation in their intensity or color upon changes in their environment. This relative insensitivity is very useful for either anisotropy or fluctuation measurements. Moreover, the extinction coefficients and quantum yields of these dyes are very high, rendering them appropriate for the study of very high-affinity interactions for which the target molecules must be used at very low concentrations, as well as single molecule applications.

3.2. Protein–ligand interactions

When confronted with determining the parameters governing the interaction of a protein with its ligand(s), fluorescence may prove to be the method of choice under certain circumstances. The main issue is the fact that one needs a fluorescent observable to make such a measurement, whereas for methods such as isothermal titration calorimetry, one simply requires a reasonable heat of reaction. However, if one suspects that the affinity is very high, or else one is interested in the kinetic parameters of the interaction, then it may be worthwhile to try to develop a fluorescence-based assay. In many cases, the binding of the ligand to the protein causes a change in the intrinsic tryptophan (or tyrosine) fluorescence of the protein. In this case, the protein is titrated with the ligand. However, one must bear in mind that the sensitivity of the intrinsic tryptophan fluorescence is not particularly high. Depending upon the instrument used and its configuration (monochromator or filter in emission) the best one can do is about 0.1 μM in tryptophan. Moreover, if the effect of the ligand binding is to quench the intrinsic fluorescence, then the sensitivity is limited even more. This is not so much of a problem if one is interested in carrying out kinetic measurements, but limits the accessible K_d values to those in the micromolar range or above. In any case, one can test the system using the protein at relatively high concentration 10 μM or above, and saturating with ligand to ascertain whether the intensity, color, or polarization of the intrinsic tryptophan emission is perturbed by the interaction. Interestingly, this is often the case since tryptophan residues are often found in binding sites or at subunit interfaces, given their versatility in forming both hydrophobic interactions and hydrogen bonds. Often ligand binding tends to limit both the solvent accessibility and the local mobility of the fluorophore, and thus the spectrum will shift to the blue, the intensity will change and the polarization will increase. In certain cases, if the ligand is close enough to the tryptophan(s) or if a conformational change is associated with ligand binding and this brings a quenching group into the vicinity of the fluorophore, then a decrease in intensity can be observed. Thus, while one is never assured to observe an intrinsic fluorescence change upon ligand binding, it turns out that this is often the case.

If no change in intrinsic fluorescence occurs when the protein interacts with its ligand, and due to other limitations such as sample availability or behavior that is incompatible with other measurement techniques, fluorescence remains the most viable choice for a binding assay, then one must consider various means employing extrinsic fluorescent probes. One can either use a fluorescent ligand, or alternatively if the structure of the protein and the binding site of the ligand are known, one can use site directed mutagenesis to place a cysteine residue near the binding site in order to specifically place a fluorescent dye close to the binding site if one seeks to monitor changes in color or intensity. On the other hand, the dye can be placed at a large distance from the binding surface if one plans to use fluorescence anisotropy or fluctuation analysis. In this latter design, the perturbation to the system is minimized. Moreover, one need not engineer a cysteine residue into the protein since specific and unique labeling of the amino terminus can be achieved using amine-reactive dyes at pH 8 for short times, since the pKa of the amino terminus is at least two log units below that of the lysine and arginine residues in the protein, and they are, therefore, much less reactive.

If the experimental design is to be based on a fluorescent ligand, there are a number of issues that should be considered. Chemically modifying a small ligand may be both difficult and costly, so this is typically only done in cases where the assay will be used for a large number of experiments. Moreover, the extrinsic dye may be as large or even larger than the ligand itself, thus introducing a significant perturbation of the system. However, this latter issue is not truly a problem. Indeed, even if the affinity of the protein for the fluorescent ligand is significantly reduced (or even enhanced) compared to the natural ligand, one can always use competition assays with unlabeled ligand to determine the true affinity.

One of the simplest situations for labeling the ligand is when the ligand is a peptide. There are a variety of chemical approaches for introducing a fluorescent label into small peptides, and most peptide synthesis companies offer the possibility of a fluorescent moiety. The most appropriate and simplest assay for titration of a fluorescent ligand with a protein is that based on changes in fluorescence anisotropy. Indeed, given that typically the ligand is much smaller than the protein, the dynamic range for the observed change in the fluorescence parameter will be quite large. In analyzing anisotropy, it is important to remember that this observable represents the sum of the anisotropy values of all of the fluorescent species detected, weighted for their fractional population and their relative quantum yields. Thus, the dyes which show little variation in intensity with changes in environment are to be preferred in this case.

In Fig. 5.1 is shown an example of the titration of a small fluorescent peptide (derived from a nuclear receptor coactivator) with its nuclear receptor partner (Pogenberg et al., 2005). In this particular case, the interaction is dependent upon a third small retinoid ligand, which was present at saturating concentrations in these titrations. The affinity of the receptor for the peptide is modulated by the structure of the bound retinoids, which cause distinct conformational changes in the receptor that influence the peptide-binding site. It can be seen from Fig. 5.1 that increasing the receptor concentration causes a large change in anisotropy (100 milli-

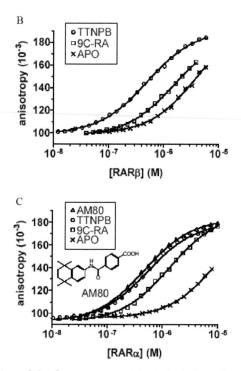

Figure 5.1 Titration of the fluorescent peptide derived from the second interacting domain (NR Box 2) of the nuclear receptor coactivator 1 (SRC-1) protein by the ligand-binding domains of the retinoid receptors (B) RARβ and (C) RARα in the absence (apo) or in the presence of various agonists. The figure was taken from Pogenberg et al. (2005). The synthetic retinoids (TTNPB or AM80) activate RARα and RARβ more efficiently than 9C-RA does, and hence the coactivator peptide-binding affinity is higher.

anisotropy units) of the target fluorescent peptide. The dissociation constants were in the 10–100 nM range, and the dye used in these experiments was fluorescein, which has a very high extinction coefficient and a high quantum yield, allowing the target fluorescent peptide concentration to be about 5 nM, well below the measured K_d values. As can be seen from the quality of the data, the signal to noise in the anisotropy values was very high, and the binding curves are very well defined. This example demonstrates that very subtle structure–function relationships can be investigated using such techniques. We note that in addition to the high quantum yield probe, the polarimeter used to collect the data is a very high sensitivity polarimeter with very low noise analogue detection (Beacon 2000, Panvera Corp., Madison, WI), and that this quality of data is not necessarily available on all fluorometers, although we have obtained similar quality data at even lower fluorescein concentrations (200 pM) using a CW argon ion laser-based excitation coupled into an ISS KOALA (ISS, Champaign, IL) with cut-on filters (rather than monochromators) in emission (Grillo and Royer, 2000; Grillo et al., 1999). In Fig. 5.2 is shown an example of the RAR/RXR heterodimeric system (Pogenberg et al., 2005), in which the fluorescent peptide derived from the sequence of the second interacting domain NR Box 2 of the coactivator, SRC-1, is used in competition assays with nonfluorescent peptides of the same and different composition, demonstrating the principle that the introduction of a fluorescent label does not necessarily preclude determination of the "true" affinity.

Fluctuation spectroscopy can also be used to study protein small ligand interactions. In Fig. 5.3 are shown the autocorrelation profiles of the fluorescent peptide used in the study above free and in presence of saturating receptor. Despite the fact that the translational diffusion time is only a weak function of the molecular weight (since it depends upon molecular volume

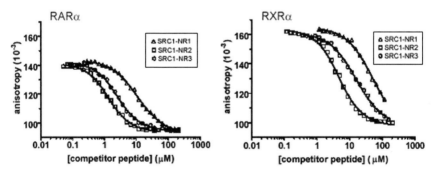

Figure 5.2 Competition assays based on the fluorescence anisotropy of a the peptide derived from the NR Box 2 sequence of the SRC-1 nuclear receptor coactivator with nonfluorescent peptides derived from the NR Box 1, 2, and 3 sequences. The figure is taken from Pogenberg et al. (2005).

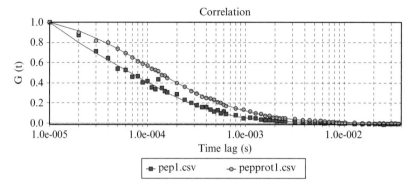

Figure 5.3 Normalized autocorrelation functions obtained from two-photon FCS measurements of the fluorescein-labeled SRC-1 NR Box 2 peptide in absence and in presence of the RAR/RXR ligand-binding domain heterodimer (V. Pogenberg, W. Bourguet, and C. A. Royer, unpublished results).

it increase with the cube root of the molecular weight), the very large change in size that occurs when the small peptide binds to a large protein is easily detected in FCS measurements. To construct the binding profile, one collects autocorrelation profiles at multiple concentrations of the protein and then the curves must be analyzed in terms of the diffusion times of the free and bound species, τ_{dF} and τ_{dB}, respectively:

$$G(\tau) = 1/N(Y_F(1 + \tau/\tau_{dF})^{-1}(1 + r_o^2\tau/z_o^2\tau_{dF})^{-1/2} \\ + Y_B(1 + \tau/\tau_{dB})^{-1}(1 + r_o^2\tau/z_o^2\tau_{dB})^{-1/2}), \quad (5.14)$$

where

$$Y_F = \eta_F^2 X_F/(\eta_F X_F + \eta_F X_F^2) \text{ and } Y_F = \eta_B^2 X_B/(\eta_B X_F + \eta_F X_F^2), \quad (5.15)$$

with η_F and η_B and X_F and X_B corresponding to the molecular brightness and the fractional population of the free and bound species, respectively. If the brightness of the free and bound species is the same, then Y_F and Y_B correspond directly to the fractional populations of the two species. In carrying out such experiments it is important to use a global analysis of the entire family of autocorrelation curves for the multiple concentrations tested to recover the diffusion times, molecular brightness values, and populations with a reasonable degree of certainty. Once the fractional populations of free and bound species as a function of concentration are known, these titration profiles can be analyzed to recover the affinities.

3.3. Protein–nucleic acid interactions

Fluorescence has been used extensively to study protein–nucleic acid interactions since the early 1970s (Helene and Dimicoli, 1972; Helene et al., 1969, 1971). These early studies were typically based on measuring the quenching of intrinsic tryptophan or tyrosine fluorescence of proteins and peptides upon their interactions with nucleic acids. In these cases the proteins were titrated by the nucleic acids. More recently, the labeling of the nucleic acid (Heyduk and Lee, 1990) has allowed for much increased sensitivity allowing the study of very high-affinity complexes. In such studies, whether detection is by FRET or by anisotropy, or more recently by FCS, the labeled DNA is titrated with the protein. This allows the target DNA to be present at much lower concentration than the dissociation constant, corresponding to true equilibrium conditions.

It should be noted however, that in protein–DNA interactions, titrating the protein with the DNA and titrating the DNA with the protein are not necessarily symmetric and equivalent operations. In simple binding, it does not matter which partner is the target and which is the titrant, but in protein–nucleic acid interactions, the binding is often not simple. Indeed, resolving multiple complexes of different stoichiometry is a major headache in the protein–nucleic acid field. Under conditions of saturating protein and low concentrations of DNA, one often observes the population of higher-order species, containing multimers of the protein bound more or less specifically to the DNA or directly to the specifically bound protein. The problem, of course, is to know that these species exist. In gel mobility shift assays, such higher-order species appear as separate bands on the gel of lower mobility. The problem with these assays is that prior to entering the gel and even during electrophoresis, a re-equilibration of the complexes under the salt, pH, and temperature conditions of the gel itself renders the observations difficult to interpret. In filter-binding assays, of course there is no indication of the size of the complex. This is also true for studies based on the detection of the quenching of intrinsic tryptophan or tyrosine fluorescence. However, whether it is the labeled nucleic acid or the labeled protein that is under study, both fluorescence anisotropy and FCS can provide information about the sizes of the protein–nucleic acid complexes.

In titrations of a labeled protein (the histone octamer labeled with Dansyl chloride) by the DNA, we have observed by anisotropy, first the formation of higher-order complexes at low concentrations of the DNA in the beginning of the titration, and then as more DNA was added, the partitioning of the proteins out of these higher-order complexes onto the free DNA (Royer et al., 1989). This sort of behavior causes a peak in the anisotropy profile and depends upon the concentration of the two partners (protein and DNA) relative to each other and to the various K_d values of the system. In the case of titration of a labeled DNA (typically fluorescein labeled at the 5′-end during oligonucleotide synthesis using a fluorescein phosphoramidite with a six carbon linker), the formation of higher-order complexes is sometimes evident as intermediate plateaus in the titration curve followed by a further increase in anisotropy as more protein is added. We have observed such behavior for the trp repressor (Brown *et al.*, 1999; Grillo and Royer, 2000; Grillo *et al.*, 1999) and for the RAR/RXR heterodimer binding to their cognate DNA sites (Poujol *et al.*, 2003). Depending upon the degree of cooperativity of the formation of these higher-order species they can be identified more or less easily, and can sometimes be differentially competed with specific and nonspecific DNA. In Fig. 5.4A, the titration of the fluorescein-labeled target DR5 oligonucleotide with the RAR/RXR heterodimer (Poujol *et al.*, 2003)

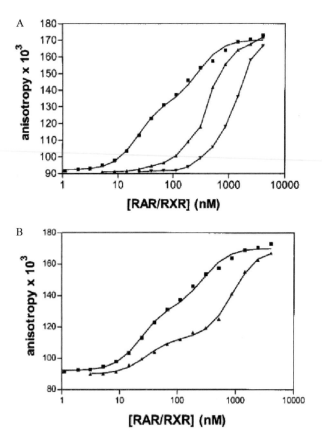

Figure 5.4 Titration by RAR/RXR of a fluorescein-labeled 32-bp oligonucleotide bearing the target sequence (DR5) for the RAR/RXR heterodimer alone and in presence of (A) 10 and 50 excess unlabeled DR5 and (B) in absence and in presence of 50-fold excess nonspecific unlabeled oligonucleotide. The figure was taken from the work of Poujol *et al.* (2003).

reveals a relatively subtle bump in the titration curve that could correspond to a first dimer bound followed by the binding of a second dimer. Competition with specific unlabeled DR5 leads to a shift of the curves to much higher concentration, and thus a disappearance of this complex behavior. However, when competition is performed with an unlabeled nonspecific sequence (Fig. 5.4B), the intermediate plateau is accentuated in the titration curve because the nonspecific DNA is more efficient at competing out the higher-order, lower-affinity complex than the specific, high-affinity heterodimeric complex.

We note also that the salt concentration in protein–nucleic acid titrations not only affects the affinity of the interaction, but can drastically modify the stoichiometry of the complexes formed, since increasing salt concentration typically affects nonspecific interactions more strongly than specific ones. Thus, while the overall affinity will decrease with increasing salt, the specificity ratio will increase. A specific complex of the estrogen receptor with a fluorescein-labeled estrogen response element shows strong salt dependence of the affinity between 150 and 300 mM KCl (Fig. 5.5A) (Boyer *et al.*, 2000). In Fig. 5.5B, one can see that at very low salt concentration the titration of a fluorescein-labeled mutant estrogen response element with the estrogen receptor leads to a very large increase in anisotropy, whereas in presence of 200 mM KCl, specific dimeric complexes, exhibiting a much lower anisotropy are formed only with the wild-type response element.

Despite the fact that fluorescence anisotropy can provide information about the size and number of the different protein–nucleic acid complexes that are populated during titrations under various conditions, the sizes of the complexes that can be studied by this technique are limited by the lifetime of the fluorophore, which is typically around 4 ns. Thus, the maximum size of complex that can be distinguished is limited to rotational correlation times of about 40 ns, which with approximately 1 ns for each 2500 Da, leads to a size limit of about 100 kDa. Moreover, the steady-state anisotropy that one measures includes depolarization due to the global tumbling of the particle, as well as the local rotational mobility of the probe. In our experience, it is often latter that is greatly constrained upon formation of the higher-order complexes, since the global rotational correlation time limit has already been reached. We note that the interaction of the dye with the oligonucleotide can cause quenching of the fluorescence that is dependent upon the sequence of the oligonucleotide at the $5'$-end, as well as the electrostatic environment that can be modified by the binding of protein. Thus, while anisotropy studies can clearly show that complexes that are different structurally and dynamically are formed, it is often difficult to be certain that these differences are due to higher-order species.

We have found that FCS can be quite useful to characterize protein–nucleic acid interactions. The autocorrelation signal that decays due to

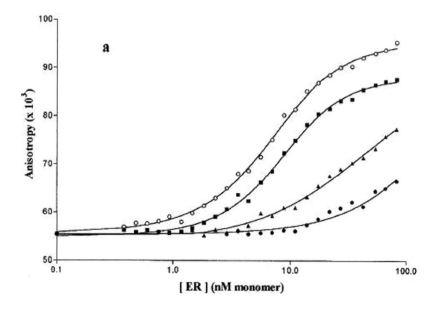

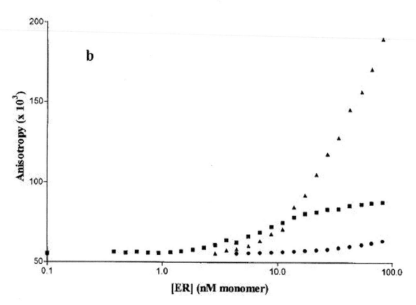

Figure 5.5 Anisotropy-binding profiles of (A) a fluorescein-labeled oligonucleotide bearing the consensus sequence of the estrogen response element F-vitERE with full-length human estrogen receptor α at salt concentrations between 150 and 300 mM KCl and (B) the titration of F-vitERE at 200 mM KCl, and that of a mutant sequence mutF at 200 mM KCl and in absence of salt. This figure was taken from the work of Boyer *et al.* (2000).

translational diffusion, is not limited by the lifetime of the fluorophore, and moreover, is not sensitive to the local rotational motions of the dye. While the translational diffusion is a weaker function of the molecular weight than is the rotational diffusion, the resolution of FCS does allow distinguish the approximate twofold difference in molecular weight that occurs when a second protein binds to a protein–nucleic acid complex. In Fig. 5.6 are shown the anisotropy profiles obtained from the titration of the fluorescein-labeled target ERE with estrogen receptor α in presence of estradiol, 4-hydroxytamoxifen and the antagonist, ICI 182780 (Margeat et al., 2003). While hydroxytamoxifen changes the affinity and cooperativity or ER–DNA interactions, the antagonist, ICI 182780 results in a much higher plateau in the anisotropy value. While we suspected that in presence of ICI, complexes of higher order were formed, time-resolved fluorescence demonstrated that the global correlation time was too long to be well defined using the fluorescein probe, and that the large difference in anisotropy plateau was due primarily to a decrease in local probe rotational mobility. While the anisotropy observable suggested that a higher-order complex was forming, the ambiguity inherent in these measurements did not allow us to be certain. FCS experiments on the other hand (Fig. 5.7), clearly show that the translational diffusion of the ER/ERE complex formed in presence of ICI 182780 is significantly slower, and hence the complex significantly larger, than in absence of ligand or in presence of the agonist estradiol (A. Bourdoncle and C. A. Royer, unpublished results). Analysis of the data are consistent with a complex involving a two dimers of the receptor bound to the ERE in presence of ICI, as opposed to a simple dimer/DNA complex in presence of estradiol. These

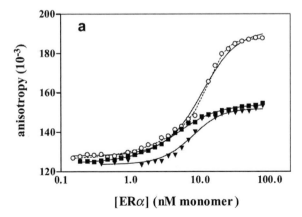

Figure 5.6 Anisotropy titrations of the F-vitERE from Fig. 5.5 with the estrogen receptor β in absence of ligand and in presence of the antagonists 4-hydroxytamoxifen and ICI 182780. This figure is taken from the work of Margeat *et al.* (2003).

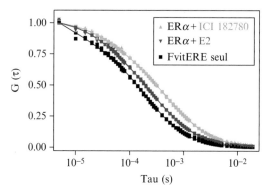

Figure 5.7 Normalized two-photon FCS autocorrelation profiles of the fluorescein-labeled F-vitERE from Fig. 5.5 in alone and in presence of saturating human estrogen receptor β and saturating agonist estradiol (E2) and in presence of the saturating receptor and saturating antagonist ICI 182780 (A. Bourdoncle and C. A. Royer, unpublished results). (See Color Insert.)

results highlight the complimentarity of anisotropy and FCS experiments. Indeed, we increasingly use both approaches in our studies of protein–nucleic acid interactions.

3.4. Aqueous phase protein–protein interactions

The same fluorescence-based techniques that are used to study protein–nucleic acid interactions can also be used to study protein–protein interactions. Since intrinsic tryptophan fluorescence is not particularly sensitive, it is best to use extrinsic fluorescent dyes for such studies. The simplest approach is to specifically label the amino terminus using isothiocyanate or NHS ester derived probes at pH 8. If for some reason, the amino terminus is not reactive, then one must consider engineering a specific cysteine residue to covalently attach the probe. To study homologous protein subunit oligomerization, of course there is only one protein involved, and one simply can measure the anisotropy of the solution as a function of protein concentration. We have learned that much better data are obtained by diluting a mixture of unlabeled protein and a low concentration of labeled protein with a buffer solution containing only the labeled protein at the same mow concentration. This low concentration should be set just above the detection limit of the system, and defines the lowest concentration to be tested. This approach ensures that the total fluorescence intensity does not change by orders of magnitude as one dilutes the sample, thus minimizing artifacts due to scatter or contaminants, since the relative contribution will remain constant throughout the dilution. This same approach can be used in FCS experiments. In this latter case, the autocorrelation curves obtained at all the tested concentrations should be analyzed globally for the fractional contribution of the diffusion times of the oligomer and monomer to obtain the best confidence in the recovered fractional populations. These, then, can be analyzed as a function of concentration to obtain the value of the dissociation constant. We note that the same limitations for the anisotropy and FCS experiments mentioned above hold for the study of protein–protein interactions. Alternately, the anisotropy decrease associated with homotransfer between the same fluorophores has been used to characterize protein subunit interactions (Runnels and Scarlata, 1995).

When the subject of interest is a heterologous protein–protein interaction, then it is wisest to label the smaller of the two partners, if this is at all possible, since in both FCS and anisotropy experiments this will afford the largest dynamic range for the signal change upon binding. It is also possible to study heterologous protein–protein interactions using FRET, although this approach is more complicated since both proteins must be labeled with a different dye. Thus, unless the system is not amenable to FCS and anisotropy (such as is the case below with proteins at surfaces), these would be the methods of choice. Titration experiments are, then, carried out with the labeled protein at low concentration, titrating with the unlabeled partner (see, e.g., Auguin et al., 2004; Margeat et al., 2001, 2003). As in the case of protein–nucleic acid interactions, the stoichiometries of the complexes can be an issue. Recently in the study of the interaction of a nuclear receptor coactivator, SRC-1, with the estrogen receptor α, we turned to the photon counting histogram as a means of characterizing the stoichiometry of the complex (Margeat et al., 2001). In PCH analysis, the intensity versus time profile obtained in a typical FCS experiment is represented as a probability distribution (the probability of observing x number of photons; Chen et al., 1999, 2005). This distribution can be analyzed in terms of the average number of molecules in the detection volume and their individual molecular brightness. Intuitively, one can imagine a molecule diffusing into the observation volume. If that molecule incorporates one fluorescent dye, then there is a probability of obtaining a certain number of photons. If on the other hand, the molecule contains two fluorescent dyes (in absence of any photophysical interaction between the two, if the molecule is 100% labeled, and assuming that the bin time for the counting is much smaller than the diffusion time), then one should observe twice as many photons for this molecule. In our work, we demonstrated that under all of the concentration conditions that we could test, the complex between a dimer of the estrogen receptor α and the labeled SRC-1 coactivator domain contained one molecule of SRC-1 (Margeat et al., 2001).

While fluorescence anisotropy is quite difficult to implement, both FRET and FCS have been used extensively in live cells to study protein–protein interactions (see, e.g., Bacia and Schwille, 2003; Bacia et al., 2004; Berland et al., 1995; Brock and Jovin, 1998). The most reliable means of ascertaining that FRET is occurring using steady-state approaches is by the acceptor photobleaching method, in which the destruction of the acceptor using a strong laser pulse results in an enhancement of the intensity in the donor channel. More reliable still is FRET–FLIM or FRET coupled to fluorescence lifetime imaging (Bastiaens and Squire, 1999; Cremazy et al., 2005; Harpur et al., 2001). And while FCS autocorrelation has been used to measure the diffusion of proteins in cells (Bacia and Schwille, 2003), the anomalous diffusion inherent to these highly inhomogeneous environments renders the interpretation of the results rather difficult (Clamme et al., 2003; Weiss et al., 2003).

More recently, a very promising approach, that of two color crosscorrelation measurements, has been applied to the study of heterologous protein–protein interactions in live cells (Bacia and Schwille, 2003; Bacia et al., 2002; Baudendistel et al., 2005; Kim et al., 2004, 2005; Saito et al., 2004; Weiss et al., 2003). The key to this approach is to be able to excite simultaneously two fluorophores of highly separated emission energy and to crosscorrelate their signals. Only if the species labeled with the two fluorophores are diffusing together in the same complex will there be a crosscorrelation signal. Exciting two fluorophores simultaneously is most easily accomplished using two-photon excitation due to the broad two-photon excitation cross sections of most dyes that arise from the different selection rules in two-photon excitation. Alternately two well-aligned lasers, at different wavelengths, can be used. For best results, the crosstalk or bleed through of one of the dyes into the other channel should be avoided, since the crosstalk itself is 100% correlated. Although microinjection of labeled material has been used (Kim et al., 2004, 2005), most often, the autofluorescent protein variants of GFP are used in such applications. However, only recently has a monomeric red fluorescent protein become available (Baudendistel et al., 2005) allowing to minimize the crosstalk inherent with the more common CFP–YFP pair. We also note that one should avoid dye pairs that can undergo FRET as well, since the FRET signal is anticorrelated. This technique has been widely applied to the study of interactions of biomolecules *in vitro* (Eigen and Rigler, 1994; Kohl et al., 2002; Koltermann et al., 1998; Rarbach et al., 2001; Rippe, 2000). In this case, the labeling is not such an issue. Although very few studies of this type have appeared to date, the two color crosscorrelation approach appears extremely promising for the study of protein–protein interactions in live cells, a field which represents one of the most important challenges in the postgenomic era.

3.5. Protein–protein interactions in or at the cell membrane

Monitoring protein–protein association on a membrane surface is more limited than in aqueous solution since changes in molecular size and diffusion cannot be distinguished from the slower movement of the bilayer, thus eliminating fluorescence anisotropy and FCS as possible methods. There are two methods that allow one to measure protein associations on membranes in real time: changes in the emission/lifetime of a fluorophore attached to a protein, and FRET. Before describing these methods, it is important to insure that the protein–protein associations being viewed are truly occurring on a membrane surface. This is done by measuring the membrane-binding affinity of both of the proteins. Unlike protein–protein associations, measurements of membrane binding can be carried out using most types of fluorescence methods, especially FCS and fluorescence anisotropy. Since the study of membrane binding using fluorescence has been recently reviewed (Scarlata, 2005), we focus here on the measurements of membrane-bound proteins occur at high enough lipid concentrations so that the proteins remain completely bound through the study.

It is almost always necessary to label at least one of the proteins with a fluorescent probe to isolate it from its partner, and in some cases, labeling occurs at a site on or close to protein–protein interaction region, or in a region that undergoes an environmental change upon association. As described above, there are many types of commercially available environmental probes that undergo shifts in emission energy, or changes in fluorescence intensity which changes in local polarity. An example of this type of study is shown in Fig. 5.8. Here, the α-subunits from a heterotrimeric

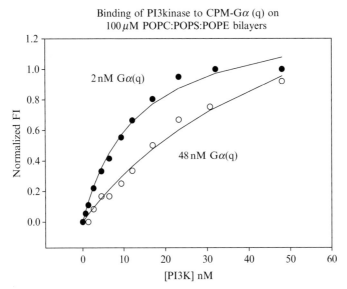

Figure 5.8 Binding of unlabeled phosphoinositol-3-kinase (PI3K) to deactivated CM-Gα(q) reconstituted on large, unilamellar vesicles. The figure shows the shift in the titration curves as the initial concentration of CM-Gα(q) is increased. Here, binding is assessed by the increase in coumarin fluorescence intensity upon association.

G protein were purified and labeled with the thiol-reactive probe 7-diethylamino-3-(4′-maleimidylphenyl)-4-methylcoumarin (CPM). Note that this study used Gα-subunits in their deactivated, GDP-bound form. CPM only becomes fluorescent after linkage to a thiol group and thus any unreactive probe bound noncovalently to the protein will be silent (see Molecular Probes Handbook). Like most membrane associating proteins, G protein subunits are stored in dialyzable detergent and a cryoprotectant (i.e., 10% glycerol) at $-70\ ^\circ$C until use. CPM can be added directly to the thawed G protein solution at a 2:1 probe:protein ratio. After allowing the probe to react at 4 $^\circ$C for 1 h, both unreacted probe and detergent can be removed by dialysis. Generally, extruded bilayers added to the sample between the second and third dialysis steps allowing for membrane reconstitution of the G protein subunit.

Membrane-associated proteins, including G proteins, generally can be reconstituted by simple addition of lipid to the protein solution and we often add extruded lipid bilayers to the sample between the second and third dialysis step when much of the detergent has been removed and before protein aggregation occurs. For G protein subunits, aggregation, as measured by fluorescence homotransfer (see below) begins approximately 12 h after detergent is removed. Reconstituting the protein onto membranes at this step, requires a fairly concentrated solution to avoid membrane crowding which can be estimated by calculating the surface area available to the proteins and the surface area that the proteins occupy. For example, if size of the membrane-binding protein area is 50×50 A^2, as estimated by the molecular weight, and assuming the area of a lipid head group is $\sim$75 A^2, then for a 300 μM lipid solution, the membrane will be completely occupied with protein at the protein concentration of $((75\ A^2) (300\ \mu M) \times 0.5)/(50\ A)^2$, or 4.5 μM where the 0.5 is included because only the outer leaflet is available for protein association. Thus, using at least a 100-fold excess of lipid will insure that membrane crowding does not occur.

In Fig. 5.8, we present the normalized change in fluorescence intensity of CPM-Gα(q) upon binding to the bilayer where the change in fluorescence intensity was normalized to increase from 0 to 1.0 in order to compare samples containing different initial concentrations of proteins. This change was taken from the area under the curve, and in this case, the total increase was 1.5-fold which significant changes in emission energy. However, the response from other probe–protein systems varies considerably.

While following the changes in the emission properties to detect protein–protein associations on membrane surfaces has the advantage of using only one labeled species, there are several important controls that should be carried out before assessing protein affinity. First, it should be verified that the location of the probe is not affecting the interactions. This idea can be tested by labeling the protein in the presence of its partner and repurifying the protein. An easier approach is to label the probe on another reactive group, such as a Lys, and show that the same apparent affinity is obtained. Alternately, one can label the protein partner and following its changes in fluorescence intensity. Second, it is important that the change in fluorescence being followed truly reflects changes due to protein–protein association. Several methods can be used for this purpose. One, presented in Fig. 5.8, is to show that the titration curve shifts appropriately when the initial concentration is changed. For CPM-Gα(q), increasing the concentration from 2 to 48 nM shifts the midpoint of the curve from ∼8 to 21 nM PI3Kinase. Other approaches often used are substituting one of the proteins with a known nonbinding partner that could be an unrelated protein, a protein with a deleted binding domain, or a denatured protein.

Since probes usually react with proteins on an external site that is not sensitive to protein interactions, they are frequently insensitive to protein–protein associations, and in this case, FRET should be used. As noted in the introductory section on FRET, donor/acceptor pairs should be selected based on spectral overlap and quantum yield and proteins should be labeled close to a 1:1 probe:protein ratio. Van der Meer *et al.* (2005) has tabulated many of these parameters including values of R for a variety of probe pairs. Experimentally, FRET is assessed by the decrease in donor fluorescence in the presence of an acceptor, or the increase in acceptor fluorescence in the presence of a donor using unlabeled proteins as control. Our laboratory frequently employs a nonfluorescent energy transfer acceptor, such as DABCYL-SE (4-((4-(dimethylamino)phenyl)azo)benzoic acid succinimidyl ester) to avoid correcting for contributions of acceptor fluorescence in the donor emission spectra.

A specialized case of FRET is homotransfer, or energy transfer between identical probes, and this method has been used to follow oligomerization of several membrane proteins (Runnels and Scarlata, 1995). Since the donor and acceptor are identical and cannot be followed by changes in spectra, homotransfer is detected by the loss in polarization that occurs when the emission energy is transferred. This method has the advantage of only following one labeled specifies, however a considerable amount of light is lost by the polarizers, making it difficult to apply this method to weakly fluorescing proteins.

While protein–protein associations in solution can be readily evaluated by their bimolecular association constants, similar calculations for membrane proteins are more complicated because these associations occur on a quasi two-dimensional surface and their concentration on the membrane surface must be considered. Thus, apparent affinities will not only depend on the protein concentrations and their interaction energies, but also the lipid concentration at which the binding curve is measured. An example of this dependence is shown in Fig. 5.9, where the association between CM-G$\beta\gamma$ to DAB-Gα(GDP) is monitored by FRET. Increasing the amount of lipid effectively dilutes the proteins on the membrane surface and shifts the

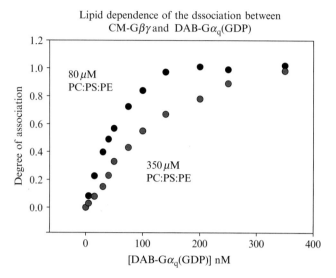

Figure 5.9 Shift in the binding curves between CM-Gβγ and DAB-Gα(q) as the concentration of lipid in which the proteins are bound is raised. Binding is assessed by FRET. (See Color Insert.)

titration curves to higher amounts of protein. This complication has prompted the development of methods to translate the apparent membrane-dependent affinities to membrane independent ones (i.e., those for freely diffusing proteins in solution; Runnels and Scarlata, 1999). We assume that the proteins interact within a surface volume (v) equal to the surface area of the bilayer with an outer radius r, multiplied by the thickness of the membrane solvent interface d, which is estimated by the diameter of the protein:

$$v = 4\pi r^2 d. \tag{5.16}$$

The effective concentration of the protein on the membrane surface, $[P]_m$ can be related to the total protein concentration in the sample $[P]$, by simply relating the total concentration of the bulk solution to this reduced membrane volume in which the proteins are confined:

$$[P]_m = (V/v), [P] = \varepsilon[P], \tag{5.17}$$

where ε can be calculated by the lipid concentration. For proteins that are roughly $d = 10$ nm in diameter associating in a solution of a lipid concentration of 350 μM, then

$$\varepsilon = (2 \times 10^{24})/([\text{lipid}] \times N_{av} \times 10 \text{ nm} \times 0.75 \text{ nm}^2) = 135, \tag{5.18}$$

where N_{av} is the Avogadro's number, 0.75 nm² is the average surfaces area per head group. Thus, confining membrane protein associations to 350 mM lipid has the effect of concentrating them 135-fold, and the apparent bimolecular dissociation constant is 135 times stronger then if the proteins were not bound to membranes but freely diffusing in solution.

While estimates of the affinities between membrane proteins in model systems can be readily made, those in biological systems, such as cells and natural membranes, are complex. These affinities are critical if one is to understand the protein pathways a signal undergoes during various cellular events. Isolation of a target protein from other cellular proteins has been overcome by the use of GFP analogues, however, problems in determining the cellular concentrations, showing that overexpression of the tagged proteins is not affecting the localization of the protein or the physiological state of the cell, and showing that the signal attributed to FRET is due to protein–protein association rather than to changes in some cellular factors, still exist. Also, many proteins undergo postsynthetic modifications which can alter their interactions with both proteins partners and membrane surfaces. While many of these factors are approachable, it is clear that more cellular methods must be developed to allow one to determine these cellular affinities.

4. Conclusions

The above examples underline the versatility of fluorescence as it is applied to the study of biomolecular interactions. We have also attempted to draw attention to some of the pitfalls and limitations of the technique and to discuss which approaches are best suited to specific applications. Clearly, it is an extremely useful and widely used technique, and given the large number of complexes for which the interaction parameters and structure–function relationships have not been determined, we expect that the applications of fluorescence in this field will continue to grow and evolve.

REFERENCES

Auguin, D., Barthe, P., Royer, C., Stern, M. H., Noguchi, M., Arold, S. T., and Roumestand, C. (2004). Structural basis for the co-activation of protein kinase B by T-cell leukemia-1 (TCL1) family proto-oncoproteins. *J. Biol. Chem.* **279**, 35890–35902.
Bacia, K., and Schwille, P. (2003). A dynamic view of cellular processes by *in vivo* fluorescence auto- and cross-correlation spectroscopy. *Methods* **29**, 74–85.
Bacia, K., Majoul, I. V., and Schwille, P. (2002). Probing the endocytic pathway in live cells using dual-color fluorescence cross-correlation analysis. *Biophys. J.* **83**, 1184–1193.
Bacia, K., Scherfeld, D., Kahya, N., and Schwille, P. (2004). Fluorescence correlation spectroscopy relates rafts in model and native membranes. *Biophys. J.* **87**, 1034–1043.

Bastiaens, P. I., and Squire, A. (1999). Fluorescence lifetime imaging microscopy: Spatial resolution of biochemical processes in the cell. *Trends Cell Biol.* **9,** 48–52.

Baudendistel, N., Muller, G., Waldeck, W., Angel, P., and Langowski, J. (2005). Two-hybrid fluorescence cross-correlation spectroscopy detects protein–protein interactions in vivo. *ChemPhysChem* **6,** 984–990.

Beechem, J. M., and Brand, L. (1986). Global analysis of fluorescence decay: Applications to some unusual experimental and theoretical studies. *Photochem. Photobiol.* **44,** 323–329.

Berland, K. M., So, P. T., and Gratton, E. (1995). Two-photon fluorescence correlation spectroscopy: Method and application to the intracellular environment. *Biophys. J.* **68,** 694–701.

Boyer, M., Poujol, N., Margeat, E., and Royer, C. A. (2000). Quantitative characterization of the interaction between purified human estrogen receptor alpha and DNA using fluorescence anisotropy. *Nucleic Acids Res.* **28,** 2494–2502.

Brock, R., and Jovin, T. M. (1998). Fluorescence correlation microscopy (FCM)–fluorescence correlation spectroscopy (FCS) taken into the cell. *Cell. Mol. Biol. (Noisy-le-grand)* **44,** 847–856.

Brown, M. P., Grillo, A. O., Boyer, M., and Royer, C. A. (1999). Probing the role of water in the tryptophan repressor–operator complex. *Protein Sci.* **8,** 1276–1285.

Chen, Y., Muller, J. D., So, P. T., and Gratton, E. (1999). The photon counting histogram in fluorescence fluctuation spectroscopy. *Biophys. J.* **77,** 553–567.

Chen, Y., Wei, L. N., and Muller, J. D. (2005). Unraveling protein–protein interactions in living cells with fluorescence fluctuation brightness analysis. *Biophys. J.* **88,** 4366–4377.

Clamme, J. P., Krishnamoorthy, G., and Mely, Y. (2003). Intracellular dynamics of the gene delivery vehicle polyethylenimine during transfection: Investigation by two-photon fluorescence correlation spectroscopy. *Biochim. Biophys. Acta* **1617,** 52–61.

Clegg, R. M. (1995). Fluorescence resonance energy transfer. *Curr. Opin. Biotechnol.* **6,** 103–110.

Cremazy, F. G., Manders, E. M., Bastiaens, P. I., Kramer, G., Hager, G. L., van Munster, E. B., Verschure, P. J., Gadella, T. J., Jr., and van, D. R. (2005). Imaging *in situ* protein–DNA interactions in the cell nucleus using FRET–FLIM. *Exp. Cell Res.* **309,** 390–396.

Eigen, M., and Rigler, R. (1994). Sorting single molecules: Application to diagnostics and evolutionary biotechnology. *Proc. Natl Acad. Sci. USA* **91,** 5740–5747.

Elson, E. L. (2004). Quick tour of fluorescence correlation spectroscopy from its inception. *J. Biomed. Opt.* **9,** 857–864.

Grillo, A. O., and Royer, C. A. (2000). The basis for the super-repressor phenotypes of the AV77 and EK18 mutants of trp repressor. *J. Mol. Biol.* **295,** 17–28.

Grillo, A. O., Brown, M. P., and Royer, C. A. (1999). Probing the physical basis for trp repressor–operator recognition. *J. Mol. Biol.* **287,** 539–554.

Harpur, A. G., Wouters, F. S., and Bastiaens, P. I. (2001). Imaging FRET between spectrally similar GFP molecules in single cells. *Nat. Biotechnol.* **19,** 167–169.

Haustein, E., and Schwille, P. (2004). Single-molecule spectroscopic methods. *Curr. Opin. Struct. Biol.* **14,** 531–540.

Helene, C., and Dimicoli, J. L. (1972). Interaction of oligopeptides containing aromatic amino acids with nucleic acids. Fluorescence and proton magnetic resonance studies. *FEBS Lett.* **26,** 6–10.

Helene, C., Brun, F., and Yaniv, M. (1969). Fluorescence study of interactions between valyl-tRNA synthetase and valine-specific tRNA's from *Escherichia coli. Biochem. Biophys. Res. Commun.* **37,** 393–398.

Helene, C., Brun, F., and Yaniv, M. (1971). Fluorescence studies of interactions between *Escherichia coli* valyl-tRNA synthetase and its substrates. *J. Mol. Biol.* **58,** 349–356.

Heyduk, T., and Lee, J. C. (1990). Application of fluorescence energy transfer and polarization to monitor *Escherichia coli* cAMP receptor protein and lac promoter interaction. *Proc. Natl Acad. Sci. USA* **87,** 1744–1748.

Kim, S. A., Heinze, K. G., Waxham, M. N., and Schwille, P. (2004). Intracellular calmodulin availability accessed with two-photon cross-correlation. *Proc. Natl Acad. Sci. USA* **101,** 105–110.

Kim, S. A., Heinze, K. G., Bacia, K., Waxham, M. N., and Schwille, P. (2005). Two-photon cross-correlation analysis of intracellular reactions with variable stoichiometry. *Biophys. J.* **88,** 4319–4336.

Knutson, J. R., Walbridge, D. G., and Brand, L. (1982). Decay-associated fluorescence spectra and the heterogeneous emission of alcohol dehydrogenase. *Biochemistry* **21,** 4671–4679.

Kohl, T., Heinze, K. G., Kuhlemann, R., Koltermann, A., and Schwille, P. (2002). A protease assay for two-photon crosscorrelation and FRET analysis based solely on fluorescent proteins. *Proc. Natl Acad. Sci. USA* **99,** 12161–12166.

Koltermann, A., Kettling, U., Bieschke, J., Winkler, T., and Eigen, M. (1998). Rapid assay processing by integration of dual-color fluorescence cross-correlation spectroscopy: High throughput screening for enzyme activity. *Proc. Natl Acad. Sci. USA* **95,** 1421–1426.

Margeat, E., Poujol, N., Boulahtouf, A., Chen, Y., Muller, J. D., Gratton, E., Cavailles, V., and Royer, C. A. (2001). The human estrogen receptor alpha dimer binds a single SRC-1 coactivator molecule with an affinity dictated by agonist structure. *J. Mol. Biol.* **306,** 433–442.

Margeat, E., Bourdoncle, A., Margueron, R., Poujol, N., Cavailles, V., and Royer, C. (2003). Ligands differentially modulate the protein interactions of the human estrogen receptors alpha and beta. *J. Mol. Biol.* **326,** 77–92.

Pogenberg, V., Guichou, J. F., Vivat-Hannah, V., Kammerer, S., Perez, E., Germain, P., de Lera, A. R., Gronemeyer, H., Royer, C. A., and Bourguet, W. (2005). Characterization of the interaction between retinoic acid receptor/retinoid X receptor (RAR/RXR) heterodimers and transcriptional coactivators through structural and fluorescence anisotropy studies. *J. Biol. Chem.* **280,** 1625–1633.

Poujol, N., Margeat, E., Baud, S., and Royer, C. A. (2003). RAR antagonists diminish the level of DNA binding by the RAR/RXR heterodimer. *Biochemistry* **42,** 4918–4925.

Rarbach, M., Kettling, U., Koltermann, A., and Eigen, M. (2001). Dual-color fluorescence cross-correlation spectroscopy for monitoring the kinetics of enzyme-catalyzed reactions. *Methods* **24,** 104–116.

Rippe, K. (2000). Simultaneous binding of two DNA duplexes to the NtrC-enhancer complex studied by two-color fluorescence cross-correlation spectroscopy. *Biochemistry* **39,** 2131–2139.

Royer, C. A., Rusch, R. M., and Scarlata, S. F. (1989). Salt effects on histone subunit interactions as studied by fluorescence spectroscopy. *Biochemistry* **28,** 6631–6637.

Runnels, L. W., and Scarlata, S. F. (1995). Theory and application of fluorescence homo-transfer to melittin oligomerization. *Biophys. J.* **69,** 1569–1583.

Runnels, L. W., and Scarlata, S. F. (1999). Determination of the affinities between heterotrimeric G protein subunits and their phospholipase C-beta effectors. *Biochemistry* **38,** 1488–1496.

Saito, K., Wada, I., Tamura, M., and Kinjo, M. (2004). Direct detection of caspase-3 activation in single live cells by cross-correlation analysis. *Biochem. Biophys. Res. Commun.* **324,** 849–854.

Scarlata, S. F. (2005). "Use of Fluorescence Spectroscopy to Monitor Protein-Membrane Associations." Plenum, New York.

van der Meer, W., Coker, G., and Chen, S. S.-Y. (2005). "Resonance Energy Transfer, Theory and Data." VCH Publishers, New York.

Weber, G. (1952a). Polarization of the fluorescence of macromolecules. I. Theory and experimental method. *Biochem. J.* **51,** 145–155.

Weber, G. (1952b). Polarization of the fluorescence of macromolecules. II. Fluorescent conjugates of ovalbumin and bovine serum albumin. *Biochem. J.* **51,** 155–167.

Weber, G. (1976). Practical applications and philosophy of optical spectroscopic probes. *Horiz. Biochem. Biophys.* **2,** 163–198.

Weber, G., and Farris, F. J. (1979). Synthesis and spectral properties of a hydrophobic fluorescent probe: 6-Propionyl-2-(dimethylamino)naphthalene. *Biochemistry* **18,** 3075–3078.

Weiss, M., Hashimoto, H., and Nilsson, T. (2003). Anomalous protein diffusion in living cells as seen by fluorescence correlation spectroscopy. *Biophys. J.* **84,** 4043–4052.

CHAPTER SIX

FÖRSTER RESONANCE ENERGY TRANSFER MEASUREMENTS OF TRANSMEMBRANE HELIX DIMERIZATION ENERGETICS

Mikhail Merzlyakov *and* Kalina Hristova

Contents

1. Introduction	108
2. Challenges in Quantitative Measurements of Interactions Between TM Helices in Bilayers	108
3. Bilayer Platforms for Measuring TM Helix Interactions Using FRET	110
3.1. Multilamellar vesicles	110
3.2. Extruded large unilamellar vesicles	111
3.3. Small unilamellar vesicles	111
3.4. Surface-supported bilayers	111
4. FRET Due to Random Colocalization of Donors and Acceptors (Proximity FRET)	114
5. FRET Efficiencies and Energetics of TM Helix Dimerization	115
5.1. Direct calculation of FRET efficiencies from donor quenching	115
5.2. The EmEx-FRET method	116
5.3. Energetics of TM helix heterodimerization	122
6. Biological Insights from FRET Measurements	124
6.1. Thermodynamics of protein homodimerization in lipid bilayer membranes: Wild-type FGFR3 TM domain, and the pathogenic Ala391Glu, Gly380Arg, and Gly382Asp mutants	124
6.2. Thermodynamics of protein heterodimerization in lipid bilayer membranes	125
7. Conclusion	125
References	125

Abstract

Lateral interactions between hydrophobic transmembrane (TM) helices in membranes underlie the folding of multispan membrane proteins and signal transduction by receptor tyrosine kinases (RTKs). Quantitative measurements of

Department of Materials Science and Engineering, Johns Hopkins University, Baltimore, Maryland 21218 dimerization energetics in membranes are required to uncover the physical principles behind these processes. Here, we overview how FRET measurements can be used to determine the thermodynamics of TM helix homo- and heterodimerization in vesicles and in supported bilayers. Such measurements can shed light on the molecular mechanism behind pathologies arising due to single-amino acid mutations in membrane proteins.

1. INTRODUCTION

The folding of membrane proteins is mediated by lateral interactions between adjacent transmembrane (TM) helices (MacKenzie, 2006; White and Wimley, 1999). The lateral interactions between the TM domains of receptor tyrosine kinases (RTKs) are important for the process of signal transduction across the plasma membrane and for the regulation of cell growth, differentiation, motility, or death (Li and Hristova, 2006). Therefore, free energy measurements of TM helix dimerization are needed for understanding the physical principles behind membrane protein folding and signal transduction. Importantly, these measurements need to be carried out in quantitative manner in the native bilayer environment. Here, we overview how FRET can be used to determine the concentrations of monomeric and dimeric TM helices in bilayers. Such measurements yield association constants and free energies of TM helix dimerization in vesicles and in supported bilayers, environments that mimic the biological membrane.

2. CHALLENGES IN QUANTITATIVE MEASUREMENTS OF INTERACTIONS BETWEEN TM HELICES IN BILAYERS

First, there are challenges due to the chemical nature of the hydrophobic TM helices. The production of pure TM helices that are labeled with FRET dyes can be expensive and labor-intensive, due to low synthesis yields, low labeling yields, and low purification yields (Iwamoto et al., 2005). Second, there are challenges associated with the incorporation of the TM helices in the bilayers, and in general, sample preparation and handling.

Each sample is prepared from stocks of lipids and donor- and acceptor-labeled proteins. Since there is no material exchange between the membranes and the aqueous medium, a new sample has to be prepared when an experimental parameter, such as peptide-to-lipid ratio, donor-to-acceptor ratio, or peptide concentration, is changed (discussed in details in You et al., 2005). Furthermore, for each FRET measurement, three different samples are needed: (1) a sample containing both donor and acceptor, (2) a "no FRET" control containing the donor only, and (3) an acceptor-only control to monitor the direct excitation of the acceptor fluorophore.

Synthesis and purification protocols for TM helices are now available in the literature (Iwamoto et al., 1994, 2005). Many hydrophobic TM domains contain aromatic residues such as Trp or Tyr, and their absorbance can be used to determine the concentration of the stock peptide solutions. Circular dichroism (CD) can be used to confirm that the peptides are helical. Some solvents, such as methanol, dissolve TM helices but do not support their helical structure. Such solvents should be avoided because they promote the misfolding and the aggregation of the TM domains into β-sheets. Furthermore, in liposomes, one must prove that the orientation of the helix is indeed transmembrane (Li et al., 2005; You et al., 2005). If the organic solvents used do not dissolve the two components (lipids and peptides) equally well, the helices may get "trapped" at the interface. Therefore, the tilt of the helices with respect to the bilayer normal should be measured using oriented circular dichroism (OCD) in oriented multilayers prior to hydration (Li et al., 2005). TM helices exhibit characteristic OCD spectra with a single minimum around 230 nm and a maximum around 200 nm.

Furthermore, when peptides and lipids are mixed in organic solvents, they can either (1) form a homogeneous mixture (i.e., a single "phase") or (2) segregate into two or more distinct lipid- and peptide-rich phases. Therefore, homogencity of peptide/lipid mixtures should be assessed using different methods such as X-ray diffraction, fluorescence microscopy, and FRET efficiency measurements (Li et al., 2005).

Phase separation in lipid systems can be observed using X-ray diffraction. A homogeneous sample gives rise to a single set of Bragg peaks. A phase-separated sample shows either (1) two sets of Bragg peaks or (2) a single set of Bragg peaks, identical to pure lipid samples, and one or several sharp lines due to protein aggregates. Since phase separation is particularly likely to occur in dry samples, X-ray diffraction of dry multilayers provides a very stringent test for possible phase separation (You et al., 2005). Aggregation of the proteins due to their dissolution from the lipid matrix can be further detected by measuring FRET as a function of acceptor fraction (Li et al., 2005, 2006; You et al., 2006). If the helices form dimers but no higher-order aggregates (such that only monomers and dimers are present), the FRET efficiency depends linearly on the acceptor ratio (Adair and Engelman, 1994). However, larger peptide aggregates will show a nonlinear dependence on acceptor fraction.

Further experimental challenges can arise due to sample-to-sample variations in protein concentrations due to the very low solubility of the hydrophobic TM helices. These variations in concentration can introduce uncertainties in the calculated dimerization free energies. This problem can be resolved by using the EmEx-FRET method (Merzlyakov et al., 2007) described below.

Before measuring FRET, one needs to investigate if the dyes affect (1) the secondary structure of the TM domains, by measuring the helicities of labeled and unlabeled peptides using CD, and (2) the dimerization propensities by running SDS-PAGE gels of labeled and unlabeled peptides (Iwamoto et al., 2005). If the dyes are attached to the termini, or close to the termini, they generally have no effect on helicity or dimerization (Iwamoto et al., 2005).

For quantitative FRET measurements, one needs to measure the labeling yields. The attachment of the fluorophores to the hydrophobic TM domains does not always change the elution time on the HPLC column. In this case, the labeled and unlabeled peptides cannot be separated, and the labeling yield is determined by comparing the concentration of the labeled peptides (determined via absorbance measurements of fluorophore concentrations) and the concentration of all peptides (measured using CD). Quantitative FRET measurements of dimerization energetics require high labeling yields (You et al., 2005).

3. Bilayer Platforms for Measuring TM Helix Interactions Using FRET

Three different bilayer platforms have proven useful in the studies of TM helix dimerization: multilamellar vesicles (MLVs), large unilamellar vesicles (LUVs), and supported bilayers (Merzlyakov et al., 2006c; You et al., 2005). In all cases, lipids and peptides are first codissolved in organic solvents to ensure complete mixing of the two components, followed by the removal of the organic solvents. We recommend that a mixture of solvents, such as HFIP/TFE/chloroform is used to codissolve peptides and lipids.

3.1. Multilamellar vesicles

The hydration of dry protein/lipid films leads to the formation of MLVs containing TM helices (You et al., 2005). MLVs can be used for FRET measurements of dimerization energetics after a single freeze–thaw cycle (You et al., 2005). Just after one cycle, the turbidity of the MLVs is greatly reduced. This decrease in turbidity is surprising, given that it does not occur for lipid MLVs that do not contain the peptides. Therefore, it appears that the TM helices promote the formation of relatively small MLVs for peptide concentrations ranging from 0.01 to 1 mol%. Thus, the CD and fluorescence spectra can be measured in MLVs (You et al., 2005).

We have observed that the TM peptides are homogeneously distributed when MLVs are prepared, and their distribution remains homogeneous over time. The fluorescence intensity and the FRET signal of such MLVs are very stable over a month (You et al., 2005). Furthermore, the fluorescence intensity measured for the MLVs is not affected by light scattering. Finally, the FRET efficiency is determined only by the protein-to-lipid ratio, not by the total peptide and lipid concentrations, such that the measurements are relevant for processes occurring in cellular membranes (You et al., 2005).

3.2. Extruded large unilamellar vesicles

The MLVs can be extruded using a 100nm pore diameter membrane (Avanti) to produce LUVs. There is no statistically significant change in FRET in the LUVs compared to MLVs. Therefore, FRET efficiencies and helix–helix interactions measured in MLVs and LUVs are comparable.

A potential problem with extrusion is the loss of peptides and lipids in the extrusion process. We, therefore, determined peptide and lipid concentrations before and after extrusion (You et al., 2005). We found that typical losses of peptides and lipids were identical; that is, between 13% and 18% of both lipids and peptides were lost during the extrusion process. The peptide-to-lipid ratio does not change during the extrusion process, and therefore, the FRET efficiency, which is determined solely by the peptide-to-lipid ratio, remains unchanged after extrusion (You et al., 2005).

3.3. Small unilamellar vesicles

We have shown that the SUVs with TM peptides aggregate and come out of solution, such that the fluorescence signal gradually decreases over time (You et al., 2005). Thus, SUVs are not an appropriate system for FRET measurements, particularly if Eq.(6.2) below is used to calculate FRET efficiencies.

3.4. Surface-supported bilayers

To study the interactions between TM peptides in supported bilayers, we have recently developed a surface-supported bilayer platform based on the so-called "directed assembly method" (Merzlyakov et al., 2006c; see Fig. 6.1). In this assembly method, the peptides are incorporated into lipid monolayers at the air/water interface, and the monolayers are then transferred onto glass substrates using Langmuir–Blodgett (LB) deposition. The bilayers are completed via lipid vesicle fusion on top of the LB monolayers. The novelty in the assembly is the incorporation of the peptides into the monolayer at the first step of the bilayer assembly, which allows control over the peptide concentration and orientation (Merzlyakov et al., 2006c). The TM orientation of the peptides in the supported bilayer was confirmed directly by OCD—both the shape and the amplitude of the OCD spectra showed that the peptides are transmembrane (Merzlyakov et al., 2006c). The lateral mobility of the peptides was assessed in fluorescence recovery after photobleaching (FRAP) experiments. Although the peptides were moving much slower than lipids, there was no immobile peptide fraction

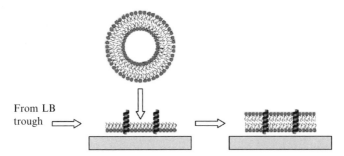

Figure 6.1 Illustration of the directed assembly method. Reprinted with permission from (Merzlyakov et al., 2006). Copyright (2006) American Chemical Society.

(Merzlyakov et al., 2006d), such that the energetics of TM helix dimerization can be studied.

The quartz slide, with the coverslip supporting the bilayer, was inserted into a home-built adapter as shown in Fig. 6.2A. The adapter was designed to fit into the standard cuvette holder of a fluorometer. The emission and excitation spectra of labeled peptides in supported bilayers were measured in a Fluorolog-3 fluorometer (Jobin Yvon, Edison, NJ). To reduce the background from stray light, the angle between the excitation beam and the quartz slide was set to 35° (Fig. 6.2B), such that the reflected excitation beam was not captured by the photodetector.

The fluorescence spectra of labeled peptides in the supported bilayers are very reproducible: the typical standard deviation in fluorescence intensity is below a few percent (Merzlyakov et al., 2006a). Furthermore, the fluorescence intensity in these bilayers is proportional to the fluorophore concentration, as expected. Bleaching during spectral collection is negligible. Thus, measurements of fluorescence spectra in surface-supported bilayers are reproducible and reliable, even for dyes with poor photostability, such as fluorescein. Furthermore, the spectra of the surface-supported bilayers (i.e., close to the surface) are very similar to spectra of suspended liposomes (Merzlyakov et al., 2006a).

We have recorded FRET spectra of fluorescein- and rhodamine-labeled peptides, while varying the donor-to-acceptor ratio and keeping the total peptide concentration fixed. The dependence of the FRET efficiency on acceptor fraction is linear, indicating that only monomers and dimers are present and that the peptides do not aggregate in the surface-supported bilayers (Merzlyakov et al., 2006a). Furthermore, we have shown that the FRET efficiencies, measured in surface-supported bilayers, are similar to the ones in liposomes (Merzlyakov et al., 2006a).

There are advantages to performing the measurements in supported bilayers, as compared to free liposomes in suspension:

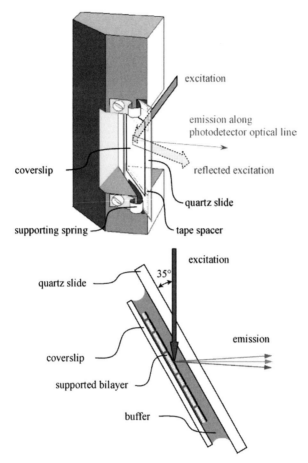

Figure 6.2 (A) Schematic drawing of the adapter used for spectral FRET measurements of TM helix dimerization in surface supported bilayers. The adapter fits in the cuvette holder of a standard fluorometer. A fluid bilayer is supported on the coverslip, and is sandwiched between the coverslip and the quartz slide. The gap between the coverslip and the quartz slide is filled with buffer, such that the bilayer is fully hydrated. (B) Top view showing the bilayer, supported on the coverslip, and the quartz slide (not drawn to scale). Reprinted with permission from (Merzlyakov *et al.*, 2006). Copyright (2006) American Chemical Society.

1. The amount of peptide required for an experiment in surface-supported bilayers is 1/100 of the peptide amount typically used in vesicle solutions, thus substantially reducing the cost of research.
2. Supported bilayer platforms could be adapted to parallel high-throughput measurements of lateral protein interactions in bilayers, paving the way for the development of novel sensing devices that utilize membrane proteins (Sackmann, 1996; Sackmann and Tanaka, 2000).

3. The directed assembly method may offer a means to achieve unidirectional orientation of the peptides (Merzlyakov et al., 2006c) because the orientation of the peptides occurs at the air–water interface, and therefore asymmetry in the sequence of the flanking residues may ensure the unidirectionality of the peptides and may prevent nonbiological lateral interactions.

4. FRET Due to Random Colocalization of Donors and Acceptors (Proximity FRET)

When measuring FRET in a fluid lipid bilayer, it is important to recognize that FRET can arise simply due to random proximity of the acceptors and donors. Therefore, it is desirable to use low peptide concentration, such that the average distances between the peptides always exceed the Förster radii for the donor/acceptor pairs. It should be further taken into account that the peptides diffuse randomly in the bilayers, such that for any concentration some acceptors will come in close contact with donors and FRET will occur even without sequence-specific dimerization.

An indicator of what portion of the measured FRET efficiency is due to dimerization rather than random colocalization is the deviation of the measured FRET from the expected FRET calculated for randomly distributed fluorophores. In this calculation, carried out first by Wolber and Hudson (1979) and later by Wimley and White (2000), FRET from randomly distributed peptides is determined by averaging the donor quenching by acceptors over a large number of acceptor configurations. The FRET efficiency (E) for a specific acceptor configuration is given by (Wolber and Hudson 1979)

$$E = \frac{1}{1 + \left(\frac{1}{\sum_i (R_0/r_i)^6} \right)}, \tag{6.1}$$

where r_i is the distance between the donor and the ith acceptor in the system and R_0 is the Förster radius for the donor/acceptor FRET pair. The simulation of FRET from random colocalization in bilayers shows that this random proximity effect can contribute to the measured FRET efficiencies even at acceptor concentrations far less than 1mol% (You et al., 2005). This nonspecific FRET is higher for fluorophores with larger R_0.

We have determined the Forster radii, R_0, of the fluorescein/rhodamine and the BODIPY-fluorescein/rhodamine pairs, when the dyes are attached to the TM helices and are likely positioned in the bilayer interface as $R_0=56$ Å using two different methods. First, R_0 was calculated by measuring the fluorescent yield of the donor, QY_{DONOR}, and the degree of overlap between donor emission and acceptor excitation, as described in Li et al.,

2006. R_0 was also calculated by fitting the measured FRET efficiencies to a sum of two contributions: a sequence-specific dimerization contribution and a random colocalization contribution, via a two-parameter fit, by simultaneously varying K (or ΔG) and R_0 (or the predicted FRET due to random colocalization of donors and acceptors) (Li et al., 2006).

A control experiment that can distinguish between random colocalization and sequence-specific dimerization is to monitor the effect of "dilution" of labeled peptides with unlabeled peptides. If sequence-specific dimerization occurs, the addition of unlabeled peptide to donor and acceptor-labeled dimers will decrease the FRET signal. FRET that is due to random colocalization will not decrease in the presence of unlabeled peptide. This control experiment has demonstrated that FGFR3 TM domains form sequence-specific dimers in POPC bilayers (Li et al., 2005). This experiment can identify sequence-specific interactions even for very weakly dimerizing helices. It should be kept in mind, however, that this experiment works only when high labeling yields can be achieved.

5. FRET Efficiencies and Energetics of TM Helix Dimerization

5.1. Direct calculation of FRET efficiencies from donor quenching

FRET efficiencies can be calculated from the emission spectra of donor- and acceptor-labeled peptides of known concentration in MLVs, LUVs, and supported bilayers. Bilayers containing only donor-labeled peptides serve as the "no FRET control." The Energy transfer, efficiency E, is calculated from the donor emission in the absence (F_D) and presence of the acceptor (F_{DA}) according to

$$E = 1 - \frac{F_{DA}(\lambda_{em}^D)}{F_D(\lambda_{em}^D)}, \qquad (6.2)$$

where λ_{em}^D is the wavelength of the donor emission maximum.

The measured FRET efficiency has two contributions, one due to random colocalization of donors and acceptors (proximity effect) and one due to sequence-specific dimerization:

$$E = E_{proximity} + E_{dimer}. \qquad (6.3)$$

The FRET efficiency due to sequence-specific dimerization E_{dimer} can be presented as

$$E_{dimer} = f_D p_D E_R, \qquad (6.4)$$

where f_D is the fraction of molecules in the dimeric state, $f_D = 2[D]/[T]$, $[D]$ is the dimer concentration, $[T]$ is the total peptide concentration, p_D is the probability for donor quenching to occur in the dimer, and E_R is the FRET efficiency in the dimer. If the distance between the FRET pair in the dimer is smaller than the Förster radius, then $E_R \approx 1$. The probability of a donor-labeled peptide to dimerize with an acceptor-labeled peptide equals the fraction of acceptor-labeled molecules:

$$x_A = \frac{[a]}{[d]/f_d + [a]/f_a}, \qquad (6.5)$$

where $[d]$ and $[a]$ are the donor and acceptor concentrations, f_d and f_a are the donor and acceptor labeling yields. A donor will be quenched if it dimerizes with an acceptor, and therefore

$$p_D = x_A. \qquad (6.6)$$

Since $[d]$ and $[a]$ are known as aliquoted (or can be determined using the EmEx-FRET method described below) for each sample, $E_{\text{proximity}}$ and x_A can be calculated too, such that the dimer concentration can be calculated as

$$[D] = \frac{E_{\text{dimer}}[T]}{2x_A}. \qquad (6.7)$$

The mole fraction association constant, K, is given by

$$K = [D]/[M]^2, \qquad (6.8)$$

where $[M] = [T] - 2[D]$ is the monomer concentration in the lipid vesicles. K is usually determined from a plot of measured dimer fractions versus total peptide concentration (Li et al., 2006; You et al., 2006, 2007). The free energy of dimerization is given by

$$\Delta G = -RT \ln K. \qquad (6.9)$$

5.2. The EmEx-FRET method

As discussed above, sample-to-sample variations in protein concentrations (due to the very low solubility of the hydrophobic TM helices) can introduce uncertainties in the measured FRET efficiencies and the calculated dimerization free energies. Here, we overview the EmEx-FRET method (Merzlyakov et al., 2007) which reduces such experimental uncertainties. In addition, this method can be useful when the concentrations of the donor- and acceptor-tagged molecules cannot be controlled, such as in cellular studies.

The EmEx-FRET method relies on the acquisition of both excitation and emission spectra for "the FRET sample" which contains both donor- and acceptor-labeled TM helices. Furthermore, both excitation and emission spectra need to be acquired for a donor-only and an acceptor-only "standards" of known donor and acceptor concentration (shown in Fig. 6.3). The EmEx-FRET method uses these standard spectra to calculate not only the FRET efficiency, but also the actual donor and acceptor concentration, and therefore the equilibrium constants and the free energy of lateral dimerization with high experimental precision.

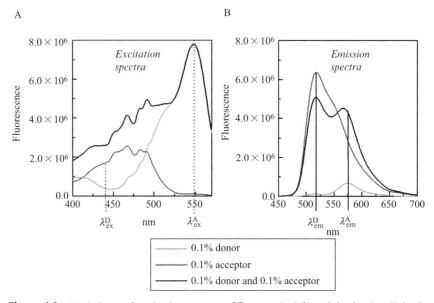

Figure 6.3 Emission and excitation spectra of fluorescein (Fl) and rhodamine (Rhod), a common FRET pair, conjugated to wild-type FGFR3 TM domain. POPC concentration was 0.25 mg/ml; the concentration of the protein is reported in moles of protein per mole of lipid. (A) Excitation spectra, collected by recording emission at 595 nm while scanning the excitation from 400 to 570 nm. (B) Emission spectra, with excitation fixed at 439 nm, and emission scanned from 450 to 700 nm. In (A) and (B), the blue lines correspond to 0.1 mol% Fl-TM$_{WT}$, while the red lines are for 0.1 mol% Rhod-TM$_{WT}$. The blue and red spectra serve as standard excitation and emission spectra, to be used with the EmEx-FRET method. These spectra are averages, derived from measurements of multiple samples. Also shown are the excitation and emission spectra of 0.1 mol% fluorescein-labeled wild-type and 0.1 mol% rhodamine-labeled wild-type (the donor/acceptor sample, or "FRET sample", black lines). The FRET spectrum (black) is the sum of three contributions: direct donor emission in the presence of the acceptor F_{DA}, direct acceptor emission F_A, and sensitized acceptor emission F_{sen}. Inspection of the excitation spectra in (A) reveals that at excitation wavelength λ_{ex}^A the acceptor excitation reaches its maximum, while the donor excitation is negligible. As a result, the excitation of the donor/acceptor sample (black) at λ_{ex}^A is contributed by the acceptor only (red). $\lambda_{ex}^D = 439$ nm is the excitation wavelength used in the acquisition of the emission spectra. In (B), the emissions of the donor and the acceptor reach their maxima at λ_{em}^D and λ_{em}^A, respectively. Note that the emission of the acceptor is low, but not negligible, at λ_{em}^D. Reprinted from (Merzlyakov et al., 2007), with permission from Springer Science and Business Media. (See Color Insert.)

5.2.1. EmEx-FRET theory

The efficiency of energy transfer E from an excited donor to an acceptor depends on the donor emission in the absence (F_D) and presence (F_{DA}) of the acceptor according to Eq. (6.2). It can be also determined from the acceptor emission in the absence (F_A) and presence (F_{AD}) of the donor. When both the donor and the acceptor are present, we can write

$$F_{DA}(\lambda_{em}^D) = S_D[\varepsilon_D(\lambda_{ex}^D)[d] - \varepsilon_D(\lambda_{ex}^D)[d]E], \quad (6.10)$$

$$F_{AD}(\lambda_{em}^A) = S_A[\varepsilon_A(\lambda_{ex}^D)[a] + \varepsilon_D(\lambda_{ex}^D)[d]E], \quad (6.11)$$

where $\varepsilon_D(\lambda_{ex}^D)$ and $\varepsilon_A(\lambda_{ex}^D)$ are the extinction coefficients of the donor and the acceptor at λ_{ex}^D (see Fig. 6.3), S_D and S_A are scaling factors which depend on the quantum yields of the donor and acceptor, the photodetector efficiencies at λ_{em}^D and λ_{em}^A, and the geometry of the experimental setup. At zero FRET efficiency ($E = 0$), the donor and the acceptor emit due to direct excitation only, such that $F_{DA} = F_D$ and $F_{AD} = F_A$. At nonzero FRET efficiency ($E \neq 0$), part of the energy, absorbed by the donor, is transferred to the acceptor. The donor fluorescence is quenched ($F_{DA} < F_D$) and the acceptor fluorescence is enhanced ($F_{AD} > F_A$). The acceptor enhancement, termed "sensitized fluorescence" F_{sen}, can be calculated as

$$\begin{aligned}F_{sen} &= F_{AD} - F_A, \\ F_{sen}(\lambda_{em}^A) &= S_A \varepsilon_D(\lambda_{ex}^D)[d]E.\end{aligned} \quad (6.12)$$

The donor quenching is given by

$$\begin{aligned}\Delta F_D &= F_D - F_{DA}, \\ \Delta F_D(\lambda_{em}^D) &= S_D \varepsilon_D(\lambda_{ex}^D)[d]E = F_{sen}(\lambda_{em}^A) \frac{S_D}{S_A}.\end{aligned} \quad (6.13)$$

Therefore, the efficiency of donor quenching at λ_{em}^D is proportional to the sensitized acceptor fluorescence at λ_{em}^A, with S_D/S_A being the scaling factor (Merzlyakov et al., 2007).

The scaling factor S_D/S_A depends on the dyes used and on the characteristics of the instrument, not on the particular dye concentration. S_D/S_A can be calculated from the FRET spectrum for well-defined dye concentrations and known FRET efficiency according to

$$\frac{S_D}{S_A} = \frac{\Delta F_D(\lambda_{em}^D)}{F_{sen}(\lambda_{em}^A)}. \quad (6.14)$$

Alternatively, if the photodetector efficiency at λ_{em}^D and λ_{em}^A is the same (or if the fluorescence signal is normalized by the photodetector efficiency), then S_D/S_A equals the ratio of donor and acceptor quantum yields, $\emptyset_D/\emptyset_A$, which can be calculated from measured absorbance and emission spectra. Both methods yield $S_D/S_A = 1 \pm 0.05$ for the fluorescein–rhodamine FRET pair (Merzlyakov et al., 2007).

Once S_D/S_A is determined, it can be used to calculate the donor fluorescence in the absence of the acceptor (i.e., the donor control, F_D) from the measured donor fluorescence in the presence of the acceptor F_{DA} according to

$$F_D(\lambda_{em}^D) = F_{DA}(\lambda_{em}^D) + F_{sen}(\lambda_{em}^A)\frac{S_D}{S_A}. \qquad (6.15)$$

5.2.2. EmEx-FRET protocol

Here, we show how to calculate the FRET efficiency E, the donor concentration $[d]$, and the acceptor concentration $[a]$ using (1) the excitation and emission spectra acquired in the presence of donors and acceptors of unknown concentrations $[d]$ and $[a]$, (2) the donor emission standard spectrum (acquired for a known donor concentration), and (3) the acceptor excitation and emission standard spectra (acquired for a known acceptor concentration).

Step 1. The excitation and emission spectra of bilayers containing donor- and acceptor-labeled peptides are acquired in a fluorometer. Each FRET spectrum has three contributions: direct donor emission in the presence of the acceptor F_{DA}, direct acceptor emission F_A, and sensitized acceptor emission F_{sen}. Furthermore, $F_{AD} = F_A + F_{sen}$.

Step 2. The acquired FRET excitation spectrum (Fig. 6.4A, black line) is compared to the standard excitation spectrum of the acceptor (Fig. 6.3A, red line). The standard spectrum is multiplied by a coefficient, ξ, to produce the red line in Fig. 6.4A, such that the amplitude of the scaled standard (Fig. 6.4A, red line) is identical with the FRET excitation spectrum (Fig. 6.4A, black line) at λ_{ex}^A. This step gives the concentration of the acceptor in the sample, $[a]$, as ξ times the standard acceptor concentration, in this case $\xi \times 0.1$ mol%.

Step 3. The emission standard of the acceptor (Fig. 6.3B, red line) is multiplied by ξ to obtain the direct emission contribution of the acceptor (Fig. 6.4B, red line). The direct acceptor contribution (red line) is subtracted from the FRET emission spectrum (Fig. 6.4B, black line), to reveal the sum of the direct donor emission and the sensitized acceptor emission, $F_{DA} + F_{sen}$. This sum is plotted in Fig. 6.4C with the black dashed line.

Step 4. The emission standard spectrum of the donor (Fig. 6.3B, blue line) is multiplied by a coefficient to produce the blue dashed line in Fig. 6.4C, such that the amplitude of the scaled standard (blue dashed line) is identical

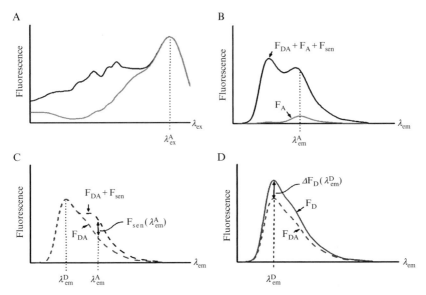

Figure 6.4 The EmEx-FRET method. (A) An acquired FRET excitation spectrum (black line) is compared to the excitation standard spectrum of the acceptor (Fig. 3A, red line.) This step gives the concentration of the acceptor in the sample. (B) The emission standard of the acceptor (Fig. 3B, red line) is scaled according to the acceptor concentration determined in (A) (red line). The difference between the red line and the FRET emission spectrum (black line) is the sum $F_{DA} + F_{sen}$. (C) Dashed black line: $F_{DA} + F_{sen}$, sum of the direct donor (Bell et al., 2000) emission in the presence of the acceptor, F_{DA}, and the sensitized acceptor emission, F_{sen}. The standard emission spectrum of the donor (Fig. 3B, blue line) is scaled, such that the amplitude of the scaled standard (blue dashed line) is identical to the amplitude of the dashed black line at λ_{em}^A. The difference between the black dashed and the blue dashed line is the sensitized acceptor emission, F_{sen}. The value of the sensitized emission at λ_{em}^A, $F_{sen}(\lambda_{em}^A)$, is related to the decrease in donor emission at λ_{em}^D, $\Delta F_D(\lambda_{em}^D)$ (Merzlyakov et al., 2007). (D) The value $\Delta F_D(\lambda_{em}^D)$, determined in (C), is used to determine the donor emission in the absence of the acceptor F_D (blue line), and the donor concentration. Reprinted from (Merzlyakov et al., 2007), with permission from Springer Science and Business Media. (See Color Insert.)

with the black dashed line in Fig. 6.4C at λ_{em}^D. The difference between the black dashed and the blue dashed line is the sensitized acceptor emission, F_{sen}. The value of the sensitized emission at λ_{em}^A, $F_{sen}(\lambda_{em}^A)$, is related to the decrease in donor emission at λ_{em}^D, $\Delta F_D(\lambda_{em}^D)$, given by Eq. (6.13).

Step 5. The value $\Delta F_D(\lambda_{em}^D)$, determined in Step 4, is added to $F_{DA}(\lambda_{em}^D)$ (Fig. 6.4D) to give the value of the donor emission in the absence of acceptor at λ_{em}^D, that is, $F_D(\lambda_{em}^D) = \Delta F_D(\lambda_{em}^D) + F_{DA}(\lambda_{em}^D)$. The donor emission standard (Fig. 6.3B, blue line) is multiplied by a coefficient ζ, such that the amplitude of the standard spectrum at λ_{em}^D equals $F_D(\lambda_{em}^D)$. This step gives the concentration of the donor in the sample, [d], as ζ times the standard donor concentration, in this case $\zeta \times 0.1$ mol%. It also gives the complete emission spectrum of the donor in the absence of the acceptor F_D (Fig. 6.4D, blue line) as the emission of the donor standard (Fig. 6.3B, blue line), multiplied by ζ.

Further data analysis is preformed using Eqs. (6.2)–(6.9).

We have studied the dimerization energetics of the Gly382Asp TM domain of fibroblast growth factor receptor 3 (FGFR3) using the EmEx-FRET method (Merzlyakov et al., 2007). The results are shown in Fig. 6.5, and each data point is derived from a single experiment. Yet, one can have high confidence in the data, because the actual donor and acceptor concentrations for each data point have been determined. Thus, errors associated with uncertainties in peptide concentrations due to low peptide solubility have been reduced. The free energy of dimerization is calculated as -2.78 ± 0.04 kcal/mol. On the other hand, the calculation of the dimerization free energy from donor quenching (i.e., not using the EmEx-FRET method) gives -2.9 ± 0.7 kcal/mol. Thus, the application of the EmEx-FRET method reduces the uncertainty in the free energy calculation from ± 0.7 to ± 0.04 kcal/mol, and therefore greatly improves the experimental precision.

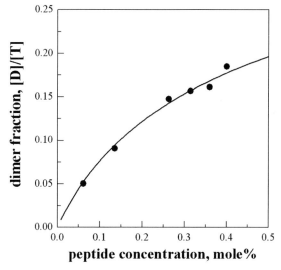

Figure 6.5 Dimer fraction [D]/[T] vs total peptide concentration [T] for TM$_{382Asp}$. The symbols represent the dimer fractions calculated using the EmEx-FRET method, for different protein concentrations. The solid line is the theoretical equilibrium curve, obtained as described in details previously (Li et al., 2005, 2006). The free energy of dimerization, as the average of the six experiments, is -2.78 ± 0.04 kcal/mol. For comparison, the average free energy value, obtained from donor quenching as previously described (You et al., 2005) (i.e. not using the EmEx-FRET method), is -2.9 ± 0.7 kcal/mol. Thus, the EmEx-FRET method greatly improves the precision of TM helix dimerization energetics measurements. Reprinted from (Merzlyakov et al., 2007), with permission from Springer Science and Business Media.

5.3. Energetics of TM helix heterodimerization

Measurements of TM helix heterodimerization energetics present a unique challenge if the helices have both homo- and heterodimerization propensities (Merzlyakov et al., 2006d). Here, we overview the thermodynamics behind heterodimer formation and outline a method for calculating the free energy of heterodimerization using FRET.

5.3.1. Theory and protocol

In a lipid bilayer with two different TM helices, X and Y, three different types of dimers will form: XX, YY, and XY. The monomer–dimer equilibrium is described by three equilibrium constants, K_X, K_Y, and K_{XY}:

$$X + X \underset{}{\overset{K_x}{\rightleftarrows}} XX,\ Y + Y \underset{}{\overset{K_y}{\rightleftarrows}} YY,\ X + Y \underset{}{\overset{K_{xy}}{\rightleftarrows}} XY, \qquad (6.16)$$

$$K_X = \frac{[XX]}{[X]^2},\ K_Y = \frac{[YY]}{[Y]^2},\ K_{XY} = \frac{[XY]}{[X][Y]}, \qquad (6.17)$$

where $[XX]$, $[YY]$, and $[XY]$ are the concentrations of the XX homodimers, YY homodimers, and XY heterodimers, respectively; $[X]$ and $[Y]$ are the concentrations of the X and Y monomers, respectively. The homodimer equilibrium constants K_X and K_Y can be measured independently for helices X and Y as described above and by Li et al. (2005, 2006) and You et al. (2005).

The total concentrations of the X and Y helices, $[T_X]$ and $[T_Y]$, are known as aliquoted and equal to

$$[T_X] = [X] + 2[XX] + [XY]\ \text{or}\ [T_X] = [X] + 2K_X[X]^2 + [XY], \qquad (6.18)$$

$$[T_Y] = [Y] + 2[YY] + [XY]\ \text{or}\ [T_Y] = [Y] + 2K_Y[Y]^2 + [XY]. \qquad (6.19)$$

Let X be labeled with a FRET donor, and Y be labeled with the appropriate acceptor.

The FRET efficiency due to sequence-specific heterodimerization is given by

$$E_{\text{dimer}} = \frac{[da]}{[d]}, \qquad (6.20)$$

where $[da]$ is the concentration of donor–acceptor dimers. $[da]$ would be equal to $[XY]$, and $[d]$ would be equal to $[T_X]$ if the labeling yield is 100% (i.e., the ratio of peptides to conjugated fluorescent dyes is 1). If the labeling yields, f_d and f_a, for the donor and the acceptor are lower than 100%, then

$$[d] = f_d[T_X], 0 < f_d < 1,$$
$$[a] = f_a[T_Y], 0 < f_a < 1. \quad (6.21)$$

Assuming that labeling does not affect the dimerization energetics, the concentration of donor–acceptor heterodimers relates to the total concentration of heterodimers XY as follows:

$$[da] = [XY] f_d f_a. \quad (6.22)$$

From Eqs. (6.20)–(6.22), we obtain

$$E_{\text{dimer}} = \frac{[XY] f_d f_a}{f_d [T_X]} = \frac{[XY]}{[T_X]} f_a. \quad (6.23)$$

Equations (6.18), (6.19), and (6.23) can be solved to determine the three unknowns: $[X]$, $[Y]$, and $[XY]$. The solution is

$$[XY] = E_{\text{dimer}}[T_X]/f_A, \quad (6.24)$$

$$[X] = \frac{\sqrt{1 + 8K_X[T_X](1 - E_{\text{dimer}}/f_A)} - 1}{4K_X}, \quad (6.25)$$

$$[Y] = \frac{\sqrt{1 + 8K_Y([T_Y] - E_{\text{dimer}}[T_X]/f_A)} - 1}{4K_Y}. \quad (6.26)$$

The heterodimerization equilibrium constant K_{XY} is calculated as

$$K_{XY} = \frac{E_{\text{dimer}}[T_X]}{[X][Y] f_A}, \quad (6.27)$$

where $[X]$ and $[Y]$ are given by Eqs. (6.25) and (6.26), respectively. Then, the free energy of heterodimerization can be calculated as

$$\Delta G_{XY} = -RT \ln K_{XY}. \quad (6.28)$$

6. Biological Insights from FRET Measurements

Using the described FRET methodology, we have studied the thermodynamic principles that underlie signal transduction across biological membranes (Li and Hristova, 2006). The application of these methods to wild-type and mutant RTK TM domains has yielded new information pertaining to the physical basis of complex biological processes, and to the mechanism of pathogenesis in humans (Li and Hristova, 2006; Li et al., 2006; You et al., 2006, 2007).

6.1. Thermodynamics of protein homodimerization in lipid bilayer membranes: Wild-type FGFR3 TM domain, and the pathogenic Ala391Glu, Gly380Arg, and Gly382Asp mutants

Mutations in the TM domain of FGFR3 have been linked to growth disorders and cancers. One example is the Ala391Glu mutation, causing Crouzon syndrome with acanthosis nigricans, as well as bladder cancer (Li and Hristova, 2006; Meyers et al., 1995; van Rhijin et al., 2002). To gain insight into the molecular mechanism behind the pathology, we have determined the free energy of dimerization of the wild-type and the Ala391Glu mutant in lipid bilayers. The change in the free energy of dimerization due to the Ala391Glu pathogenic mutation is -1.3 kcal/mole (Li et al., 2006). This seemingly modest value can lead to a large increase in receptor dimer fraction and thus, profoundly affect FGFR3-mediated signal transduction (Li et al., 2006). The pathogenic stabilization is most likely due to Glu-mediated hydrogen bonds (Li and Hristova, 2006).

We have also determined the energetics of dimerization of the Gly380Arg FGFR3 variant linked to achondroplasia, the most common form of human dwarfism. The molecular mechanism of pathogenesis in achondroplasia is under debate, and two different mechanisms have been proposed to contribute to pathogenesis: (1) Arg380-mediated FGFR3 dimer stabilization (Webster and Donoghue, 1996) and (2) slow downregulation of the activated mutant receptors (Cho et al., 2004). We have shown that the Gly380Arg mutation does not alter the dimerization energetics of FGFR3 TM domain in detergent micelles or in lipid bilayers (You et al., 2006). This result indicates that pathogenesis in achondroplasia cannot be explained simply by a higher dimerization propensity of the mutant FGFR3 TM domain, thus highlighting the potential importance of the observed slow downregulation in phenotype induction (Cho et al., 2004).

We have also investigated the effect of the Gly382Asp mutation, identified in the KSM-18 myeloma cell line (Otsuki et al., 1998), on the dimerization propensity of FGFR3 TM domain. The measured free energy of dimerization, -2.78 ± 0.04 kcal/mol (Merzlyakov *et al.*, 2007), is the same as the value determined for wild-type, -2.8 ± 0.1 kcal/mol (Li *et al.*, 2006), indicating that the Gly382Asp mutation does not stabilize the FGFR3 TM domain dimer. Thus, the mechanism of pathogenesis in multiple myeloma is not associated with dimer stabilization due to Asp-mediated hydrogen bonds.

6.2. Thermodynamics of protein heterodimerization in lipid bilayer membranes

We have determined the propensity for heterodimer formation between wild-type FGFR3 TM domain and the Ala391Glu mutant, linked to the autosomal dominant Crouzon syndrome with acanthosis nigricans (Merzlyakov *et al.*, 2006d). The cells of the affected heterozygotes express both wild-type and mutant receptors, and thus, measurements of heterodimer stabilities are critical for understanding the induction of the phenotype. The free energy of heterodimerization was determined as -3.37 ± 0.25 kcal/mol. Comparison of this value to the homodimerization free energies for the wild-type, -2.8 ± 0.2 kcal/mol, and the mutant, -4.1 ± 0.2 kcal/mol, reveals that the heterodimer stability is the average of the two homodimer stabilities. This finding may indicate that the mutant homodimer is stabilized by two hydrogen bonds, while a single hydrogen bond forms in the heterodimer.

7. CONCLUSION

Quantitative FRET measurements of TM helix dimerization energetics in lipid bilayers can be carried out to determine the homo- and heterodimerization propensities with high precision. Such experiments have yielded new insights into the role of RTK TM domains in cell signaling and human pathologies.

REFERENCES

Adair, B. D., and Engelman, D. M. (1994). Glycophorin A helical transmembrane domains dimerize in phospholipid bilayers—A resonance energy transfer study. *Biochemistry* **33**, 5539–5544.

Cho, J. Y., Guo, C. S., Torello, M., Lunstrum, G. P., Iwata, T., Deng, C. X., and Horton, W. A. (2004). Defective lysosomal targeting of activated fibroblast growth factor receptor 3 in achondroplasia. *Proc. Natl Acad. Sci. USA* **101**, 609–614.

Iwamoto, T., Grove, A., Montal, M. O., Montal, M., and Tomich, J. M. (1994). Chemical synthesis and characterization of peptides and oligomeric proteins designed to form transmembrane. *Int. J. Pept. Protein Res.* **43**, 597–607.

Iwamoto, T., You, M., Li, E., Spangler, J., Tomich, J. M., and Hristova, K. (2005). Synthesis and initial characterization of FGFR3 transmembrane domain: Consequences of sequence modifications. *Biochim. Biophys. Acta* **1668,** 240–247.

Li, E., and Hristova, K. (2006). Role of receptor tyrosine kinase transmembrane domains in cell signaling and human pathologies. *Biochemistry* **45,** 6241–6251.

Li, E., You, M., and Hristova, K. (2005). SDS-PAGE and FRET suggest weak interactions between FGFR3 TM domains in the absence of extracellular domains and ligands. *Biochemistry* **44,** 352–360.

Li, E., You, M., and Hristova, K. (2006). FGFR3 dimer stabilization due to a single amino acid pathogenic mutation. *J. Mol. Biol.* **356,** 600–612.

MacKenzie, K. R. (2006). Folding and stability of alpha-helical integral membrane proteins. *Chem. Rev.* **106,** 1931–1977.

Merzlyakov, M., Li, E., Casas, R., and Hristova, K. (2006a). Spectral Forster resonance energy transfer detection of protein interactions in surface-supported bilayers. *Langmuir* **22,** 6986–6992.

Merzlyakov, M., Li, E., Gitsov, I., and Hristova, K. (2006b). Surface-supported bilayers with transmembrane proteins: Role of the polymer cushion revisited. *Langmuir* **22,** 10145–10151.

Merzlyakov, M., Li, E., and Hristova, K. (2006c). Directed assembly of surface-supported bilayers with transmembrane helices. *Langmuir* **22,** 1247–1253.

Merzlyakov, M., You, M., Li, E., and Hristova, K. (2006d). Transmembrane helix heterodimerization in lipids bilayers: Probing the energetics behind autosomal dominant growth disorders. *J. Mol. Biol.* **358,** 1–7.

Merzlyakov, M., Chen, L., and Hristova, K. (2007). Studies of receptor tyrosine kinase transmembrane domain interactions: The EmEx-FRET method. *J. Membrane Biol.* **215,** 93–103.

Meyers, G. A., Orlow, S. J., Munro, I. R., Przylepa, K. A., and Jabs, E. W. (1995). Fibroblast-growth-factor-receptor-3 (Fgfr3) transmembrane mutation in Crouzon-syndrome with acanthosis nigricans. *Nat. Genet.* **11,** 462–464.

Otsuki, T., Nakazawa, N., Taniwaki, M., Yamada, O., Sakaguchi, H., Wada, H., Yawata, Y., and Ueki, A. (1998). Establishment of a new human myeloma cell line, KMS-18, having t(4;14)(p16.3;q32.3) derived from a case phenotypically transformed from Ig A-lambda to BJP-lambda, and associated with hyperammonemia. *Int. J. Oncol.* **12,** 545–552.

Sackmann, E. (1996). Supported membranes: Scientific and practical applications. *Science* **271,** 43–48.

Sackmann, E., and Tanaka, M. (2000). Supported membranes on soft polymer cushions: Fabrication, characterization and applications. *Trends Biotechnol.* **18,** 58–64.

van Rhijin, B., van Tilborg, A., Lurkin, I., Bonaventure, J., de Vries, A., Thiery, J. P., van der Kwast, T. H., Zwarthoff, E., and Radvanyi, F. (2002). Novel fibroblast growth factor receptor 3 (FGFR3) mutations in bladder cancer previously identified in non-lethal skeletal disorders. *Eur. J. Hum. Genet.* **10,** 819–824.

Webster, M. K., and Donoghue, D. J. (1996). Constitutive activation of fibroblast growth factor receptor 3 by the transmembrane domain point mutation found in achondroplasia. *EMBO J.* **15,** 520–527.

White, S. H., and Wimley, W. C. (1999). Membrane protein folding and stability: Physical principles. *Annu. Rev. Biophys. Biomol. Struct.* **28,** 319–365.

Wimley, W. C., and White, S. H. (2000). Designing transmembrane α-helices that insert spontaneously. *Biochemistry* **39,** 4432–4442.

Wolber, P. K., and Hudson, B. S. (1979). An analytic solution to the Förster energy transfer problem in two dimensions. *Biophys. J.* **28,** 197–210.

You, M., Li, E., Wimley, W. C., and Hristova, K. (2005). FRET in liposomes: Measurements of TM helix dimerization in the native bilayer environment. *Anal. Biochem.* **340,** 154–164.

You, M., Li, E., and Hristova, K. (2006). The achondroplasia mutation does not alter the dimerization energetics of FGFR3 transmembrane domain. *Biochemistry* **45,** 5551–5556.

You, M., Spangler, J., Li, E., Han, X., Ghosh, P., and Hristova, K. (2007). Effect of pathogenic cysteine mutations on FGFR3 transmembrane domain dimerization in detergents and lipid bilayers. *Biochemistry* **46,** 11039–11046.

CHAPTER SEVEN

Application of Single-Molecule Spectroscopy in Studying Enzyme Kinetics and Mechanism

Jue Shi,*,§ Joseph Dertouzos,† Ari Gafni,*,‡ *and* Duncan Steel*,†

Contents

1. Introduction	130
2. Experimental Considerations	134
3. Sample Preparation	136
4. Instrumentation	138
5. Data Analysis of Single-Molecule Trajectories	141
5.1. Static heterogeneity	145
5.2. Dynamic heterogeneity	146
5.3. Kinetics of a single oligomeric enzyme molecule	147
6. Additional Considerations	151
7. Summary	154
References	155

Abstract

This chapter reviews recent developments in the application of single-molecule spectroscopy (SMS) to studies of enzyme kinetics and mechanism. Protocols for conducting single-molecule experiments on enzymes, based largely on the experience in our laboratory, are provided, including methods of sample preparation, instrumentation, and data analysis. We also address general issues related to the design of meaningful single-molecule experiments and include specific examples of the application of SMS to enzyme studies, which reveal new and intriguing aspects of enzyme behavior, including static and dynamic heterogeneity, as well as subunit cooperativity. Finally, we discuss the advantages of employing single-molecule approach in obtaining new information beyond ensemble studies.

---

* Biophysics Research Division, University of Michigan, Ann Arbor, Michigan
† Department of Physics and EECS, University of Michigan, Ann Arbor, Michigan
‡ Department of Biological Chemistry, University of Michigan, Ann Arbor, Michigan
§ Current address: Department of Physics, Hong Kong Baptist University, Hong Kong

ABBREVIATIONS

AFM	atomic force microscopy
APD	avalanche photodiode
CCD	charge-coupled device
DHOD	dihydroorotate dehydrogenase
CO_x	cholesterol oxidase
FCS	fluorescence correlation spectroscopy
FRET	Forster resonance energy transfer
NA	numerical aperture
NSOM	near-field scanning optical microscopy
PMT	photomultiplier tube
SMS	single-molecule spectroscopy
TIR	total internal reflection

1. Introduction

The past 20 years have witnessed a growing interest in studying biological processes and interactions at the single-molecule level. Beginning with patch-clamp studies, the opening and closing of individual ion channels can be followed by measuring the stepwise change of ion currents across the membrane (Sakmann and Neher, 1995). The advances in optical instrumentation and theory have expanded the single-molecule capability to encompass a wide variety of biomolecules.

For example, near-field scanning optical microscopy (NSOM) overcomes the optical diffraction limit, giving a spatial resolution of 50–100 nm (de Lange et al., 2001). It provides optical images and information about the specific spectroscopic properties of single molecules on a sample surface by placing an optical probe tip within 5–10 nm (~probe diameter/π) of the sample, which serves as a restriction for light passage with a small aperture.

Molecular structures of biological surfaces can be also probed by atomic force microscopy (AFM), a nonoptical detection technique that gives nanometer to subnanometer lateral resolution. It uses a sharp probe tip to interact with atoms on the surface. Forces (e.g., Van der Waals, electrostatic, magnetic, etc.) generated between the tip and the sample are transmitted to an attached flexible cantilever, causing it to bend. The bending of the cantilever is monitored by the deflection of a laser beam reflected by the cantilever (Santos and Castanho, 2004). Although AFM provides higher resolution than NSOM, it does not provide information about the chemical identity of the sample because the tip interacts indiscriminately with the sample surface. This problem can be solved by specifically labeling the probe tip with biomolecules, such as antibodies (Willemsen et al., 2000).

In another approach to single-molecule studies, mechanical properties of biomolecules and biological interactions can be conveniently measured at the single-molecule level by using optical tweezers. The basic concept of optical tweezers is to manipulate individual biomolecules by attaching them to micron-sized dielectric beads (e.g., polystyrene beads) that can be trapped at the center of a focused laser beam (Ashkin, 2000; Bustamante et al., 2000). Variations of the force acting on the bead are measured by its displacement or by the deflection of the laser beam. This gives insight into the mechanics of the attached biomolecules.

Fluorescence correlation spectroscopy (FCS) also has the capability to offer single-molecule resolution. It measures the temporal fluorescence fluctuations from a small number of molecules (sample concentration of 0.01–100 nM in a small volume of 0.01–100 pl). Temporal autocorrelation of these fluorescence fluctuations decays with time. The rate and the shape of the decay provide information about the rates of the processes that lead to the fluorescence fluctuations, while the magnitude of the autocorrelation function provides information about the composition of the fluorescent species in the sample volume (Elson, 2004; Elson and Magde, 1974). Therefore, FCS can be applied to study diffusion, chemical reactions, and molecular association/dissociation.

Although all the techniques discussed above are useful for diverse biological systems, one of the most rapidly developing areas of single-molecule study is single-molecule spectroscopy (SMS) based on laser-induced fluorescence (Weiss, 1999). This technique can reveal new information not observable in ensemble studies. At a more fundamental level it provides empirical means to evaluate the applicability of the Ergodic theorem to a specific enzymatic system, which asserts that ensemble-averaged behavior is the same as the temporal average of the behavior of a single molecule. SMS also enables studies to determine if the models describing the reaction kinetics are Markovian.

A broad spectrum of SMS studies have already shed light on various biological processes, such as molecular motions (Ariga et al., 2002; Forkey et al., 2003; Sosa et al., 2001), nucleic acids and protein folding (Ha et al., 1999; Margittai et al., 2003; Rhodes et al., 2003; Schuler et al., 2002; Yang et al., 2003), nucleic acid–protein interactions (Blanchard et al., 2004; Ha et al., 2002), and enzyme kinetics (Edman et al., 1999; Lu et al., 1998; Rajagopalan et al., 2002; Shi et al., 2004; Zhuang et al., 2002). For example, Sosa et al. (2001) determined the orientation and mobility of kinesin bound to microtubule in the presence or absence of different nucleotides by monitoring the polarization of single kinesin molecules labeled with bis-((N-iodoacetyl)piperazinyl) sulfoerhodamine (BSR).

Forster resonance energy transfer (FRET) measurement is a useful tool in probing molecular distance and has been successfully employed in single-molecule studies. FRET is energy transfer from an excited donor fluorophore to an acceptor fluorophore, resulting in quenching of the donor fluorescence. The rate of transfer depends on the sixth power of the distance between the donor and acceptor. Therefore, FRET is a useful for measuring distance, especially for studies of biomolecules because FRET typically occurs in the range of 20–60 Å, which is comparable to the dimension of biomolecules (Lakowicz, 1999). For instance, Schuler et al. (2002) studied the conformational distribution of a cold-shock protein and derived the free-energy surface for folding by measuring the FRET efficiency of a single cold-shock protein labeled with Alexa 488 and Alexa 594 dyes. Also by following FRET, Ha et al. examined single protein–nucleic acid interactions during the unwinding of DNA by Rep helicase (Ha et al., 2002). It was shown that Rep monomers bind DNA but do not initiate unwinding, suggesting that the functional complex of Rep helicase is a dimer. This complex appears to be relatively unstable and its frequent dissociation causes DNA unwinding to stall, explaining the low unwinding processivity of Rep helicase observed in ensemble experiments.

In addition to the successful *in vitro* single-molecule experiments, *in vivo* trafficking and protein interactions have been visualized at the single-molecule level by imaging molecules labeled with organic dyes or fused with fluorescent proteins (Harms et al., 2001; Itoh et al., 2002; Murakoshi et al., 2004; Sako et al., 2000). SMS is thus facilitating research in various areas. In this review we focus on applying SMS to study the kinetics and mechanisms of enzymes, and describe the general protocols for conducting such single-molecule measurements.

The unique advantage of following enzyme kinetics one molecule at a time is that it provides new information for elucidating the catalytic mechanism, such as static and dynamic heterogeneity, nonpopulated intermediates, as well as subunit activities and cooperativity of oligomeric enzymes. When the activity of one molecule is monitored, different reaction processes can be distinguished and kinetic properties of individual molecules can be obtained. Therefore, single-molecule measurement can distinguish between different populations of molecules (static heterogeneity). Molecular heterogeneity plays an important role in various biological processes, for example, early intermediates in amyloid aggregation are highly heterogeneous and some aggregates eventually cause neurodegenerative diseases such as Alzheimer (Klein et al., 2004). Obtaining information on static heterogeneity with the single-molecule approach can help elucidate the reaction mechanism of these processes.

A key distinguishing feature of SMS is seen by recalling that ensemble experiments follow reactions of a large number of molecules simultaneously. The lack of synchrony among different molecules makes the ensemble-averaged results uninformative in terms of the temporal dynamics of individual molecules in reaction. In contrast, single-molecule experiments follow the sequential reaction cycles of individual molecules, thereby enabling dynamic correlation analysis of the time evolution of the reaction kinetics. Such correlation can provide insight into the dynamic fluctuation of an enzyme, possibly arising from conformational fluctuations and environmental modulation and resulting in a non-Markovian process.

Single-molecule studies can also help identify intermediates that are difficult to observe in ensemble studies. Ensemble studies essentially follow concentration changes over time, therefore short-lived intermediate states that are sparsely populated are not observable. In contrast, a single-molecule measurement has the potential to observe all intermediates because each molecule has to go through all the reaction steps sequentially to complete the reaction cycle. The same applies to the study of subunit kinetics and cooperativity. For oligomeric enzymes consisting of multiple catalytic subunits, ensemble stopped-flow studies give rate constants that are the ensemble-averaged results of all subunits, thus it is difficult to separate and determine the contribution from each subunit and study the kinetics of an individual subunit. Because single-molecule measurements follow the sequential reactions of the individual subunits of one oligomer over time, subunit kinetics can be studied separately if subunit activity can be distinguished by distinct fluorescent states. Moreover, interactions between subunits that affect catalytic efficiency (cooperativity) can be studied by analyzing catalytic turnovers at the single-molecule level.

Flavoproteins are particularly amenable to SMS studies because the flavin cofactor is an intrinsic fluorescent reporter. The redox-active 7,8-dimethylisoalloxazine moiety of the flavin is generally fluorescent in the oxidized state and nonfluorescent in the reduced state. This signature fluorescence change of flavin upon reductive and oxidative reactions has been utilized to study the turnover of single flavoenzymes with SMS (Lu et al., 1998; Shi et al., 2004). Flavoenzymes most frequently use flavin mononucleotide (FMN) or flavin adenine dinucleotide (FAD) as noncovalently bound prosthetic groups. As the flavin bound in a single flavoenzyme molecule cycles between the oxidized and the reduced state, the kinetics of catalysis can be followed as the fluorescence signal alternates between the fluorescent and nonfluorescent state. The probability distribution of the fluorescence on-states provides the lifetimes of the oxidized state, while the probability distribution of the off-states provides the lifetimes of the reduced state. Because the on- and off-times are the waiting time for flavin reduction and oxidation, respectively, they are directly related to the rate constants of the reactions, enabling kinetic analysis using single-molecule data. Kinetics of a single enzyme can also be measured with substrates fluorescence change as they convert into products (Dyck and Craig, 2002; Tan and Yeung, 1997). While this approach reports on a single enzyme molecule, the fluorescence signal is from many substrate (or product molecules), hence it is not considered SMS. In this review we focus exclusively on SMS studies based on fluorescence changes of the single enzyme molecule itself.

2. Experimental Considerations

To record a reasonable single-molecule fluorescence signal, the experiment needs to be designed with several technical issues taken into consideration:

1. When the protein does not contain an intrinsic fluorophore like a flavin (tryptophan and tyrosine are too photolabile to be of practical use), it is necessary to label the protein with a dye or FRET pair, whose fluorescence is modulated by the reaction steps of interest.
2. Irreversible photobleaching of the fluorophore limits the number of available photons for single-molecule detection. In general, a relatively photostable fluorophore, such as Rhodamine, can emit between 10^5 and 10^6 photons before it is bleached (Nie and Zare, 1997). Dyes like AlexaFluor, Cy3, and Cy5 also have good photostability comparable to Rhodamine, while fluorescein and intrinsic fluorescent amino acids, such as tyrosine and tryptophan, are photolabile and thus not appropriate for single-molecule detection. Because oxygen free radicals generated by light excitation can react with fluorophore in the excited state, deoxygenation of the sample can retard the process of photobleaching and prolong the lifetime of fluorophores. This is especially important for the Cy3 and Cy5 dyes. An oxygen scavenging system that is commonly used includes 0.1 mg/ml glucose oxidase, 0.02 mg/ml catalase, 1% 2-mercaptoethanol, and 3% (w/v) glucose (Ha, 2001). However, deoxygenation is not always effective. For example, Texas Red and AlexaFluor 488 do not exhibit extended lifetime after oxygen removal. And in the case of fluorescein, deoxygenation has a negative effect on the number of photons it emits before it is bleached (Ha, 2001).
3. Due to photobleaching one cannot follow kinetics continuously by dye fluorescence for more than several minutes typically, and therefore we need to optimize the rate of data acquisition. To do this, the approach that is employed in recording slowly evolving events on video (time-lapsed photography) can be adapted for SMS measurements. The idea is to turn the excitation light on for short periods of time (*ca.* 10 ms or shorter, depending on excitation rate and detector noise) separated by time intervals when there is no light on the dye (the length of the off periods is adjusted to allow observation of the full reaction, while the time resolution is sacrificed). This approach enables creation of a series of still "snapshots" that can be joined together to make a relatively detailed

time history. Specifically, the laser-on time τ_{on} is determined by the requirements to obtain an adequate signal to noise. The time interval $\tau_{off}(t)$ between data samplings is determined by the needed time resolution. The time interval $\tau_{off}(t)$ needed to monitor a nonlinear chemical reaction is expected to be time dependent where, for example, the reaction may proceed slowly, and then increase in time such as in nucleation dependent oligomerization.

4. For a fluorophore that is noncovalently bound, as in most flavoenzymes, dissociation of the fluorophore also limits the length of single-molecule traces. Our study of several flavoenzymes showed that the flavin dissociated from the holoenzyme and diffused away from the field of view within seconds, before the flavin was photobleached. Therefore, dissociation can be the limiting factor for measuring single-molecule fluorescence time traces, especially considering the low concentration (subnanomolar) typical of these measurements.

   Note: The appropriate concentration in single-molecule measurements is set by the need for the microscope to spatially resolve individual sites of fluorescence. For example, assuming dye-labeled molecules immobilized in a thin film of agarose gel (see below) of thickness (~5–10 μm) and a spatial resolution on the order of 500 nm (diffraction limit), single molecules can be resolved at concentrations up to 0.1 nM. Due to significant flavin dissociation, the concentration of flavoenzyme has to be on the order of 1–5 nM to achieve the same density of single-molecule signals as that observed for molecules with covalently linked chromophore.

5. If we assume no background or electronic detection noise, and shot noise limited photon emission (common for most cases of interest), and further assume a signal-to-noise ratio of $\sqrt{N}$ in each time bin (corresponding to the shot noise for an average of N photons detected per time bin), we can derive that the length of a single-fluorophore trajectory as $T = \xi^{-1}(N_0/N)$, where ξ is the fluorescence duty cycle (ratio of average fluorescence on-times to average time between on-times), N_0 is the total number of detected photons (limited by photobleaching or fluorophore dissociation as discussed above) and N is the number of photons detected in each time bin. $N = \varepsilon \Delta t$, where ε is the detection rate of emission and Δt is the chosen width of the time bin. For $\varepsilon = 10^3$ s^{-1}, $\Delta t = 100$ ms (achieving a signal-to-noise ratio of $\sqrt{N} = 10$) and $\xi = 0.5$, we find $T = 2000$ s for $N_0 = 10^5$.

6. Assuming that the laser intensities used for excitation are below the fluorophore saturation intensity, the detection rate of the emission, ε, is determined by $\varepsilon = (I\sigma/\hbar\omega)q\eta$, where I is the excitation intensity at the focal plane, σ is the absorption cross section (in chemistry and biology a similar coefficient termed extinction coefficient, ε_0, is more commonly used than σ. ε_0 is related to σ by $\varepsilon_0 = 4.3 \times 10^{-3} A_0 \sigma$, where A_0 is the Avogadro number), $\hbar$ is the Planck constant, ω is the excitation

frequency, q is the quantum efficiency, and η is the microscope detection efficiency. With a high numerical aperture (NA) objective (NA ≥ 1.4) and a single-photon-counting avalanche photodiode (APD, quantum efficiency typically 70%), the detection efficiency of a confocal microscope, η, is around 10% (Sandison et al., 1995). To calculate appropriate laser power (I) for exciting the single fluorophore, we first consider the desired detection rate, ε, determined by the required photon counts in each time bin (N) and the width of the time bin Δt (the time resolution of the experiment). In this case, the desired rate is then given by $\varepsilon = N/\Delta t$. Using this result and the above expression for ε, the laser intensity I can be determined.

7. Background noise can be reduced optically by exciting a small area of the sample, such as applying total internal reflection (TIR) (Axelrod, 1989) or using multiphoton excitation (Williams et al., 1994) to illuminate a very thin layer of the sample, or by using a pinhole at the emission side (confocal geometry) to reduce detection volume (Sandison et al., 1995). In essence, TIR, multiphoton excitation and confocal detection all decrease the out-of-focus background noise by achieving a reduced depth of field (the depth of the sample that is effectively in-focus). In a TIR setup, the depth of field is small because the "evanescent field" resulted from TIR light incident beyond the critical angle only excites fluorophores within about 100 nm of the surface. As for multiphoton excitation, because the probability of fluorophore absorbing multiple photons is proportional to the square (two-photon absorption) or higher order of the excitation light intensity, only the area proximal to the focal point, where the photon density is high, can be excited, resulting in a small depth of field. By placing a small pinhole at the focus of the image plane, confocal microscopy suppresses the passage of emission from above and below the focal plane, thus also achieving a small depth of field.

8. Detectors with high quantum efficiency and low dark counts are preferred. Therefore, single-photon-counting APD should be used rather than photomultiplier tube (PMT). As for imaging, cooled charge-coupled device (CCD) camera with photointensifier or on-chip electronic gain provides a better signal-to-noise ratio than ordinary CCD camera.

3. Sample Preparation

SMS requires immobilizing the single molecules in the focal volume, since the diffusion-limited transit time for focal spot with diameter on the order of an optical wavelength is only 0.01 s. The least intrusive way to immobilize a single enzyme molecule for fluorescence detection is to use agarose gel. In general, enzyme molecules with a molecular weight larger than 20 kD can be effectively trapped in the pores of a 1% agarose gel (Shi et al., 2004).

The procedure for single-molecule sample preparation used in our lab is as follows. Low-gelling point agarose (Sigma, type 4) is melted in the desired buffer with gentle stirring and heating. Solution of enzyme molecules is diluted into the agarose solution to a concentration of 1 nM just above the gelling temperature ($\sim$30 °C), and the mixture is spun on a coverslip at 2000 RPM for 10 s, yielding a smooth thin layer of gel containing the enzyme molecules. After the slide is mounted on the microscope stage, a small volume of buffer ($\sim$120 μl) is applied on the gel sample to keep it moist, which also helps to reduce background noise. In enzyme turnover experiments, the substrates are premixed with the agarose solution. Because the substrate molecules are generally small, they freely diffuse in and out of the pores of the agarose gel. On average, at enzyme concentrations on the order of 1 nM, no more than one molecule will reside in the confocal field of view. It should be noted that the process of dissociation might become significant in single-molecule experiments that work with sample concentrations in the nanomolar and subnanomolar range. For example, dissociation of flavin from the holoflavoenzyme essentially limits the length of the single-molecule trajectory in our study of DHOD (Shi et al., 2004). Also, subunit dissociation conceivably poses a problem for measurement of the kinetics of a single oligomeric enzyme.

Other common methods for immobilizing biomolecules for single-molecule detection include trapping the molecules in liposomes, and covalently tethering the molecules to the glass surface. Individual molecules can be encapsulated in lipid vesicles that typically range in size from 100 nm to 1 μm. The biotin–streptavidin interaction can tether the biotinylated vesicles to a membrane surface immobilized on the glass slide (Boukobza et al., 2001). One caveat of using lipid vesicles is that even small molecules cannot diffuse in or out after the vesicles are formed. Therefore, this approach is not suitable for experiments that require freely diffusing substrates. However, it provides an advantage to studies of oligomers and molecules with noncovalently bound cofactors. Due to the small volume of the vesicle, the effective sample concentration inside the lipid vesicle is on the order of micromolar so dissociation is minimal.

Molecular immobilization by directly tethering a molecule on a glass slide is another option, though more challenging. The glass slides are generally first silanized and then the molecule is covalently linked to the functionalized surface through different bifunctional linkers or biotin–streptavidin interaction. The most difficult problem of tethering biomolecules onto a glass surface is nonspecific binding, especially for protein molecules. Coating the glass surface with bovine serum albumin (BSA) or polyethylene glycol (PEG) generally helps decrease such nonspecific binding. Detailed protocols for tethering molecules onto a glass surface can be found in Heyes et al. (2004). Some less common methods of preparing single-molecule samples include trapping the single molecules inside femtoliter-size vials (Tan and Yeung, 1997), which act as individual nanoscopic reactors, or inside capillaries (Dyck and Craig, 2002).

Rapid mixing and exchange of single-molecule samples can be achieved using commercially available perfusion imaging chamber (ICP-25T, Dagan Corporation, Minneapolis, MN). Placing the imaging chamber in a heating/cooling thermal stage (HE-100 series, Dagan Corporation) with temperature controller (HCC-100A, Dagan Corporation) provides temperature control capacity. Encapsulating single-molecule samples in microfluidic chips, a microsize device that can manipulate nanoliters of material, potentially offers high-throughput processing and analysis in single-molecule measurements (Hong and Quake, 2003).

Because the signal from a single molecule is generally small, it is important to avoid contaminants that increase the level of background fluorescence in the sample. The glass coverslips have to be extensively cleaned. We find the following cleaning procedure is reliable in removing impurity. (1) Soak coverslips in 30% detergent (Aquinox) overnight. (2) Sonicate coverslips sequentially in the following solvents for 30 min each: (a) 30% detergent (Aquinox), (b) acetone, (c) 3 M potassium hydroxide, and (d) ethanol. Between changing solvent, the coverslips are sonicated in MilliQ water for 5 min and rinsed extensively with MilliQ water individually. (3) Soak coverslips in Aqua Regis (hydrochloric acid and nitric acid (3:1)) for 15 min and then sonicate for 20 min. (4) Rinse coverslips extensively with MilliQ water and spin-dry.

To further reduce "optical contaminants," all reagents and organic solvents should be of spectroscopic grade. All beakers, bottles and cylinders should be cleaned extensively. The buffers should be made freshly and filtered through a 0.2 μm filter. Before every single-molecule measurement, a blank slide without a fluorescent sample should be examined under the microscope to ensure that there is no optical contaminant on either the coverslip or in the buffers.

In confocal scanning mode, typical background noise in our experiment with 50 mM sodium phosphate buffer is 1–2 counts per 10 ms at laser power of 1 μW with the 457 nm line of the Argon laser.

4. Instrumentation

Single-molecule fluorescence trajectories can be collected using an inverted confocal microscope. As illustrated in Fig. 7.1, laser light at the appropriate wavelength is passed through a circular polarizer and expanded. It is then reflected by a dichroic mirror and focused by a high-NA oil-

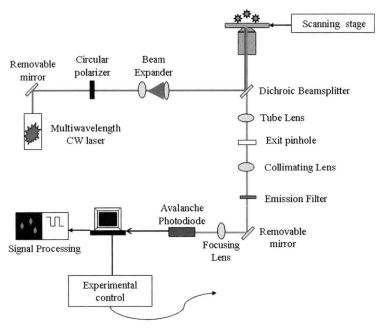

Figure 7.1 Block diagram of a basic inverted confocal scanning microscope.

immersion objective on a glass slide coated with the sample. The fluorescence light is collected by the same high-NA objective and transmitted through the dichroic mirror. It is filtered by a band-pass filter and then detected by a single-photon-counting APD. The sample slide is mounted on a piezoscanning stage, which precisely moves the sample stage for scanning different locations of the sample. In this setup, the image of the sample is reconstructed by raster scanning.

The main components of our confocal microscope include:

1. *Laser source.* A variety of CW lasers can be used to provide the appropriate excitation wavelength, such as line tunable helium–neon, solid-state lasers, dye lasers, ion lasers, etc. We typically use a line-tunable Argon ion laser (532R-AP-AO1, Melles-Griot, Carlsbad, CA) because it provides the suitable range of laser lines to excite the fluorophores we are interested in observing. For experiments that involve multiphoton excitation or measurement of single-molecule fluorescence lifetime and anisotropy, an ultrafast pulse laser source is required. Our lab is equipped with a standard ultrafast Ti:Sapphire laser (Coherent, Inc.).

2. *Fluorescence filters.* Dichroic mirrors, excitation filters, and emission filters used in our lab are all purchased from Chroma technology (Brattleboro, VT). The cutoff wavelength of the filters is selected based on the optical properties of the fluorophore. For example, the set of filters we used for

measuring flavin fluorescence is dichroic mirror (470/DCLP), excitation filter (D457/10×), and emission filter (D520/60M). For studying tetramethylrhodamine (TMR), we used dichroic mirror (Z514rdc), excitation filter (Z514/10×), and emission filter (HQ580/60m).
3. *Microscope objective.* (a) APO 100× oil-immersion objective, NA = 1.65, working distance (WD) = 0.15 mm (APO100XO-HR-SP, Olympus). Special quartz coverslips (9-U991, Optical Analysis, Nashua, NH) and immersion oil (168M×1780, McCrone, Westmont, IL) with refractive index 1.78 are required for this high-NA objective. (b) Plan APO 60× oil-immersion objective, NA = 1.45, WD = 0.17 mm (PLAPO60XO/TIRFM-SP, Olympus). The matching immersion oil with refractive index 1.479 (type FF) is purchased from Cargille Laboratory (Cedar Grove, NJ). Normal glass slides can be used with this objective. Because both objective (a) and (b) are infinity corrected and focus at infinity, a matching tube lens shown in Fig. 7.1 is used to bring the image into focus at the intermediate image plane for detection. The tube lens is purchased with the specific objectives.
4. *Scanning stage.* XYZ piezoflexure nanopositioner (P-517.3CL) with a high-speed digital piezoelectric controller (E-710.4CL, Polytech PI, Germany). This scanning apparatus can precisely move the sample stage with a resolution of 1 nm and a linear travel range to $100 \times 100 \times 20$ μm in the *XYZ*-axis.
5. *Exit pinhole.* The size of the pinhole can be roughly calculated as the product of the total magnification of the microscope and the size of the diffraction-limited spot (Airy disk), although optimization of the pinhole size requires more elaborate considerations (Sandison *et al.*, 1995). In practice, a pinhole is usually not necessary because the APD serves as a point detector with its small active area (0.2–0.4 mm in diameter), effectively acting as a light-restricting aperture.
6. *Photon-counting unit.* Single-photon-counting APD (SPCM-AQ 161, PerkinElmer Optoelectronics, Canada). The "dead time" between pulses is 50 ns and single photon can be detected with a timing resolution of 350 ps FWHM.
7. *Data acquisition devices.* The pulsed signal output of the APD is sent to a digital I/O block (CB-68LPR, National Instruments) in a CA-1000 box (National Instruments) and then counted by the Counter/Timers board (PCI-6602, National Instruments). The maximal timebase of the PCI-6602 counter is 80 MHz.

It is potentially more efficient to use a CCD camera to image a large area of the sample, thereby recording the fluorescence time evolution of multiple single molecules simultaneously. However, in this case the setup is no longer confocal and loses the advantage of excluding out-of-focus light. Moreover, as a point detector, the APD has better time resolution, lower noise, and better quantum efficiency than a CCD camera. Thus the APD is more suitable for single-molecule detection when the signal-to-noise ratio of the sample is small and the reaction kinetics are fast. If wide-field imaging by a CCD camera is chosen as the detection mode, an extended area of the sample needs to be illuminated. This can be achieved by adding a focusing lens at the excitation side of the microscope in front of the dichroic mirror (Fig. 7.1) to focus the excitation light at the back focal plane of the objective. Moreover, by moving this lens laterally, the excitation beam can be focused at the critical angle with a high-NA objective (NA $\geq$ 1.4), achieving objective-type TIR (Tokunaga *et al.*, 1997). This decreases background noise significantly by limiting the excitation to a very thin layer, usually on the order of the excitation wavelength. The CCD camera used in our lab is a back-illuminated EMCCD with on-chip electronic gain that can be cooled to $-90\ °C$ (iXon DV887, Andor Technology, Allentown, NJ).

Various microscope models are commercially available, such as Zeiss Axiophot and Nikon TE2000. There are several advantages of using ready-to-use commercial microscopes: they provide both binocular and electronic detection; they have rugged platforms; and technical support is available to facilitate usage. We chose to custom build our microscope mainly because it provides more flexibility in internal modifications to improve optical performance.

By pairing the confocal microscope discussed above with the standard setup for ensemble measurement of fluorescence lifetime, anisotropy and emission spectrum, these fluorescence parameters can be probed at the single-molecule level. For example, for a basic single-molecule anisotropy measurement, the sample is excited by a pulsed laser and the single-molecule signal collected by the inverted confocal microscope is split into parallel and perpendicular components by the polarization cube shown in Fig. 7.2. The two-channel signals are then detected by two APDs and standard electronics for typical time-resolve single-photon counting (Hu and Lu, 2004). Combining the confocal microscope with ensemble spectroscopic measurement components has been applied in measuring single-molecule fluorescence lifetime (Yang and Xie, 2002) and spectrum (Lu *et al.*, 1998).

5. Data Analysis of Single-Molecule Trajectories

Single-molecule behavior can be followed through changes in fluorescence intensity, lifetime, anisotropy, or emission spectrum. The most commonly used variable in single-molecule studies is the time evolution of fluorescence intensity. In general, observation of a one-step fluorescence loss due to photobleaching or fluorophore dissociation is a good indication

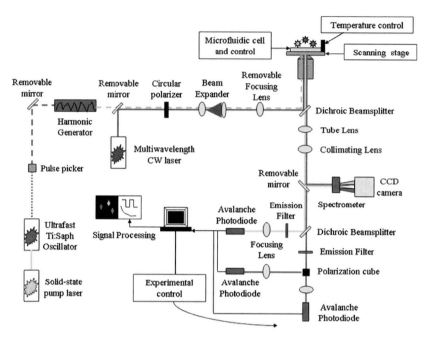

Figure 7.2 Diagram of a single-molecule spectrometer system with a microfluidic reaction cell, which is capable at measuring single-molecule intensity, emission spectrum, lifetime, anisotropy, and FRET. (See Color Insert.)

that a single molecule is being followed, not an aggregate of several molecules. This is shown in Fig. 7.3, which presents trajectories of single dihydroorotate dehydrogenase (DHOD, a flavoenzyme) molecules (Shi et al., 2004). Because photobleaching increases with excitation intensity while fluorophore dissociation is photo-independent, the two processes can be distinguished by examining the power dependence of the total fluorescence time. Variation of the vertical displacement of the molecules in the agarose gel from the focal point of the illumination may account for the difference in fluorescent intensities of different molecules.

The observed rapid fluctuation in fluorescence intensity of each trajectory in Fig. 7.3 can be attributed to photon-counting noise (shot noise) and possibly also to minor instrumental noise. The variance due to shot noise has a Poisson distribution and is equal to $N^{1/2}$, where N is the total number of detected photons. Fluorescence fluctuations (quantified by standard deviation of the fluorescence intensity) significantly larger than the Poisson noise indicate the presence of dynamic processes that cause changes in fluorescence, that is, chemical reactions or conformational changes, assuming no comparable noise in excitation intensity, detection electronics, etc.

Application of SMS in Enzyme Kinetics and Mechanism

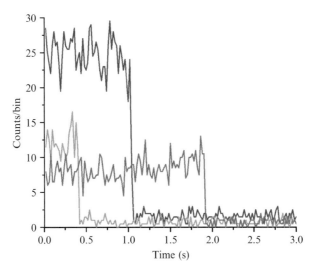

Figure 7.3 Fluorescence trajectories of single molecules of a dihydroorotate dehydrogenase (DHOD) mutant. DHOD is a monomeric flavoenzyme with FMN bound as the prosthetic group. The fluorescence loss is the result of FMN dissociating from the enzyme and rapidly diffusing away from the field of observation. The one-step fluorescence loss indicates that a single DHOD molecule is being observed. (See Color Insert.)

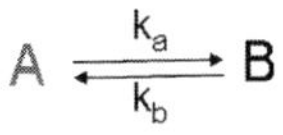

$$A \underset{k_b}{\overset{k_a}{\rightleftharpoons}} B$$

Figure 7.4 Reaction scheme of a simple first-order reaction. State A is fluorescent and converts to state B with rate k_a. State B is not fluorescent and converts to state A with rate k_b.

To introduce the analysis for extracting kinetic information from single-molecule data, we begin by considering a simple reaction scheme as shown in Fig. 7.4. The molecule leaves state A for state B with a forward reaction rate, k_a, and converts back to state A with a backward reaction rate, k_b. States A and B are assumed to be distinguishable spectroscopically, that is, state A is fluorescent (on) and state B is nonfluorescent (off). In single-molecule experiments the molecule is observed as randomly turning over between the fluorescent state A and nonfluorescent state B. Therefore, single-molecule trajectories of the reaction alternate between the fluorescent and nonfluorescent state ("blinking") in a repetitive fashion, as shown in Fig. 7.5A. The fluorescence fluctuation is clearly much larger than the shot noise, indicating the presence of a dynamic process that affects single-molecule fluorescence.

Reaction kinetics of the single-molecule turnovers can be analyzed in terms of the time durations that the molecule resides in the different states. Considering the lifetime that the molecule resides in the fluorescent state A, $k_a\Delta t$ is the probability that the molecule converts from A to B during Δt,

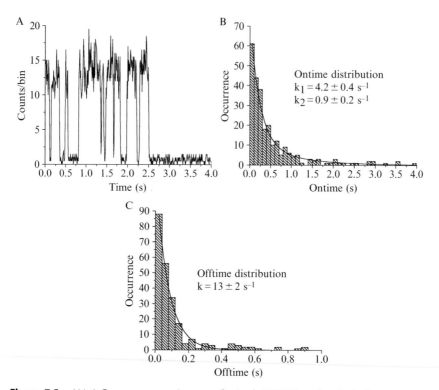

Figure 7.5 (A) A fluorescence trajectory of a single DHOD molecule during turnover (50 µM dihydroorotate (DHO) and 10 µM dichlorophenolindophenol (DCIP) in 50 mM phosphate buffer, pH 7 at 22 °C). At 2.5 s, FMN dissociated from the enzyme and diffused away from the field of view, resulting in the irreversible loss of fluorescence. (B) On-time and (C) off-time distributions of 50 individual molecules. Two phases were observed in the on-time distribution, which, by virtue of single-molecule analysis, were resolved as the result of two populations of DHOD molecules with distinctive reductive rates.

while $(1 - k_a \Delta t)$ is the probability that the molecule remains in state A. Therefore, $P_A(t + \Delta t)$, the probability of a molecule being in state A for $t + \Delta t$, is related to $P_A(t)$, the probability of a molecule being in A for time t, by the following equation:

$$P_A(t + \Delta t) = P_A(t)(1 - k_a \Delta t). \tag{7.1}$$

In the limit of $\Delta t \to 0$, this defines the differential equation for $P_A(t)$:

$$dP_A(t)/dt = -k_a P_A. \tag{7.2}$$

The solution decays as $\exp(-k_a t)$. Similar rationale can be applied to obtain $P_B(t)$, the probability that the molecule resides in the nonfluorescent state B for time t, as $\exp(-k_b t)$. Therefore, for the simple first-order reaction shown in Fig. 7.4, the lifetime histograms of both the on-times and off-times fit to single exponential decays and the rate constants are equal to the forward reaction rate k_a and the backward rate k_b, respectively. Note that while the equations for the probability are similar to the rate equations, they do not include the back-reaction terms. This reflects a fundamental difference in the half-reaction type of kinetic information that can be obtained in single studies, as compared to that obtained in ensemble studies, which is the sum of the forward and backward reactions.

From a single-molecule reaction trajectory the lifetime distribution histogram can be constructed by counting the number of on-times (off-times) that occur between t and $t + \Delta t$, where Δt is the bin size selected based on the kinetics of the reaction and the time resolution of the single-molecule measurement, as shown in Fig. 7.5C. The on-time and off-time distribution histograms correspond to the probability function $P_A(t)$ and $P_B(t)$, respectively. The rate constants observed in the decay of the on-time and off-time distribution thus are equal to the forward and backward reaction rate k_a and k_b.

Ensemble transient kinetic data of the first-order reaction shown in Fig. 7.4 appear as a single exponential decay with an observed rate constant, $k_a + k_b$, which is the sum of the individual forward and backward reaction rate constant. In contrast, single-molecule data directly provides kinetic information for the forward and backward reaction separately, giving individual rate constant of k_a and k_b because the two half-reactions are distinguished by their unique fluorescence characteristic in the single-molecule measurement.

5.1. Static heterogeneity

Characterization of different populations of enzymes with distinct reaction behaviors and kinetics (static heterogeneity) facilitates understanding of the reaction mechanism and structural stability of the enzyme. When distributions of time durations in different states are constructed for an individual molecule, the reaction rates of individual molecules can be obtained with no ensemble averaging. The distribution of reaction rates obtained for different molecules reveals the degree of heterogeneity in the system. By applying such a single-molecule approach, we determined that the two distinct kinetic phases exhibited in the reductive half-reaction of a Tyr318leu DHOD mutant from *E. coli*, could be attributed to two populations of enzyme molecules with two different reductive rates (Fig. 7.6) (Shi *et al.*, 2004). The Tyr318Leu mutant was studied instead of the wild-type because the FMN in wild-type *E. coli* DHOD is heavily quenched by contact with Tyr 318 and therefore is not fluorescent, preventing single-molecule studies on the wild-type enzyme. By substituting the tyrosine residue to leucine,

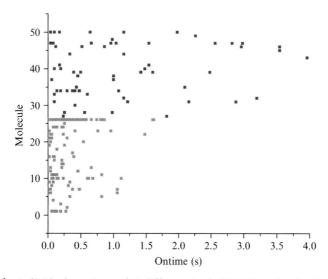

Figure 7.6 Individual on-times of 50 different single DHOD molecules in turnovers. Slow molecules, shown in blue, have less than 10% on-times shorter than 0.1 s, while fast molecules, shown in red, have few on-time longer than 1 s. Different molecules were grouped according to their kinetic behavior and then numbered. (See Color Insert.)

the mutant enzyme was highly fluorescent in the oxidized state. Identification of two populations of the DHOD molecules in our single-molecule study suggests that the enzyme is present in distinct conformations that result in different catalytic efficiencies. A number of studies on the wild-type DHOD did not show any evidence of heterogeneity in its catalysis, therefore the mutation of Tyr318Leu in DHOD mostly likely accounts for the alternate conformations possibly because the loss of van der Waals or other interactions between Tyrosine 318 and FMN in the catalytic site disrupts the native structure of wild-type DHOD.

5.2. Dynamic heterogeneity

Temporal fluctuation of catalytic rates, termed dynamic heterogeneity, is an important indicator of environmental modulation or conformational changes of an enzyme (Schenter *et al.*, 1999; Yang and Cao, 2002). Correlation analysis of consecutive turnovers in the single-molecule trajectories provides a direct measure of such fluctuations of reaction rates. If there is no dynamic disorder, the consecutive reactions are not correlated and the long and short time durations are randomly distributed across the trajectories. However, if dynamic disorder exists, the reaction rate fluctuates and the time durations in a given molecular state would show a dependence on previous ones, making the process non-Markovian. Our computer simulation of single-molecule trajectories with dynamic heterogeneity showed that the most sensitive way to characterize the degree of dynamic heterogeneity is to calculate an autocorrelation parameter $r(m)$ defined as

$$r(m) = \frac{\frac{n^2}{n-m}\sum_{i=1}^{n-m} t_i t_{i+m} - \left(\sum_{i=1}^{n} t_i\right)^2}{n\sum_{i=1}^{n} t_i^2 - \left(\sum_{i=1}^{n} t_i\right)^2}, \qquad (7.3)$$

where t_i is the on-time for the ith reaction event, m is the number of reaction events separating the pairs of on-times in the sequence, and n is the total number of on-times. Figure 7.7 displays the autocorrelation function, $r(m)$, for the on-times of a single cholesterol oxidase (CO_x) molecule [19] and a mutant DHOD molecule [20], respectively. With saturating amount of substrate, the CO_x reaction shows a clear dynamic correlation, while with a lower substrate concentration the CO_x catalysis is temporally uncorrelated. The random characteristics of $r(m)$ for DHOD indicates the absence of dynamic correlation on the time scale of the reaction.

Changes in the single-molecule emission spectrum as a function of time can also indicate dynamic fluctuations of the enzyme (Ambrose and Moerner, 1991; Basche and Moerner, 1992; Zumbusch et al., 1993). In the single-molecule study of the flavoenzyme CO_x, slow fluctuation of the emission spectrum was observed in the absence of substrate (Fig. 7.8) (Lu et al., 1998). Further autocorrelation analysis of the spectral-mean trajectory showed that the correlation decay was independent of excitation (i.e., not photoinduced), providing strong evidence that the spectral fluctuation was the result of an intrinsic conformational change around the flavin, which led to variation of the enzymatic reaction rate.

5.3. Kinetics of a single oligomeric enzyme molecule

Determination of subunit activity and cooperativity is critical for understanding the reaction mechanism of oligomeric enzymes. However, the lack of synchrony (as discussed above) in the reaction of different subunits and different molecules poses difficulty in interpreting the ensemble results of transient kinetics in terms of subunit activity. The decay rate constants observed in ensemble measurements are complex functions of the rates of both different subunits and the multiple reaction steps. Therefore, interpretation of the decay rate constants in terms of subunit activity requires knowledge of the reaction model and ingenious manipulation of the different variables that contribute to the distinct phases in the exponential decay. In contrast to the ensemble approach, which follows concentration change over time, a single-molecule study follows the sequential catalytic turnover

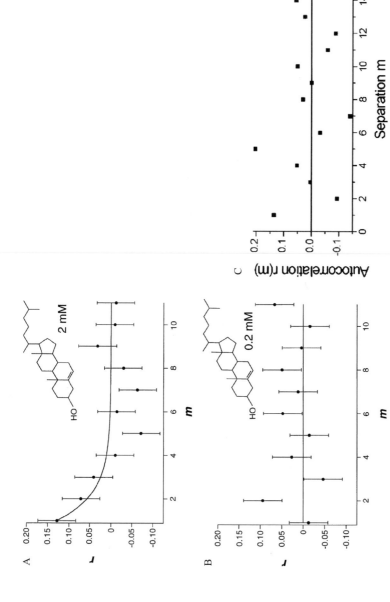

Figure 7.7 The autocorrelation function $r(m)$, where m is the separation between the pairs of on-time being correlated. (A) The $r(m)$ for a CO_x molecule with 2 mM cholesterol. The solid line is a single exponential fitting with a decay rate constant of 1.2 ± 0.5 turnovers. With an averaged turnover cycle of 500 ms, we deduce a correlation time of 0.6 ± 0.3 s. (B) The $r(m)$ for a CO_x molecule with 0.2 mM cholesterol. The averaged turnover cycle of the trajectory is 900 ms. Under this condition, there is no dynamic disorder. [(A) and (B) are taken from Fig. 7.5B and C by Lu et al. (1998).] (C) The $r(m)$ of on-times of a single DHOD molecule during catalytic turnovers. The random distribution suggests the absence of dynamic disorder.

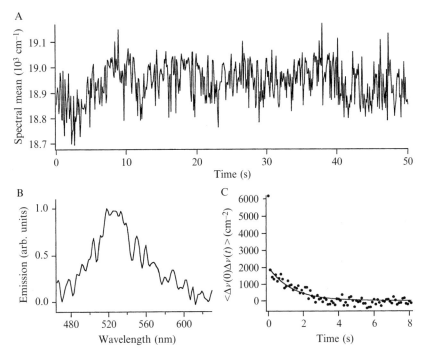

Figure 7.8 (A) A trajectory of the spectral mean of a single cholesterol oxidase (CO_x) molecule in the absence of cholesterol. Each data point is the first moment of an FAD emission spectrum sequentially recorded with a 100-ms collection time. The spectral fluctuations are attributed to spontaneous conformational fluctuations. (B) Emission spectrum of a CO_x molecule collected in 100 ms. The spectral mean of the spectrum is the first data point in (A). (C) The correlation function (dots) of the spectral mean for the trajectory in (A). The first data point, a spike at zero time, is due to uncorrelated measurement noises and spectral fluctuations faster than the 100-ms time resolution. The solid line is a fit by a single exponential with a decay time of 1.3 ± 0.3 s, which coincides with the correlation time for fluctuation of turnover rate. (Taken from Fig. 7.5 by Lu et al. (1998).)

of one oligomer. Individual subunit kinetics can then be distinguished by distinct fluorescent states, enabling analysis of each subunit as well as interactions between the subunits that affect catalytic efficiency (cooperativity). In the following we use the example of a homodimer to discuss the basic concepts of deriving subunit kinetics from a single-molecule data set. A similar rationale can be applied to analyze the single-molecule trajectory of higher oligomers with nonidentical subunits.

A simple first-order reaction scheme for a homodimeric enzyme is illustrated in Fig. 7.9, where A is the fluorescent state and B is the nonfluorescent state.

The single-molecule trajectory of this homodimer during turnovers would exhibit interconversion between three fluorescence states: fully

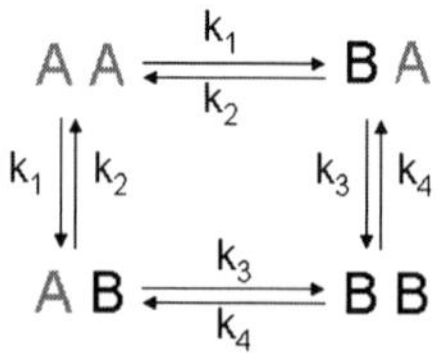

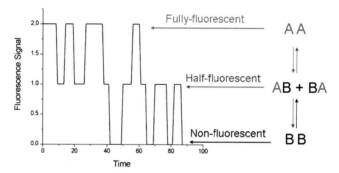

Figure 7.9 Reaction scheme of a homodimeric enzyme. Each of the two subunits can be in either the fluorescent state A or the nonfluorescent state B. k_1, k_2, k_3, and k_4 are the reaction rate constants for conversion between states A and B.

Figure 7.10 Schematic depiction of a fluorescence trajectory of a single homodimer enzyme during turnover. The forward and backward reactions of the two subunits can result in three fluorescent states: fully fluorescent, half-fluorescent and nonfluorescent. These correspond to the catalytic state of AA (A in both subunit), AB or BA (A in one subunit and B in the other subunit), and BB (B in both subunits).

fluorescent, half-fluorescent, and nonfluorescent, as shown in Fig. 7.10. These correspond to the catalytic state of AA (both subunits in state A), AB or BA (one subunit is state A and the other subunit in state B), and BB (both subunits in state B), respectively.

The length of time that the homodimer stays in a particular state is determined by the reaction kinetics and can be described by a probability distribution. When the homodimer is in state AA, $k_1 \Delta t$ is the probability that a subunit converts from A to B during Δt. Therefore, the probability that the subunit remains in state A during Δt is $(1 - k_1 \Delta t)$. In order for the molecule to stay in state AA for an interval of $t + \Delta t$, it must first remain in AA for an interval of t and still remain oxidized in both subunits during Δt. Therefore, $P_{AA}(t + \Delta t)$, the probability of the homodimer being in AA for $t + \Delta t$, is related to $P_{AA}(t)$, the probability of a molecule being in AA for time t, by the following equation:

$$P_{AA}(t + \Delta t) = P_{AA}(t)(1 - k_1 \Delta t)(1 - k_1 \Delta t) \\ = P_{AA}(t)\left(1 - 2k_1 \Delta t + (k_1 \Delta t)^2\right). \quad (7.4)$$

In the limit of $\Delta t \to 0$, this defines the differential equation for $P_{AA}(t)$:

$$dP_{AA}(t)/dt = -2k_1 P_{AA}. \tag{7.5}$$

Similar rationale can be applied to derive the differential equations for $P_{AB}(t)$, $P_{BA}(t)$, and $P_{BB}(t)$, which are the probability of residing in state AB, BA, and BB for time t, respectively. Solutions of the differential equations give

$$\begin{aligned} P_{AA}(t) &= \exp(-(2k_1)t), \\ P_{AB}(t) &= \exp(-(k_2 + k_3)t), \\ P_{BA}(t) &= \exp(-(k_3 + k_2)t), \\ P_{BB}(t) &= \exp(-(2k_4)t). \end{aligned} \tag{7.6}$$

From the solutions above, the distributions of time durations that a homodimer resides in the fully fluorescent state (AA), the half-fluorescent state (AB + BA), and the nonfluorescent state (BB) are found to decay exponentially with rate constant $2k_1$, $k_2 + k_3$, and $2k_4$, respectively.

In addition, when the homodimer is in the half-fluorescent state, the probability of it going into either the fully fluorescent or the nonfluorescent state is proportional to $k_2/(k_2 + k_3)$ and $k_3/(k_2 + k_3)$, respectively. Therefore, the fraction of the half-fluorescent state that partitions into the fully fluorescent state is $k_2/(k_2 + k_3)$ and the fraction that partitions into the nonfluorescent state is $k_3/(k_2 + k_3)$. The individual values of k_2 and k_3 can be determined using the experimentally derived rate constant for the lifetime distribution of the half-fluorescent state and the branching ratio of transitions into the fully fluorescent state and nonfluorescent state.

In the absence of cooperativity, the two subunits react independently, irrespective of each other's catalytic state. So $k_3 = k_1$ (reduction) and $k_2 = k_4$ (oxidation) would be derived from the lifetime distributions. However, if the lifetime distributions derived from the single-molecule trajectories give $k_3 > k_1$, there is positive cooperativity in the reductive half-reaction because the reduction of one subunit facilitates the reduction of the other subunit, while $k_3 < k_1$ suggests negative cooperativity, because the reduction of one subunit impedes the reduction of the other subunit. Similarly, inequality between k_2 and k_4 provides information on subunit cooperativity in the oxidative half-reaction (Table 7.1).

6. ADDITIONAL CONSIDERATIONS

Single-molecule kinetic studies of enzymes are conceptually different from ensemble steady-state assays and stopped-flow measurements. In general, steady-state assays measure the rates of substrate consumption or

Table 7.1 The relationship between the rate constants depicted in Fig. 7.9 and the measured decay rates derived from the lifetime distributions of the single-molecule trajectories for four different reaction models

Lifetime distribution	Fully fluorescent state (forward)	Half-fluorescent state (forward)	Half-fluorescent state (backward)	Nonfluorescent state (backward)
(1) No cooperativity	$2k_1$	$k_1 + k_4$	$k_1 + k_4$	$2k_4$
(2) Positive cooperativity	$2k_1$	$k_2 + k_3$ ($k_3 > k_1$)	$k_2 + k_3$ ($k_2 > k_4$)	$2k_4$
(3) Negative cooperativity	$2k_1$	$k_2 + k_3$ ($k_3 < k_1$)	$k_2 + k_3$ ($k_2 < k_4$)	$2k_4$

The four columns show the rate constants derived from the lifetime distribution of the fully fluorescent state, half-fluorescent state to the nonfluorescent state, half-fluorescent state to the fully fluorescent state, and the nonfluorescent state to the half-fluorescent state, respectively.

product dissociation when an enzyme is turning over. Steady-state results do not provide information on transient kinetics because the concentration of intermediates stays the same under the steady-state condition. The derived kinetic parameters, k_{cat} and K_m, depend on many or all of the steps in the catalytic cycle and are therefore usually complex algebraic functions of many rate constants. Stopped-flow measurements usually follow individual half-reactions in the reaction cycle under conditions when the enzyme is not turning over, giving observed rate constants that can be related to individual reaction steps. In contrast, single-molecule study provides transient kinetics of the whole reaction cycle under steady-state or equilibrium conditions when the enzyme is turning over. By looking at repetitive reaction cycles, single-molecule experiments follow individual enzyme molecules as they interconvert between different reactive states. The observed time duration that the molecule resides in each of the reactive states is determined by the reaction rate of the enzyme leaving each state. Therefore, these three different experimental approaches are done under significantly different conditions and report different aspects of the reaction kinetics. Because of the difference in experimental conditions, extra caution should be taken when comparing the kinetic results obtained by the different assays.

Studying enzyme kinetics using single-molecule data collected as a time series is still a relatively new concept. The limited number of single-molecule data sets available from experimental studies makes it difficult to test the basis for applying different statistical parameters to interpret the single-molecule data. Therefore, it is useful to simulate single-molecule reaction trajectories and analyze the simulated data against known models to identify appropriate and practical statistical indicators for interpretation of the single-molecule data with experimental limitations. For simulating the simple first-order reaction shown in Fig. 7.4, the rate constants are first used to calculate the probability, α, of the occurrence of a reaction leaving the respective state. For example, the probability of leaving state A is $\alpha_a = (1 - \exp(-k_a \Delta t))$, when Δt is the time step of data collection. α is then compared to a uniformly distributed random number R, generated by the Random function in the standard library of any programming software, whose value ranges between 0 and 1. Reactions occur when $R < \alpha$; if $R \geq \alpha$, the molecule remains unperturbed and another value of R is generated. The number of iterations of random number generation before the occurrence of a reaction, N (iteration counter), is a random variable and has a geometric probability distribution as $\alpha(1 - \alpha)^N$. In the discrete analysis the lifetimes in the different fluorescent states is given by $N \times \Delta t$.

The simulated data sets enable evaluation of different statistical parameters in revealing the kinetics and mechanism of the system faithfully. For example, our simulation analysis showed that the autocorrelation parameter $r(m)$ defined in Eq. (7.3) appropriately quantifies dynamic fluctuation of enzyme conformations and can identify different reaction channels with a relatively small data set, as is true for single-molecule data derived in experiments. In contrast, although parameters such as joint distribution function and difference distribution function, theoretically contains more information in terms of conformational dynamics (Yang and Cao, 2002), the small signature is only observable for a very large single-molecule data set (>20,000 turnovers), making them inappropriate for analyzing experimental data. Readers can find discussion of several other statistical parameters derived from intensity and event correlation analysis in a simulation study by Witkoskie and Cao (2004) in terms of their strength and the difficulties of employing them to determine possible reaction schemes.

7. SUMMARY

Single-molecule spectroscopy is a useful new tool for monitoring the catalytic behavior of individual enzyme molecules, providing new kinetic and mechanistic information about short-lived intermediates, static and dynamic heterogeneity, and subunit cooperativity. The potential of SMS goes beyond the enzymology discussed in this review. SMS can be employed to study kinetics of the interactions between multiple biological molecules, such as transcription factors and DNA, and between microtubules, motor proteins, and cargo vesicles. SMS is being developed to follow the behavior of individual biomolecules in live cell. Such *in vivo* single-molecule studies provide intriguing details about cellular trafficking and compartmentalization. Incorporating highly integrated microfluidic device in the microscope offers high-throughput capability in SMS studies. Using microfluidic chips to manipulate single-molecule samples enables multiple mixing and exchange for measurements of a series of reactions of a single sample and also parallel reactions of multiple samples.

Although there are great expectations for SMS studies, challenges need to be overcome in order to extend SMS to study a wider variety of biomolecules and processes. First of all, engineering of more photostable dyes is essential for obtaining longer single-molecule trajectories, therefore bigger single-molecule data sets. Because most of the biomolecules are not fluorescent and thus have to be labeled with fluorescent marker, more efficient labeling methods with high specificity need to be developed. As for analyzing the single-molecule data, our simulation results and others' indicate that multiple statistical parameters should be evaluated to extract meaningful kinetics from the single-molecule trajectories and establish a reliable reaction model. More theoretical and simulation work is still needed to improve our understanding of the temporal sequence of single-molecule data.

REFERENCES

Ambrose, W. P., and Moerner, W. E. (1991). Fluorescence spectroscopy and spectral diffusion of single impurity molecules in a crystal. *Nature* **349**, 225–227.
Ariga, T., Masaike, T., Noji, H., and Yoshida, M. (2002). Stepping rotation of F(1)-ATPase with one, two, or three altered catalytic sites that bind ATP only slowly. *J. Biol. Chem.* **277**(28), 24870–24874.
Ashkin, A. (2000). History of optical trapping and manipulation of small-neutral particle, atoms, and molecules. *IEEE J. Sel. Top. Quantum Electron.* **6**, 841–856.
Axelrod, D. (1989). Total internal reflection fluorescence microscopy. *Methods Cell Biol.* **30**, 245–270.
Basche, T., and Moerner, W. E. (1992). Optical modification of a single impurity molecule in a solid. *Nature* **355**, 335–337.
Blanchard, S. C., Kim, H. D., Gonzalez, R. L., Jr., Puglisi, J. D., and Chu, S. (2004). tRNA dynamics on the ribosome during translation. *Proc. Natl Acad. Sci. USA* **101**(35), 12893–12898.
Boukobza, E., Sonnenfeld, A., and Haran, G. (2001). Immobilization in surface-tethered lipid vesicles as a new tool for single biomolecule spectroscopy. *J. Phys. Chem. B* **105**(48), 12165–12170.
Bustamante, C., Smith, S. B., Liphardt, J., and Smith, D. (2000). Single-molecule studies of DNA mechanics. *Curr. Opin. Struct. Biol.* **10**, 279–285.
de Lange, F., Cambi, A., Huijbens, R., de Bakker, B., Rensen, W., Garcia-Parajo, M., van Hulst, N., and Figdor, C. G. (2001). Cell biology beyond the diffraction limit: Near-field scanning optical microscopy. *J. Cell Sci.* **114**(23), 4153–4160.
Dyck, A. C., and Craig, D. B. (2002). Individual molecules of thermostable alkaline phosphatase support different catalytic rates at room temperature. *Luminescence* **17**, 15–18.
Edman, L., Foldes-Papp, Z., Wennmalm, S., and Ridgler, R. (1999). The fluctuating enzyme: A single molecule approach. *Chem. Phys.* **247**, 11–22.
Elson, E. L. (2004). Quick tour of fluorescence correlation spectroscopy from its inception. *J. Biomed. Opt.* **9**(5), 857–864.
Elson, E. L., and Magde, D. (1974). Fluorescence correlation spectroscopy. *Biopolymers* **13**, 1–61.
Forkey, J. N., Quinlan, M. E., Shaw, M. A., Corrie, J. E., and Goldman, Y. E. (2003). Three-dimensional structural dynamics of myosin V by single-molecule fluorescence polarization. *Nature* **422**, 399–404.
Ha, T. (2001). Single-molecule fluorescence resonance energy transfer. *Method* **25**, 78–86.
Ha, T., Zhuang, X., Kim, H., Orr, J. W., Williamson, J. R., and Chu, S. (1999). Ligand-induced conformational changes observed in single RNA molecules. *Proc. Natl Acad. Sci. USA* **96**, 9077–9082.
Ha, T., Rasnik, I., Cheng, W., Babcock, H. P., Gauss, G. H., Lohman, T. M., and Chu, S. (2002). Initiation and re-initiation of DNA unwinding by the *Escherichia coli* Rep helicase. *Nature* **419**, 638–641.
Harms, G. S., Cognet, L., Lommerse, P. H., Blab, G. A., Kahr, H., Gamsjager, R., Spaink, H. P., Soldatov, N. M., Romanin, C., and Schmidt, T. (2001). Single-molecule imaging of l-type Ca(2+) channels in live cells. *Biophys. J.* **81**(5), 2639–2646.
Heyes, C. D., Kobitski, A. Yu., Amirgoulova, E. V., and Nienhaus, G. U. (2004). Biocompatible surfaces for specific tethering of individual protein molecules. *J. Phys. Chem. B* **108**(35), 13387–13394.
Hong, J. W., and Quake, S. R. (2003). Integrated nanoliter systems. *Nat. Biotechnol.* **21**(10), 1179–1183.
Hu, D., and Lu, H. P. (2004). Single-molecule nanosecond anisotropy dynamics of tethered protein motions. *J. Phys. Chem. B* **107**(2), 618–626.

Itoh, R. E., Kurokawa, K., Ohba, Y., Yoshizaki, H., Mochizuki, N., and Matsuda, M. (2002). Activation of rac and cdc42 video imaged by fluorescent resonance energy transfer-based single-molecule probes in the membrane of living cells. *Mol. Cell. Biol.* **22**(18), 6582–6591.

Klein, W. L., Stine, W. B., Jr., and Teplow, D. B. (2004). Small assemblies of unmodified amyloid β-protein are the proximate neurotoxin in Alzheimer's disease. *Neurobiol. Aging* **25**, 569–580.

Lakowicz, J. R. (1999). Principles of Fluorescence Spectroscopy. 2nd edn. Kluwer Academic/Plenum Publishers, New York (Chapter 13).

Lu, H. P., Xun, L., and Xie, X. S. (1998). Single-molecule enzymatic dynamics. *Science* **282**, 1877–1882.

Margittai, M., Widengren, J., Schweinberger, E., Schroder, G. F., Felekyan, S., Haustein, E., Konig, M., Fasshauer, D., Grubmuller, H., Jahn, R., and Seidel, C. A. (2003). Single-molecule fluorescence resonance energy transfer reveals a dynamic equilibrium between closed and open conformations of syntaxin 1. *Proc. Natl Acad. Sci. USA* **100**(26), 15516–15521.

Murakoshi, H., Iino, R., Kobayashi, T., Fujiwara, T., Ohshima, C., Yoshimura, A., and Kusumi, A. (2004). Single-molecule imaging analysis of Ras activation in living cells. *Proc. Natl Acad. Sci. USA* **101**(19), 7317–7322.

Nie, S., and Zare, R. N. (1997). Optical detection of single molecules. *Annu. Rev. Biophys. Biomol. Struct.* **26**, 567–596.

Rajagopalan, P. T. R., Zhang, Z., McCourt, L., Dwyer, M., Benkovic, S. J., and Hammes, G. G. (2002). Interaction of dihydrofolate reductase with methotrexate: Ensemble and single-molecule kinetics. *Proc. Natl Acad. Sci. USA* **99**(21), 13481–13486.

Rhodes, E., Gussakovsky, E., and Haran, G. (2003). Watching proteins fold one molecule at a time. *Proc. Natl Acad. Sci. USA* **100**(6), 3197–3202.

Sakmann, B., and Neher, E., (1995). "Single-Channel Recording". Plenum Press, New York.

Sako, Y., Minoguchi, S., and Yanagida, T. (2000). Single-molecule imaging of EGFR signalling on the surface of living cells. *Nat. Cell Biol.* **2**(3), 168–172.

Sandison, D. R., Williams, R. M., Wells, K. S., Strickler, J., and Webb, W. (1995). Quantitative fluorescence confocal laser scanning microscopy. In "The Handbook of Biological Confocal Microscopy" (J. B. Pawley, ed.), pp. 39–53. Plenum Press, New York.

Santos, N. C., and Castanho, M. A. (2004). An overview of the biophysical applications of atomic force microscopy. *Biophys. Chem.* **107**(2), 133–149.

Schenter, G. K., Lu, H. P., and Xie, X. S. (1999). Statistical analyses and theoretical models of single-molecule enzymatic dynamics. *J. Phys. Chem. A* **103**, 10477–10488.

Schuler, B., Lipman, E. A., and Eaton, W. A. (2002). Probing the free-energy surface for protein folding with single-molecule fluorescence spectroscopy. *Nature* **419**, 743–747.

Shi, J., Palfey, B., Dertouzos, J., Jensen, K. F., Gafni, A., and Steel, D. (2004). Multiple states of the Tyr318Leu mutant of Dihydroorotate Dehydrogenase revealed by single molecule kinetic. *J. Am. Chem. Soc.* **126**(22), 6914–6922.

Sosa, H., Peterman, E. J. G., Moerner, W. E., and Goldstein, L. S. B. (2001). ADP-induced rocking of the kinesin motor domain revealed by single-molecule fluorescence polarization microscopy. *Nat. Struct. Biol.* **8**(6), 540–544.

Tan, W., and Yeung, E. S. (1997). Monitoring the reactions of single enzyme molecules and single metal ions. *Anal. Chem.* **69**(20), 4242–4248.

Tokunaga, M., Kitamura, K., Saito, K., Iwane, A. H., and Yanagida, T. (1997). Single molecule imaging of fluorophores and enzymatic reactions achieved by objective-type total internal reflection fluorescence microscopy. *Biochem. Biophys. Res. Commun.* **235**(1), 47–53.

Weiss, S. (1999). Fluorescence spectroscopy of single biomolecules. *Science* **283,** 1676–1683.
Willemsen, O. H., Snel, M. M., Cambi, A., Greve, J., De Grooth, B. G., and Figdor, C. G. (2000). Biomolecular interactions measured by atomic force microscopy. *Biophys. J.* **79**(6), 3267–3281.
Williams, R. M., Piston, D. W., and Webb, W. W. (1994). Two-photon molecular excitation provides intrinsic 3-dimensional resolution for laser-based microscopy and microphotochemistry. *FASEB J.* **8**(11), 804–813.
Witkoskie, J. B., and Cao, J. (2004). Single molecule kinetics. I. Theoretical analysis of indicators. *J. Chem. Phys.* **121**(33), 6361–6372.
Yang, S., and Cao, J. (2002). Direct measurement of memory effects in single-molecule kinetics. *J. Chem. Phys.* **117**(24), 10996–11009.
Yang, H., and Xie, X. S. (2002). Probing single-molecule dynamics photon by photon. *J. Chem. Phys.* **117**(24), 10965–10979.
Yang, H., Luo, G., Karnchanaphanurach, P., Louie, T., Rech, I., Cova, S., Xun, L., and Xie, X. S. (2003). Protein conformational dynamics probed by single-molecule electron transfer. *Science* **302,** 262–266.
Zhuang, X., Kim, H., Pereira, M., Babcock, H., alter, N., and Chu, S. (2002). Correlating structural dynamics and function in single ribozyme molecules. *Science* **296,** 1473–1476.
Zumbusch, A., Fleury, L., Brown, R., Bernard, J., and Orrit, M. (1993). Probing individual two-level systems in a polymer by correlation of single molecule fluorescence. *Phys. Rev. Lett.* **70,** 3584–3587.

CHAPTER EIGHT

Ultrafast Fluorescence Spectroscopy via Upconversion: Applications to Biophysics

Jianhua Xu *and* Jay R. Knutson

Contents

1. Introduction	160
2. Basic Concepts	162
2.1. Phase-matching angle	163
2.2. Spectral bandwidth	164
2.3. Acceptance angle	165
2.4. Quantum efficiency for upconversion	166
2.5. Group velocity mismatch	166
2.6. Polarization	167
2.7. Sample handling	167
2.8. Crystal choice	168
3. Upconversion Spectrophotofluorometer and Experimental Considerations	168
4. Ultrafast Photophysics of Single Tryptophan, Peptides, Proteins, and Nucleic Acids	171
4.1. Solvent relaxation of tryptophan in water	173
4.2. QSSQ of Trp in dipeptides	174
4.3. Ultrafast fluorescence dynamics of proteins	175
4.4. Ultrafast dynamics in DNA	177
5. Summary and Future Directions	179
References	179

Abstract

This chapter reviews basic concepts of nonlinear fluorescence upconversion, a technique whose temporal resolution is essentially limited only by the pulse width of the ultrafast laser. Design aspects for upconversion spectrophotofluorometers are discussed, and a recently developed system is described. We discuss applications in biophysics, particularly the measurement of

Optical Spectroscopy Section, Laboratory of Molecular Biophysics, National Heart, Lung and Blood Institute, National Institutes of Health, Bethesda, Maryland 20892-1412 time-resolved fluorescence spectra of proteins (with subpicosecond time resolution). Application of this technique to biophysical problems such as dynamics of tryptophan, peptides, proteins, and nucleic acids is reviewed.

1. INTRODUCTION

Fluorescence spectroscopy is one of the most widely used techniques for studying the structure and function of macromolecules in biology/chemistry, especially protein interactions. For example, fluorescence (esp. lifetime) measurement reveals ligand-induced conformational changes in proteins, as fluorescence is often sensitive to subtle environmental changes of chromophores such as tryptophan and tyrosine (Gregoire *et al.*, 2007). Sensitivity to protonation or deprotonation reactions (Espagne *et al.*, 2006; Zelent *et al.*, 2006), solvent relaxation (Dashnau *et al.*, 2005; Toptygin *et al.*, 2006), local conformational changes (May and Beechem, 1993; Pan *et al.*, 2006), and (via anisotropy decay) processes coupled to translational or rotational motion (Schroder *et al.*, 2005) is available. Most fluorescence decay changes occur in the time window of a few picoseconds to nanoseconds; measurements thus employ very short light pulses. Fortunately, in recent years, a range of advanced ultrafast lasers (esp. the Ti:sapphire laser) and associated optoelectronic instruments have emerged into the general scientific and commercial marketplace. This enables spectroscopic studies of biological and chemical systems on a subpicosecond timescale. We recently reported the development of an instrument with a 150 fs full-width half-maximum response function that uses fluorescence upconversion to obtain the lifetimes of biological molecules with ultraviolet excitation (Shen and Knutson, 2001a).

Many different techniques have been proposed to obtain ultrafast time resolution in fluorescence spectroscopy. A number of relevant reviews have appeared. Recently, Andrews and Demidov (2002) have edited a multiauthor volume on laser spectroscopy, and a comprehensive discussion of techniques has been given by Fleming (1986). Before that, Topp (1979) provided a comprehensive review of pulsed laser spectroscopy, Ippen and Shank (1978) gave a general review on techniques in subpicosecond spectroscopy, and techniques and applications of fast spectroscopy in biological samples have been reviewed by Holten and Windsor (1978). Given the availability of these broader reviews, we will focus only upon certain recent developments (Fiebig *et al.*, 1999; Kennis *et al.*, 2001a; Rubtsov and Yoshihara, 1999; Schanz *et al.*, 2001).

The most direct method to study fluorescent transients is to use a fast photomultiplier (PMT) or photodetector in conjunction with fast electronics. The operating principles, design, and performance of PMTs have been reviewed by Zwicker (1981). The best time resolution that can currently be obtained by direct PMT readout is about 40 ps. Although photodetectors (e.g., PIN diodes) can be made with much faster response time (<5 ps), their low sensitivity and small active area (needed to keep RC low) restrict their use to detection of strong signals, such as direct signals from lasers.

Two photocathode-based techniques that are usually used for the detection of weaker time-resolved fluorescence are time-correlated single photon counting (TCSPC) and streak cameras. TCSPC is a relatively easy means to measure fluorescence decay times and has been well described by O'Conner and Phillips (1984), Birch and Imhof (1991), and Becker (2005). Using TCSPC together with a fast PMT, it is possible to improve the exponential time resolution to around 10 ps (Murao *et al.*, 1982), but very careful deconvolution is needed to get such good time resolution. At the shortest extreme, Holzwarth and his group used their TCSPC system to detect lifetimes on the order of 1-2 ps (with Ti:sapphire laser) (Muller *et al.*, 1996). The preferred direct technique for obtaining time resolution better than 10 ps, however, has been a streak camera. The design and application of streak cameras as detectors in the picosecond time domain have been reported and discussed by Bradley *et al.* (1980, 1983), Barbara *et al.* (1980), Ihalainen *et al.* (2005), and Van Stokkum *et al.* (2006). "Single shot" streak cameras can offer time resolution better than 1 ps in certain cases, and a synchroscan streak camera has become available recently with time resolution under a picosecond with wide spectral response (200-1600 nm) (e.g., Hamamatsu, C6860, C6138) (Van Stokkum *et al.*, 2006). Streak cameras are, however, both expensive and labile; system maintenance for day to day reproducibility in the subpicosecond regime has been said to be taxing.

The best hope to achieve even better fluorescence time resolution, comparable to laser pulse width (~100 fs), depend on nonlinear optical sampling. The use of an optical Kerr (OK) shutter was proposed by Duguay and Hansen (1968). The OK shutter makes use of the transient birefringence induced in a nonlinear medium by an intense laser pulse to create an ultrafast shutter (like a Pockels cell driven by light instead of high voltage). Several different liquid media have been used as the nonlinear shutter materials. The time resolution is determined by the recovery time of the medium. Most of the work has been done using CS_2 as the optical gate, and the time resolution can reach less than 0.5 ps by using benzene (Arzhantsev and Maroncelli, 2005), or even glass as the nonlinear shutter (Yu *et al.*, 2003). We note, however, recent reports using fused silica as a gate to get an instrument response function of ~200 fs (Gu and Shi, 2005). Unfortunately, the shutter contrast was small due to a birefringence relaxation component with rather long recovery time, and spectral-temporal correction needs to be done very carefully. Low sensitivity and (usually) restriction to a visible range (400-675 nm) has restricted the range of Kerr shutter applications in biophysics.

Another nonlinear technique, first used by Mahr and Hirch (1975), involves frequency mixing. In this technique, the fluorescence excited by an ultrafast laser pulse is mixed with another (delayed) portion of the laser pulse in a nonlinear optical crystal such as KDP or BBO to generate sum or difference frequency radiation. Since this mixing process takes place only during the presence of the second laser pulse, it provides time resolution comparable to the pulse width; delaying the gate pulses with a mechanical stage leads to an "optical boxcar approach." Fluorescence upconversion (also called sum frequency generation) was experimentally first reported in biochemical research by Halliday and Topp (1977). In their system, the fundamental 1.06 μm pulse of 7 ps from a branched mode-locked Nd^{3+} glass laser was used as the probe beam and the upconversion was accomplished in a KDP crystal with type II phase matching. The sampling gate width was about 10 ps. By changing the angle (a span of 15°) of the incident 1.06 um and fluorescence beams with respect to the optical axis in the crystal, the phase-matching conditions were changed to select a different fluorescence wavelength (250-1100 nm) for upconversion (Halliday and Topp, 1978). Kahlow et al. (1988) provided a thorough discussion about upconversion. Because of the higher intrinsic signal-to-background ratio over the Kerr-cell method and the availability of excellent PMT detectors in the deep UV regions, it is an extremely attractive technique for time-resolved fluorescence spectroscopy in biology. This technique has been exploited by other researchers in biophysics (Changenet et al., 1998; Kennis et al., 2001b; Xu et al., 2006).

This review is organized as follows. Basic concepts of upconversion are reviewed in Section 2. In Section 3, we discuss our recently developed system (which provides <400 fs FWHM gating, excellent sensitivity, and a large dynamic and wide spectral range). Section 4 discusses some recent experiments on subpicosecond time-resolved fluorescence spectroscopy in tryptophan and proteins using this technique. Section 5 presents a summary and some thoughts on future directions.

Note also that we do not review the frequent application of upconversion to problems in photosynthesis (Kennis et al., 2001b). This would be redundant, as many reviews have covered that topic extensively.

2. Basic Concepts

The time resolution mechanism underlying the upconversion technique is illustrated in Fig. 8.1. Upconversion is actually a *cross-correlation* between the fluorescence and the probe laser pulse. At time $t = 0$, the sample is electronically excited by, for example, the second or third harmonic of an ultrafast laser pulse with frequency ω_p. The collected incoherent

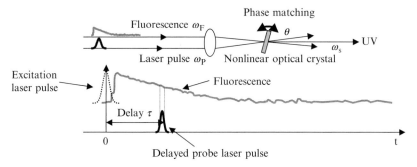

Figure 8.1 Schematic diagram of upconversion.

fluorescence (ω_F) and the probe laser pulse (ω_p) arriving at time $t = \tau$ are cofocused into a nonlinear optical crystal, such as KDP, BBO, etc., which is oriented at an appropriate angle with respect to the fluorescence and laser beams. Sum frequency photons are generated only during the time that the probe laser pulse is present in the crystal, acting as a "light gate," and thus time resolution is within the laser pulse width. The time evolution of fluorescence may then be traced by varying the delay τ of the probe laser beam. An analysis of sum frequency generation shows the intensity of the sum frequency signal at a given delay time τ is proportional to the correlation function of the fluorescence intensity with the probe laser intensity (Shen, 1984).

The basic concepts of sum frequency generation are well known and have been fully discussed (Shen, 1984; Zernike and Midwinter, 1973). We only briefly summarize previously published basic equations and discuss some of the main features for upconversion in this section.

2.1. Phase-matching angle

The sum frequency generation process is efficient only if conditions for *phase matching* are satisfied; this happens only for a narrow band of wavelengths centered at a wavelength determined by the phase-matching angle θ_m. For simplicity, we consider the case of collinear phase matching, and the appropriate equations are (Shen, 1984)

$$\omega_F + \omega_p = \omega_s, \quad h\nu_F + h\nu_p = h\nu_s, \tag{8.1}$$

$$\vec{k}_F + \vec{k}_p = \vec{k}_s \Rightarrow \frac{n_F(\lambda)}{\lambda_F} + \frac{n_p(\lambda)}{\lambda_p} = \frac{n_s(\lambda)}{\lambda_s}, \tag{8.2}$$

where ω (or ν), h, $\vec{k}$, and n are the photon frequency, Planck's constant, wave vector, and refractive index, respectively. The subscripts F, p, and s denote the fluorescence beam, probe laser beam, and sum frequency generation, respectively. Assuming θ is the angle between z-direction and $\vec{k}$; n_o and n_e are the ordinary (O) and extraordinary (E) indices, respectively. For an uniaxial crystal where the optic axis is along the z-direction, the index $n(\theta)$ satisfies

$$\frac{1}{n^2(\theta,\lambda)} = \frac{\sin^2(\theta)}{n_e^2(\lambda)} + \frac{\cos^2(\theta)}{n_o^2(\lambda)}. \quad (8.3)$$

The crystal index n_o and n_e varies with wavelength and can be obtained from Sellmeier's equations in the literature (Eimerl, 1987; Ghosh, 1995; Liu and Nagashima, 1999). In this chapter, we concentrate on the type "I" phase-matching condition (O + O → E), so the angle is given by

$$\sin^2(\theta_m) = \frac{[n_s(\theta_m)]^{-2} - (n_{o,s})^{-2}}{(n_{e,s})^{-2} - (n_{o,s})^{-2}}, \quad (8.4)$$

where $n_s(\theta_m)$ is

$$n_s(\theta_m) = n_{o,F}\frac{\lambda_s}{\lambda_F} + n_{o,p}\frac{\lambda_s}{\lambda_p}. \quad (8.5)$$

For example, β-barium borate (BBO) crystal is well suited for upconversion, its phase-matching angle is about 45° for Trp fluorescence (350 nm) with the probe pulses at 885 nm. One can easily buy a commercial type I BBO crystal with ~45° cut at different thickness.

2.2. Spectral bandwidth

If the phase-matching condition is not precisely satisfied, that is, different fluorescence wavelengths λ_F make $\Delta k \neq 0$, then the quantum efficiency falls off with increasing Δk as

$$\eta(\Delta k) = \eta(0)\frac{\sin^2(\Delta k L)}{(\Delta k L)^2}. \quad (8.6)$$

The spectral bandwidth is estimated by the place when the quantum efficiency drops to 50% of $\eta(0)$ (Shen, 1984).

For the case discussed above,

$$\Delta(h\nu_F)(\text{meV}) = \frac{3.66 \times 10^{-12}}{L(\text{cm})[\gamma_s(\text{s/cm}) - \gamma_F(\text{s/cm})]}, \quad (8.7)$$

where

$$\gamma_s = \frac{1}{c}\left|n_s(\theta_m) - \lambda_s\frac{\partial n_s(\theta_m)}{\partial \lambda}\bigg|_{\lambda=\lambda_s}\right|, \quad \gamma_F = \frac{1}{c}\left|n_{o,F} - \lambda_F\frac{\partial n_{o,F}}{\partial \lambda}\bigg|_{\lambda=\lambda_F}\right|. \tag{8.8}$$

Equation (8.7) describes the interplay of crystal thickness and fluorescence spectral bandwidth. If we concentrate on Trp fluorescence (peak emission wavelength 350 nm) using 885 nm probe pulses, then the spectral bandwidth for 0.2 mm BBO crystal is about 0.4 nm (instead of 0.08 nm for a 1 mm crystal). Upconversion is, therefore, an intrinsically high-resolution spectroscopy.

2.3. Acceptance angle

The angle of acceptance of fluorescence is another important factor in upconversion experiments. Since fluorescence is emitted in all directions from the excited spot in the sample, the collected fluorescence is refocused into the crystal in a broad cone. The larger the input angle that can be phase matched by the crystal, the larger the upconversion efficiency. The acceptance angle is defined (generously) as the angle where the phase mismatch is less than 90° (Zernike and Midwinter, 1973):

$$\Delta\theta = \frac{\pi}{L}\left(\frac{\partial k_s}{\partial \theta}\right)^{-1}. \tag{8.9}$$

Normally the acceptance angle in the plane containing the optic axis is smaller than the perpendicular plane (for collinear phase matching). For noncollinear geometry, the acceptance angle is (Zernike and Midwinter, 1973)

$$\Delta\theta = \frac{2.78 n_{o,F}\lambda_F}{L[1 - (n_{o,F}\lambda_s)/(n_s(\theta_m)\lambda_F)]}. \tag{8.10}$$

Equations (8.9) and (8.10) show that the acceptance angle increases inversely with the crystal length L, that is, thinner crystals generally have a larger angle of acceptance. Thus as the crystal is thinned, the focus can be tightened because the acceptance angle is larger, so the total upconversion signal intensity should remain relatively constant. From Eq. (8.10), one can estimate an acceptance angle of ~1.7° for Trp fluorescence studies in a 2 mm BBO crystal. This quite narrow cone angle is incompatible (for Gaussian beams) with the small spot size needed to create sufficient mixing. Thus, incoming cone angle is usually much larger than acceptance angle, and only a small fraction of available fluorescence is converted.

2.4. Quantum efficiency for upconversion

The quantum efficiency for phase matched sum frequency generation can be estimated under appropriate boundary conditions. For the "small signal" condition (no depletion-only a few percent of the power of the probe laser beam is transferred to sum frequency generation), quantum efficiency may be expressed (Shen, 1984; Zernike and Midwinter, 1973):

$$\eta(\Delta k = 0) = \frac{2\pi^2 d_{\text{eff}}^2 L^2 P_p}{c A \varepsilon_0^3 \lambda_F \lambda_s n_{o,F} n_{o,p} n_s(\theta_m)}. \qquad (8.11)$$

where P_p and A are the peak power and focus area of the probe laser beam (assuming the area of the fluorescence beam is not bigger than this area), respectively; d_{eff} is the effective nonlinear coefficient of the crystal; c is the light velocity; and ε_0 is the free-space permittivity. Note that quantum efficiency is proportional to probe pulse peak power and square of crystal thickness, but is reduced if the focal spot becomes bigger. For example, if the average power of 300 fs probe pluses is 1 W at 885 nm and 5 KHz repetition rate, and the focus spot is about 0.1 mm in diameter, for Trp fluorescence, the quantum efficiency should be about 0.1% in a 1 mm BBO crystal.

2.5. Group velocity mismatch

In linear optics, the group velocity $v_g = (|\partial k/\partial \omega|)^{-1}$ does not lead to pulse broadening. However, in nonlinear processes such as sum frequency generation (here acting as fluorescence upconversion), the mismatch between the group velocity of probe and fluorescence pulses may lead to a temporal broadening of the generated sum frequency pulse. In most cases, this restriction is more severe than the one imposed by phase mismatching. For O + O → E, the broadening introduced by the mismatched group velocity is given by (Shen, 1984)

$$\Delta t(s) = L(\text{cm})[\gamma_p(\text{s/cm}) - \gamma_F(\text{s/cm})]. \qquad (8.12)$$

And γ_p is obtained by

$$\gamma_{\rm p} = \frac{1}{c}\left|n_{\rm o,p} - \lambda_{\rm p}\frac{\partial n_{\rm o,p}}{\partial \lambda}\big|_{\lambda=\lambda_{\rm p}}\right|. \qquad (8.13)$$

In upconversion experiments, this group velocity mismatch between the two input beams results in a time broadening in the sum frequency generation, which means that the probe laser beam sweeps out an area of the fluorescence with a sampling time period wider than the pulse width itself. Unlike group velocity dispersion, this broadening effect cannot be corrected by any techniques, it can only be minimized by a proper choice of elements. Group velocity dispersion usually is only considered for the subpicosecond laser pulses, however, this effect can be pre-compensated by either grating pair, prism pair or both (Fork et al., 1987; Inchauspe and Martinez, 1997; Kafka and Baer, 1987). In our case, we can adjust the final grating compressor in the amplifier to partially compensate the dispersion of the probe pulses.

In addition to the characteristics discussed above, "walk-off" angle must be considered in some cases. For an extraordinary beam, the direction of energy flow is different from the wave direction. Walk-off angle is that between wave and energy vectors. In some cases, this walk-off will lead to a reduced overlap of the probe and fluorescence beams (reduce the upconversion efficiency in thick crystals). Fortunately it has little effect on the overlap for the O + O → E case in a thin crystal.

2.6. Polarization

The upconversion process is intrinsically a polarization selection process, only O + O → E (not E + O → E) occurs in a practical upconversion system, and the collection of different polarized emission components is usually accomplished by rotating the excitation polarization with a thin half-wave plate. This is much simpler than rotating both the entire crystal mount and the probe polarization.

2.7. Sample handling

It is critical to maintain linear spectroscopy conditions even in this higher flux experiment; thus the sample must be concentrated enough to provide a large population in the focal volume (i.e., under 1% of ground-state population should be excited, to avoid excitonic effects and saturation). Further, any photophysical events that might persist between pulses (~200 μs), such as triplet generation, obligate the removal of exposed sample from the excitation spot. Hence, either free standing jets (to avoid flow cell "wall effects") or rapidly spinning sample holders are needed. We constructed a disk to mount inexpensive thin cuvettes on a spinning mount, achieving m/s displacement rates on an easily demounted and cleaned carrier (containing an internal Raman standard).

2.8. Crystal choice

Different nonlinear optical crystals have different "phase-matchable" wavelength ranges, quantum efficiency, acceptance apertures, etc. There are several common crystals used in sum frequency generation, potassium dihydrogen phosphate (KDP) (Eimerl, 1987), lithium niobate (LiNbO$_3$) (Schlarb et al., 1995), lithium triborate (Liu and Nagashima, 1999), and β-barium borate (BBO) (Ghosh, 1995). Although the conversion efficiency of BBO is less than LiNbO$_3$, it has a much wider phase-matching wavelength range for mixing with 800 nm: from 250 to 2000 nm. It also has higher damage thresholds and lower group velocity mismatch. A BBO crystal of 1/3 the length of a similar KDP crystal has higher efficiency with roughly the same acceptance angle and group velocity mismatch. Therefore, among currently available crystals, we have chosen BBO for UV upconversion experiments for subpicosecond pulses.

3. UPCONVERSION SPECTROPHOTOFLUOROMETER AND EXPERIMENTAL CONSIDERATIONS

Upconversion is a derivative of the pump probe technique, and we developed one layout especially for studying tryptophan photophysics in proteins, which is shown in Fig. 8.2. A mode-locked Ti:sapphire laser (Tsunami, Spectra Physics) was pumped by an argon ion laser with the power of 6 W (Beamlock 2060, Spectra Physics), which generated a ~400 mW pulse train with a typical pulse duration of 120 fs at a repetition rate of 82 MHz. It was then sent to seed a Ti:sapphire regenerative amplifier (Spitfire, Spectra Physics). The output amplified pulses at 885 nm typically had a pulse energy of ~0.2 mJ and an autocorrelation pulse width of 350 fs at a repetition rate of 5 kHz. The amplifier could be tuned only between 865 and 905 nm because of the restriction caused by cavity mirrors chosen to quench self-Q switching below 860 nm and the gain pulling above 900 nm. Ultraviolet excitation pulses with an average power up to 30 mW (e.g., 295 nm) was obtained from nonlinear harmonic generation using a 1 mm BBO crystal and 0.5 mm BBO crystal for doubling and tripling, respectively. This UV beam was separated from the infrared beam (fundamental), and visible beam (doubled) by two dichroic mirrors, and spatially filtered using a pin hole, the power was carefully attenuated before excitation of the sample to avoid photodegradation, hole burning, and other undesirable effects. A zero-order half-wavelength plate (HWP) was used to change the polarization of the UV beam. A circular array of thin cells (T-20, NSG Precision Cells) with a path length of 1 mm in a delrin stacked slotted disk was used to hold samples. The disk was spun continuously (several m/s)

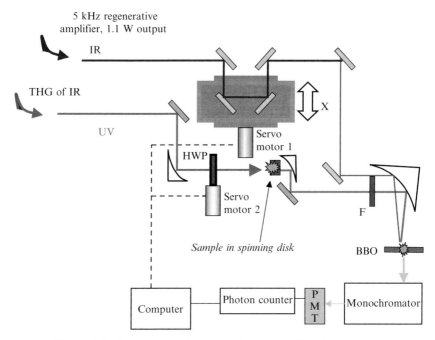

Figure 8.2 A schematic of upconversion spectrophotofluorometer.

to make each UV pulse excite a different sample spot. The infrared fundamental pulse was reflected from a broadband hollow retroreflector on a computer-controlled precision stage, and used as a gate pulse for the upconversion process. The fluorescence emission was collected by parabolic mirrors, passed through a long pass filter (F), and focused into a 0.2 mm thick BBO mixing crystal. The upconversion signal was produced via type I sum frequency generation with the gate pulse in the crystal. Here the combination of a very thin BBO crystal and no lens is selected for reducing the limitations of time resolution imposed by group velocity mismatch and dispersion. The time evolution of fluorescence at any given wavelength can be obtained by setting the BBO crystal angle, monochromator wavelength (at $\omega_F + \omega_p$), and scanning delay stage.

To reject the strong background signals (infrared laser and its second harmonic generation, remnant UV, and unconverted fluorescence) accompanying the upconverted signal, a noncollinear configuration was arranged between infrared probe laser and fluorescence. The use of a monochromator helps in reducing these unwanted signals at the PMT, also improving the spectral bandwidth. A double monochromator may be necessary if this problem is severe. Polarizations of gated fluorescence were determined by the orientation of nonlinear crystals, so no extra linear polarizer was needed (for anisotropy calculation, $G \sim 1$). By angle tuning the mixing crystal, the upconverted fluorescence signal, with a wavelength in the range 230–280 nm, always polarized in the same direction, was directed into a monochromator (Triax 320, Jobin Yvon, Inc. with a bandwidth of 0.5 nm) and a solar blind photomultiplier tube (R2078, Hamamatsu, dark rate < 1 cps). Amplified SBPMT signals were discriminated and then recorded by a gated single photon counter (994, EG&G Ortec). Photon arrival events were held to less than 5% of the repetition rate to minimize "pileup." One should mention that there are two main sources of noise for the detection. If the signal is sufficiently greater than the PMT dark current, the power fluctuations of the infrared laser beam may be the primary noise source. If the signal is very low, the shot noise from the dark current of the PMT will be the limiting factor, and in this case, cooling the PMT is advisable.

A precise determination of the bandwidth and zero time delay is essential in all ultrafast experiments. The best way of determining the zero in upconversion system is to get a cross-correlation trace between the scattered laser light from the sample and a delayed laser beam in the crystal. Such a trace provides not only an accurate zero but also an accurate measurement of the system response time. In our system, the "lamp" (AKA "apparatus" or "instrument response") function was determined by measuring the cross-correlation either between the UV and infrared pulses or UV-generated spontaneous Raman scattering in water and the infrared. In both ways, the lamp function was found to be around 400 fs (FWHM), with a timing jitter of less than 30 fs. The use of unamplified pulses can reduce this FWHM to <150 fs, but at much lower sensitivity (Fig. 8.3).

When collecting on longer timescales, we often found "detuning" to ~1 ps FWHM optimized other geometric parameters. Instrument calibration was verified with indoles and, for example, linear fluorophores like p-terphenyl, which yielded $r_0 = 0.40 \pm 0.01$ and a single rotational correlation time φ of 41 ps in cyclohexane.

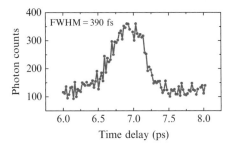

Figure 8.3 A typical response time profile of upconversion spectrophotofluorometer. (See Color Insert.)

4. Ultrafast Photophysics of Single Tryptophan, Peptides, Proteins, and Nucleic Acids

While the conjugation of small molecules such as environment-sensitive fluorescent probes to protein has been a remarkably useful strategy for studying biological structure and function, nagging concerns about probes causing perturbation have always provided impetus for the study of naturally occurring reporters. Much attention has been paid to the time-resolved fluorescence decay of tryptophan (Beechem and Brand, 1985), the most important protein fluorophore, as it can be used to study physical and dynamic properties (not only the local environment of the indole ring but also global changes of the tertiary structure of proteins). This amino acid (in aqueous solution) has been long known to be at least biexponential on the 100 ps–100 ns timescale and the principle components (usually about 0.6 and 3.1 ns) have different decay-associated spectra (DAS) (Beechem and Brand, 1985; Lakowicz, 1999). Many peptides and proteins containing single Trp residues have wavelength-dependent multiexponential decays, and each exponential term has a different DAS (Chowdhury *et al.*, 2003; Fukunaga *et al.*, 2007; Li *et al.*, 2007; Qiu *et al.*, 2006; Zhang *et al.*, 2006).

Generally two distinct explanations have been given to explain the various "lifetimes" (exponential decay terms): one is the excited population loss, that is, "population decay"; another one is the equilibration of the new excited state with surroundings, resulting in spectral shifts without change in overall excited-state population, which is termed solvent relaxation (although "solvent" may also include protein contributions). "Ground-state heterogeneity" is the most basic explanation for multiple population decay rates; it suggests the presence of several conformers, each with indole experiencing distinct environments and yielding different fluorescence lifetimes (Mcmahon *et al.*, 1997; Ross *et al.*, 1992; Szabo and Rayner, 1980). Decay amplitude correspondence with NMR-derived rotamer population data has supported such a view (Clayton and Sawyer, 1999; Mcmahon *et al.*, 1997). In addition to different true rotamers of the Trp side chain, *microconformational* states of proteins, with different local environments for the indole ring, could confer ground-state heterogeneity. Relaxation of the protein matrix (not just solvent water) would also produce a complex decay (Lampa-Pastirk *et al.*, 2004; Toptygin *et al.*, 2006).

Generalized solvent relaxation yields short-lived blue and long-lived red DAS on the nanosecond timescale (Alcala *et al.*, 1987; Toptygin *et al.*, 2001; Vincent *et al.*, 1995). Relaxation as a cause for the observed multiexponentiality is suggested by the fact that mean decay times usually increase with increasing observation wavelength and that DAS are generally strictly ordered: longer decay times usually (but not always!) go with longer wavelength components (Lakowicz, 2000).

If one carefully surveys the decay surface during the process of solvent relaxation, a continuous change in the decay shapes—from rapid decay at the short wavelength of the spectrum, to rising behavior at the longer part of the spectrum—should be observed. This is usually manifested as a *negative amplitude* associated with a short lifetime ("short negative DAS") on the red side of the surface, diagnostic of excited-state reaction (Grinvald and Steinberg, 1974).

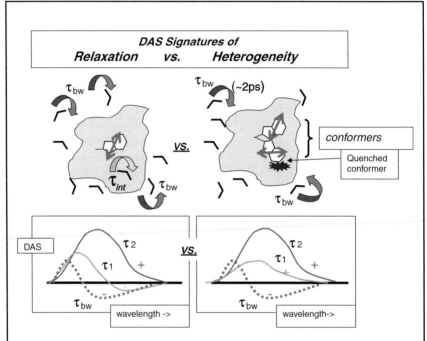

DAS are pre-exponential terms fitting $I(\lambda, t) = \sum DASi(\lambda)\exp(-t/\tau_i)$.

Homogeneous relaxation leads to negative DAS not only for ~2 ps water relaxation, but also for other short lifetimes. Heterogeneous quenching, however, yields *positive* DAS for all lifetimes longer than the water relaxation.

Time-resolved emission spectra (TRES) rising with time on red side correspond to the negative portion of the short-lived DAS terms. Positive DAS (TRES decline after 5 ps even in long wavelength regions) reveal the dominance of heterogeneity.

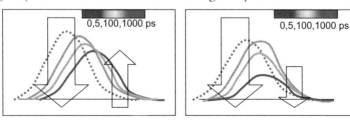

Computer simulations can also guide the modeling process. It was shown that 3-methyindole in water had bimodal character (Muino and Callis, 1994; Jimenez et al., 1994), an ultrafast Gaussian relaxation of ~15 fs (from trajectrory changes "between water collisions") was followed by an exponential response of ~400 fs due to quasi diffusive motion.

Before the 21th century, all but a handful of Trp photophysical studies had been made with 100 + picosecond and nanosecond time resolutions. Recently, however, several groups have developed upconversion instruments to broach the subpicosecond events, and all have found ~1.2 ps bulk water diffusive relaxation (Lu et al., 2004; Shen and Knutson, 2001b).

In this section, first we will revisit the picosecond solvent relaxation, complete within 5 ps, that occurs for tryptophan in water. Second, we discuss how a previously suggested quasistatic self-quenching mechanism (QSSQ; Chen et al., 1991; sub-100 ps decay terms that reduce yield and can be mistaken for scattered light or obscured by deconvolution noise) has been confirmed by upconversion fluorescence measurements. We review ps spectral and lifetime data for Trp residues in proteins (e.g., monellin and IIA^{Glc}), and finally we examine published fluorescence upconversion experiments on nucleic acids.

4.1. Solvent relaxation of tryptophan in water

Trp has an extraordinarily large Stokes' shift in physiological buffers. At times, a portion of the shift can be seen in nanosecond generalized relaxation, but most of the shift is completed before the fastest PMT can detect it. In early studies, Ruggiero et al. (1990) noted ps transients and related anisotropy decay that was, at that time, attributed to level crossing. Shen and Knutson (2001b) have measured "magic angle" (rotation invariant) TRES for Trp in water, from 400 fs to 20 ps. Figure 8.4 shows fluorescence curves of Trp in water collected at several emission wavelengths (Shen and

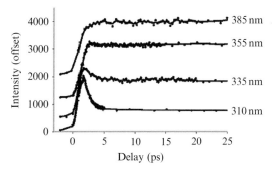

Figure 8.4 Trp emission at very early times in water. Reproduced with permission from *J. Phys. Chem. B.* **105** (2001) 6260. Copyright © 2007 American Chemical Society.

Knutson, 2001b). The fluorescence surface clearly contains a fast decay at short emission wavelengths and a fast rise at longer emission wavelengths. Further, this behavior was nearly independent of excitation wavelength, which is indicative of excited-state reaction. Both the fast decay and fast rise gave essentially the same time constant, around 1.0-1.6 ps, with an average value near 1.2 ps that matched the global analysis value gleaned from all curves. Importantly, the picosecond transient had amplitude with opposite signs at blue and red sides of the spectrum. The 1.2 ps spectral dynamics were demonstrably not from the dynamics of internal conversion (sub-100 fs events yielding anisotropy changes) (Shen and Knutson, 2001a) and the most direct explanation is solvent relaxation.

4.2. QSSQ of Trp in dipeptides

In the late 1980s, Chen *et al.* (1991) found that dipeptides had "yield defects" (lower quantum yield than expected from mean lifetimes) and postulated short "QSSQ" lifetimes. Recently, time-resolved fluorescence decay profiles of N-acetyl-L-tryptophan-namide (NATA) and tryptophan dipeptides of the form Trp-X and X-Trp, where X is another aminoacyl residue, have been investigated using our ultraviolet upconversion spectrophotofluorometer with time resolution better than 150 fs, together with a TCSPC apparatus on the 100 ps-20 ns timescale (Xu and Knutson, 2008). The set of fluorescence decay profiles has been analyzed using the global analysis technique. Nanosecond (conventional TCSPC) experiments all show the multiexponential decay of Trp dipeptides, while NATA shows monoexponential decay of 3 ns, independent of pH value. For example, the results of Leu-Trp at pH 5.2 can be fitted using three lifetimes 0.33, 1.57, and 3.68 ns. The mean lifetime $\langle \tau \rangle$ (Chen *et al.*, 1991) is around 1.18 ns. The results of Leu-Trp at pH 9.3 can be fitted using two lifetimes: 1.53 and 4.09 ns. The mean lifetime $\langle \tau \rangle$ is around 3.54 ns. All *upconversion* transients for Leu-Trp clearly exhibited a fast decay at shorter emission wavelengths and a fast rise at longer emission wavelengths (importantly, these rise terms are seen *only* in the initial 5 ps). NATA and Leu-Trp in water only displayed the fast water relaxation ~1.2 ps term and a slow component of 3 ns (and 1.2 ns for Leu-Trp). Surprisingly, another fast component was found in the corresponding Trp-X dipeptide, which is shown in Fig. 8.5. Clearly, the ultrafast data of Trp-Leu cannot be fitted satisfactorily without the use of the third fast component, while this is not necessary for NATA or Leu-Trp in water. Further, this third decay component (~30 ps) has positive amplitude even at redder wavelengths (data not shown). Solvent relaxation thus cannot be the sole source, because a homogeneous shift should result in a positive/negative DAS (or a very distorted and blue-shifted DAS). Internal conversion between ^1L_a and ^1L_b is also a poor suspect for this term, as internal conversion is faster than 50 fs (Shen and Knutson, 2001a).

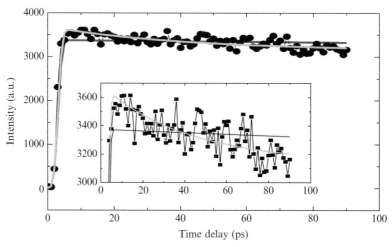

Fluorescence decay fit of trp-leu in pH 9.3
Red line: alpha 1 = 0.1, tau 1 = 30 ps; alpha 2 = 0.75, tau 2 = 6 ns
Black line: tau = 6 ns

Figure 8.5 Representative fluorescence intensity decay for dipeptides in water (Trp-Leu) within 100 ps. (See Color Insert.)

Therefore, this lifetime must originate from a quenched subpopulation (likely rotameric) of Trp. It verifies the previous logic behind QSSQ—the loss of quantum yield (but not mean lifetime detected by ns instrumentation) to a sub-100 ps decay process.

4.3. Ultrafast fluorescence dynamics of proteins

In upconversion experiments of Trp alone, the femtosecond curves showed a fast decay at "short" emission wavelengths and a fast rise at "longer" emission wavelengths—*but only during the initial 5 ps*. Trp in water displayed only this fast water relaxation (~1.2 ps) and a very slow (ns) component. Trp in the "sweet protein" monellin, however, cannot be fit without another *ultrafast* component. Figure 8.6A gives subpicosecond resolution fluorescence decays (20 ps full range) at three representative wavelengths for the protein monellin solution (black lines) and Trp solution (red lines) (Xu et al., 2006). The data in this figure were offset and peak normalized for easy comparison. Obviously, the protein monellin displays a decay process different from Trp alone, except that it is difficult to distinguish between Trp and monellin at 390 nm. In Fig. 8.6B, the temporal window was increased to 100 ps. The 16 ps term is thus more visible in Fig. 8.6B and the fit including a *positive* 16 ps exponential is also shown. The "bulk" water relaxation term (~1.2 ps) has both positive and negative regions (blue squares) below and above 365 nm, respectively. The unique (~16 ps) component (green squares) has a *significant positive amplitude* even above

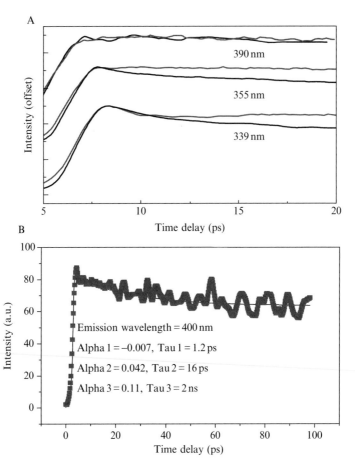

Figure 8.6 (A) Representative upconverted fluorescence intensity decay for Trp in water (red lines) and in monellin (black lines) at wavelengths 339, 355, and 390 nm using excitation at 295 nm. Reproduced with permission of *J. Am. Chem. Soc.* **128** (2006) 1214. Copyright © 2007 American Chemical Society. (See Color Insert.)

390 nm (Fig. 8.7), and the spectral shape is not significantly narrowed (Xu et al., 2006). Normal solvent relaxation on the 16 ps timescale should have resulted in a positive/negative DAS, or at least a narrowed and blue-shifted DAS without a significant positive amplitude near 400 nm. Note again, we do not include model terms for internal conversion between ^1L_b and ^1L_a since that is faster than 50 fs.

Single transients of the E21W version of the IIAGlc protein also yielded the characteristic blue positive/red negative 1.2 ps exponential previously seen for Trp solvation in bulk water. Neither wavelength region exhibited terms slower than 2 ps, however, except for previously measured terms well over 50 ps (e.g., in the TCSPC work of Toptygin and Brand) (Toptygin et al., 2001).

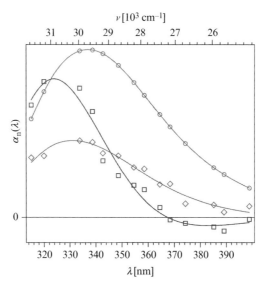

Figure 8.7 Raw decay-associated spectra of Trp in monellin extracted from upconversion data (circles) superposed with their polynomial fits (lines) (1.2 ps, blue; 16 ps, green; and 2 ns, red). Reproduced with permission of *J. Am. Chem. Soc.* **128** (2006) 1214. Copyright © 2007 American Chemical Society. (See Color Insert.)

The exposed location of Trp in the mutant E21W might lead us to think of it as neat Trp. Although the fs response of this protein looks like Trp in water, steady state iodide quenching of IIAGlc-E21W yielded a Stern-Volmer constant only ~2× that of monellin (and much less than free Trp). Thus, the "rules" for whether subpopulations will experience rapid quenching by, for example, electron transfer, appear to be specific to local constraints. QM-MM calculations (Kurz *et al.*, 2005), while not directly predictive of lifetimes in every case, provide theoretical correlates that appear to rationalize these new ~20 ps terms. Of the handful of proteins we have examined on this timescale, only GB1 shows evidence of homogeneous relaxation.

Meanwhile, it should be noticed that some research groups, for example, Abbyad *et al.*, (2007) and Cohen *et al.* (2002) have characterized the solvation kinetics of proteins using synthetic fluorescence amino acids like Aladan. Unfortunately, only TRES (not DAS) are presented, so we cannot judge the role of heterogeneity/fast quenching versus solvation there.

4.4. Ultrafast dynamics in DNA

Nucleic acids are well known to have ultrafast internal conversion after photoexcitation in the UV-visible region. In particular, fluorescence upconversion experiments on monomeric DNA constituents have shown that the fluorescence decays are extremely fast (<1 ps) and cannot be described by a single exponential, indicating complex nonradiative processes occurring in the excited state(s) (Andreatte et al., 2006; Gustavsson et al., 2006a,b; Schwalb and Temps, 2007). For DNA fluorescence upconversion experiments (Gustavsson et al., 2006a), Gustavsson, etc., used the third harmonic of a mode-locked Ti:sapphire laser as the excitation source. The 267 nm pulses are generated in a frequency-tripling system using two 0.5 mm type I BBO crystals. Typically, the average excitation power at 267 nm was 40 mW. The typical fluorescence decay curves for uracil, 6-methyluracil, 1,3-dimethyluracil, 5-methyluracil (thymine), and 5-fluorouracil are shown in Fig. 8.8. In Fig. 8.8 there is also the 330 fs (FWHM) Gaussian apparatus function. The fluorescence decays of the first three compounds are extremely fast, barely longer than the apparatus function (<1 ps). The fluorescence decays of two 5-substituted compounds, on the other hand, are much longer. This is only a taste of the work done in DNA, a problem that has been largely refractory to fluorescence analysis on longer timescale. Most recently, this work was extended (Miannay et al., 2007). Recently, an upconversion system similar to ours was used to probe the sequence and H-bonding dependence of natural base lifetimes in DNA oligonucleotides (both single and double stranded) (Schwalb and Temps, 2008).

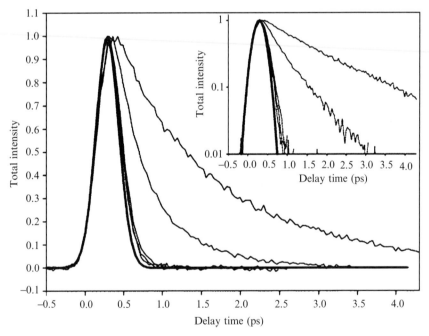

Figure 8.8 Fluorescence decays of five uracils in room-temperature aqueous solutions ($\sim 2.5 \times 10^{-3}$ mol/dm^3) at 330 nm: (in increasing order) uracil, 6-methyluracil, 1,3-dimethyluracil, 5-methyluracil (thymine), and 5-fluorouracil. Also shown is the 330 fs (FWHM) Gaussian apparatus function. The inset shows the same curves on a semilog scale. Reproduced with permission of *J. Am. Chem. Soc.* **128** (2006) 607. Copyright © 2007 American Chemical Society.

Clearly, upconversion is a suitable tool for the early dynamics of nucleic acids.

5. Summary and Future Directions

We have reviewed the basic concepts of the upconversion technique and demonstrated how it can be used in measuring time-resolved fluorescence spectra in biophysics with subpicosecond time resolution. An upconversion SPF (spectrophotofluorometer) based on these concepts has been built and design considerations have been discussed. This technique opens a window to a lot of unexplored territory in biophysics, especially studies of the previously invisible "dark" subpopulation of tryptophan in proteins and other heavily quenched chromophores. It should be possible to improve the performance features of upconversion systems by further effort, employing thinner (or composite) crystals and shorter laser pulses. More complex optics (multiplexing) can also be used to obtain TRES more rapidly. A direct fluorescence upconversion spectrograph seems promising (Zhao et al., 2005). If probe pulses are properly tilted by a prism, and then mixed with the fluorescence in the nonlinear crystal, it is possible to gate a wide wavelength range simultaneously. Improvement will be needed to translate the system to the UV for use in biophysics, such as dispersion correction, array detector UV sensitization, etc. Developments in nonlinear materials involving "poling" and nanocrystalline alignment may soon bring order of magnitude sensitivity increases. If pulse power, repetition rate, and optical properties all combine their current growth, upconversion will become as common as TCSPC, and time resolution will be selected by "detuning" (dispersing in time) the probe pulse for a more rapid, coarse collection. The ultimate limit for shorter timing will, of course, be the width of the transitions themselves, as the "transform-limited" spectral width of pulses below ~50 fs begins to excite multiple electronic bands (unavoidably). This bar to earlier times is a rare example of the uncertainty principle in action.

In all, upconversion is a technology now primed for growth in biophysics.

REFERENCES

Abbyad, P., Shi, X., Childs, W., McAnaney, T., Cohen, B., and Boxer, S. (2007). Measurement of solvation responses at multiple sites in a globular protein. *J. Phys. Chem. B* **111,** 8269.

Alcala, J. R., Gratton, E., and Prendergast, F. G. (1987). Interpretation of fluorescence decays in proteins using continuous lifetime distributions. *Biophys. J.* **51,** 925.

Andreatta, D., Sen, S., Lustres, J., Kovalenko, S., Ernsting, N., Murphy, C., Coleman, R., and Berg, M. (2006). Ultrafast dynamics in DNA: Fraying at the end of the helix. *J. Am. Chem. Soc.* **128,** 6885.

Andrews, D. L., and Demidov, A. A. (2002). An introduction to laser spectroscopy. 2nd ed. Kluwer Academic/Plenum Publisher, New York.

Arzhantsev, S., and Maroncelli, M. (2005). Design and characterization of a femtosecond fluorescence spectrometer based on optical Kerr gating. *Appl. Spectrosc.* **59,** 206.

Barbara, P. F., Brus, L. E., and Reutzepis, P. P. (1980). Picosecond time-resolved fluorescence study of s-tetrazine vibrational-relaxation in solution. *Chem. Phys. Lett.* **69,** 447.

Becker, W. (2005). Advanced time-correlated single photon counting techniques. Berlin, Springer.

Beechem, J. M., and Brand, L. (1985). Time-resolved fluorescence of proteins. *Annu. Rev. Biochem.* **54,** 43.

Bradley, D. J., Bryant, S. F., and Sibbett, W. (1980). Intensity dependent time resolution and dynamic-range of photochron picosecond streak cameras to linear photoelectric recording. *Rev. Sci. Instrum.* **51,** 824.

Bradley, D. J., Mcinerney, J., Dennis, W. M., and Taylor, J. R. (1983). A new synchroscan streak-camera readout system for use with CW mode-locked lasers. *Opt. Commun.* **44,** 357.

Birch, D. J., and Imhof, R. E. (1991). Time-domain fluorescence spectroscopy using time-correlated single-photon counting. *In* "Topics in fluorescence spectroscopy" (J. R. Lakowicz, ed.), Vol. 1, pp. 1–95. Plenum Press, New York.

Changenet, P., Zhang, H., and Van der Meer, M. J. (1998). Subpicosecond fluorescence upconversion measurements of primary events in yellow proteins. *Chem. Phys. Lett.* **282,** 276.

Chen, R. F., Knutson, J. R., Ziffer, H., and Porter, D. (1991). Fluorescence of tryptophan dipeptides-correlations with the rotamer model. *Biochemistry* **30,** 5184.

Chowdhury, P., Gondry, M., Genet, R., Martin, J. L., Menez, A., Negrerie, M., and Petrich, J. W. (2003). Picosecond dynamics of a peptide from the acetylcholine receptor interacting with a neurotoxin probed by tailored tryptophan fluorescence. *Photochem. Photobiol.* **77,** 151.

Clayton, A., and Sawyer, W. (1999). Tryptophan rotamer distributions in amphipathic peptides at a lipid surface. *Biophys. J.* **76,** 3235.

Cohen, B., McAnaney, T., Park, E., Jan, Y., Boxer, S., and Jan, L. (2002). Probing protein electrostatics with a synthetic fluorescent amino acid. *Science* **296,** 1700.

Dashnau, J. L., Zelent, B., and Vanderkooi, J. M. (2005). Tryptophan interactions with glycerol/water and trehalose/sucrose cryosolvents: Infrared and fluorescence spectroscopy and ab initio calculations. *Biophys. Chem.* **114,** 71.

Duguay, M. A., and Hansen, J. W. (1968). Optical sampling of subnanosecond light pulses. *Appl. Phys. Lett.* **13,** 178.

Eimerl, D. (1987). Electrooptic, linear, and nonliner optical-properties of KDP and its isomorphs. *Ferroelectrics* **72,** 397.

Espagne, A., Paik, D. H., Changenet-Barret, P., Martin, M. M., and Zewail, A. H. (2006). Ultrafast photoisomerization of photoactive yellow protein chromophore analogues in solution: Influence of the protonation state. *Chem. Phys. Lett.* **7,** 1717.

Fiebig, T., Chachisvilis, M., Manger, M., Zewail, A. H., Douhal, A., Garcia-Ochoa, I., and de La Hoz Ayuso, A. (1999). Femtosecond dynamics of double proton transfer in a model DNA base pair: 7-azaindole dimers in the condensed phase. *J. Phys. Chem. A* **103,** 7419.

Fleming, G. R. (1986). Chemical application of ultrafast spectroscopy. Oxford University Press, New York.

Fork, R. L., Brito, C. H., Becker, P. C., and Shank, C. V. (1987). Compression of optical pulses to 6 femtoseconds by using cubic phase compensation. *Opt. Lett.* **12,** 483.

Fukunaga, Y., Nishimoto, E., Yamashita, K., Otosu, T., and Yamashita, S. (2007). The partially unfolded state of beta-momorcharin characterized with steady state and time-resolved fluorescence studies. *J. Biochem.* **141,** 9.

Ghosh, G. (1995). Temperature dispersion of refractive-indexes in beta-bab2o4 and lib3o5 crystals for nonlinear-optical devices. *J. Appl. Phys.* **78**, 6752.

Gregoire, G., Jouvet, C., Dedonder, C., and Sobolewski, A. (2007). Ab initio study of the excited-state deactivation pathways of protonated tryptophan and tyrosine. *J. Am. Chem. Soc.* **129**, 6223.

Grinvald, A., and Steinberg, I. Z. (1974). Analysis of fluorescence decay kinetics by method of least-squares. *Anal. Biochem.* **59**, 583.

Gu, J. L., Shi, J. L., You, G. J., Xiong, L. M., Qian, S. X., Hua, Z. L., and Chen, H. R. (2005). Incorporation of highly dispersed gold nanoparticles into the pore channels of mesoporous silica thin films and their ultrafast nonlinear optical response. *Adv. Mater.* **17**, 557.

Gustavsson, T., Banyasz, A., Lazzarotto, E., Markovitsi, D., Scalmani, G., Frisch, M., Barone, V., and Improta, R. (2006a). Singlet excited-state behavior of uracil and thymine in aqueous solution: A combined experimental and computational study of 11 uracil derivatives. *J. Am. Chem. Soc.* **128**, 607.

Gustavsson, T., Sarkar, N., Lazzarotto, E., Markovitsi, D., Scalmani, G., Frisch, M., and Improta, R. (2006b). Singlet excited state dynamics of uracil and thymine derivatives: A femtosecond fluorescence upconversion study in acetonitrile. *Chem. Phys. Lett.* **429**, 551.

Halliday, L., and Topp, M. (1977). Picosecond luminescence detection using type-2 phase-matched frequency-conversion. *Chem. Phys. Lett.* **46**, 8.

Halliday, L., and Topp, M. (1978). Picosecond optical pulse sampling by frequency-conversion - studies of solvent-induced molecular relaxation. *J. Phys. Chem.* **82**, 2273.

Holten, D., and Windsor, M. W. (1978). Picosecond flash-photolysis in biology and biophysics. *Annu. Rev. Biophys.* **7**, 189.

Ihalainen, J. A., Croce, R., and Morosiuotto, T. (2005). Excitation decay pathways of Lhca proteins: A time-resolved fluorescence study. *J. Phys. Chem. B* **109**, 21150.

Inchauspe, C. M. G., and Martinez, O. E. (1997). Quartic phase compensation with a standard grating compressor. *Opt. Lett.* **22**, 1186.

Ippen, E. P., and Shank, C. V. (1978). Sub-picosecond spectroscopy. *Phys. Today* **31**, 41.

Jimenez, R., Fleming, G. R., Kumar, P. V., and Maroncelli, M. (1994). Femtosecond solvation dynamics of water. *Nature* **369**, 471.

Kafka, J. D., and Baer, T. (1987). Prism-pair dispersive delay-lines in optical pulse-compression. *Opt. Lett.* **12**, 401.

Kahlow, M. A., Jarzeba, W., Dubruil, T. P., and Barbara, P. F. (1988). Ultrafast emission-spectroscopy in the ultraviolet by time-gated upconversion. *Rev. Sci. Instrum.* **59**, 1098.

Kennis, J. T. M., Gobets, B., van Stokkum, I. H. M., Dekker, J. P., van Grondelle, R., and Fleming, G. R. (2001). Light harvesting by chlorophylls and carotenoids in the photosystem I core complex of synechococcus elongatus: A fluorescence upconversion study. *J. Phys. Chem. B* **105**, 4485.

Kurz, L. C., Fite, B., Jean, J., Park, J., Erpelding, T., and Callis, P. (2005). Photophysics of tryptophan fluorescence: Link with the catalytic strategy of the citrate synthase from thermoplasma acidophilum. *Biochemistry* **44**, 1394.

Lakowicz, J. R. (1999). Principles of fluorescence spectroscopy. 2nd edn. Kluwer Academic/Plenum Publishers, New York.

Lakowicz, J. R. (2000). On spectral relaxation in proteins. *Photochem. Photobiol.* **72**, 421.

Lampa-Pastrk, S., Lafuente, R. C., and Beck, W. F. (2004). Excited-state axial-ligand photodissociation and nonpolar protein-matrix reorganization in zn(II)-substituted cytochrome c. *J. Phys. Chem. B* **108**, 12602.

Li, T., Hassanali, A. A., Kao, Y., Zhong, D., and Singer, S. J. (2007). Hydration dynamics and time scales of coupled water-protein fluctuations. *J. Am. Chem. Soc.* **129**, 3376.

Liu, L. Q., and Nagashima, K. (1999). Optimum phase matching and effective nonlinear coefficients of lbo and ktp for continuously changing wavelength. *Opt. Laser Technol.* **31**, 283.

Lu, W., Kim, J., Qiu, W., and Zhong, D. (2004). Femtosecond studies of tryptophan solvation: Correlation function and water dynamics at lipid surfaces. *Chem. Phys. Lett.* **388**, 120.

Mahr, H., and Hirsch, M. D. (1975). Optical up-conversion light gate with picosecond resolution. *Opt. Commun.* **13**, 96.

May, J. M., and Beechem, J. M. (1993). Monitoring conformational change in the human erythrocyte glucose carrier - use of a fluorescent-probe attached to an exofacial carrier sulfhydryl. *Biochemistry* **32**, 2907.

Mcmahon, L. P., Yu, H. T., Vela, M. A., Morales, G. A., Shui, L., Fronczek, F. R., McLaughlin, M. L., and Barkley, M. D. (1997). Conformer interconversion in the excited state of constrained tryptophan derivatives. *J. Phys. Chem. B* **101**, 3269.

Miannay, F., Bányász, A., Gustavsson, T., and Markovitsi, D. (2007). Ultrafast excited-state deactivation and energy transfer in guanine-cytosine DNA double helices. *J. Am. Chem. Soc.* **129**, 14574.

Muino, P. L., and Callis, P. R. (1994). Hybrid simulations of solvation effects on electronic-spectra - indoles in water. *J. Chem. Phys.* **100**, 4093.

Muller, M. G., Drews, G., and Holzwarth, A. (1996). Primary charge separation processes in reaction centers of an antenna-free mutant of Rhodobacter capsulatus. *Chem. Phys. Lett.* **258**, 194.

Murao, T., Yamazaki, I., and Yoshihara, K. (1982). Applicability of a microchannel plate photo-multiplier to the time-correlated photon-counting technique. *Appl. Opt.* **21**, 2297.

O'Connor, D. V., and Phillips, D. (1984). Time correlated single photon counting. Academic Press, New York.

Pan, C. P., Callis, P. R., and Barkley, M. D. (2006). Dependence of tryptophan emission wavelength on conformation in cyclic hexapeptides. *J. Phys. Chem. B* **110**, 7009.

Qiu, W., Zhang, L., Okobiah, O., Yang, Y., Wang, L., Zhong, D., and Zewail, A. H. (2006). Ultrafast solvation dynamics of human serum albumin: Correlations with conformational transitions and site-selected recognition. *J. Phys. Chem. B* **110**, 10540.

Ross, J. A., Wyssbrod, H. R., Porter, R. A., Schwartz, G. P., Michaels, C. A., and Laws, W. R. (1992). Correlation of tryptophan fluorescence intensity decay parameters with H-1 NMR-determined rotamer conformations-[tryptophan2]oxytocin. *Biochemistry* **31**, 1585.

Rubtsov, I. V., and Yoshihara, K. (1999). Vibrational coherence in electron donor-acceptor complexes. *J. Phys. Chem. A* **103**, 10202.

Ruggiero, A. J., Todd, D. C., and Fleming, G. R. (1990). Subpicosecond fluorescence anisotropy studies of tryptophan in water. *J. Am. Chem. Soc.* **112**, 1003.

Schanz, R., Kovalenko, S. S., Kharlanov, V., and Ernsting, N. P. (2001). Broad-band fluorescence upconversion for femtosecond spectroscopy. *Appl. Phys. Lett.* **79**, 566.

Schlarb, U., Reichert, A., and Betzler, K. (1995). SHG phase matching conditions for undoped and doped lithium niobate. *Radiat. Eff. Defects Solids* **136**, 1029.

Schroder, G. F., Alexiev, U., and Gubmuller, H. (2005). Simulation of fluorescence anisotropy experiments: Probing protein dynamics. *Biophys. J.* **89**, 3757.

Schwalb, N., and Temps, F. (2007). Ultrafast electronic relaxation in guanosine is promoted by hydrogen bonding with cytidine. *J. Am. Chem. Soc.* **129**, 9272.

Schwalb, N., and Temps, F. (2008). Base sequence and higher-order structure induce the complex excited-state dynamics in DNA. *Sciences* **322**, 243.

Shen, Y. R. (1984). The principles of nonlinear optics. Wiley-Interscience, New York.

Shen, X., and Knutson, J. R. (2001a). Femtosecond internal conversion and reorientation of 5-methoxyindole in hexadecane. *Chem. Phys. Lett.* **339**, 191.

Shen, X., and Knutson, J. R. (2001b). Subpicosecond fluorescence spectra of tryptophan in water. *J. Phys. Chem. B* **105**, 6260.

Szabo, A. G., and Rayner, D. M. (1980). Fluorescence decay of tryptophan conformers in aqueous-solution. *J. Am. Chem. Soc.* **102,** 554.

Topp, M. R. (1979). Pulsed laser spectroscopy. *Appl. Spectrosc. Rev.* **14,** 1.

Toptygin, D., Savtchenko, R., Meadow, D., and Brand, L. (2001). Homogeneous spectrally- and time-resolved fluorescence emission from single-tryptophan mutants of IIA(Glc) protein. *J. Phys. Chem. B* **105,** 2043.

Toptygin, D., Gronenborn, A. M., and Brand, L. (2006). Nanosecond relaxation dynamics of protein GB1 identified by the time-dependent red shift in the fluorescence of tryptophan and 5-fluorotryptophan. *J. Phys. Chem. B* **110,** 26292.

Van Stokkum, I. H. M., Gobets, B., and Gensch, T. (2006). (Sub)-picosecond spectral evolution of fluorescence in photoactive proteins studied with a synchroscan streak camera system. *Photochem. Photobiol.* **82,** 380.

Vincent, M., Gallay, J., and Demchenko, A. P. (1995). Solvent relaxation around the excited-state of indole - analysis of fluorescence lifetime distributions and time-dependence spectral shifts. *J. Phys. Chem.* **99,** 14931.

Xu, J., and Knutson, J. R. (2008). Femtosecond fluorescence studies of tryptophan dipeptides in water: Explanation of quasi static self quenching. (submitted for publication).

Xu, J. H., Toptygin, D., Graver, K. J., Albertini, R. A., Savtchenko, R. S., Meadow, N. D., Roseman, S., Callis, P. R., Brand, L., and Knutson, J. R. (2006). Ultrafast fluorescence dynamics of tryptophan in the proteins monellin and IIA(Glc). *J. Am. Chem. Soc.* **128,** 1214.

Yu, B. L., Bykov, A. B., and Qiu, T. (2003). Femtosecond optical Kerr shutter using lead-bismuth-gallium oxide glass. *Opt. Commun.* **215,** 407.

Zelent, B., Vanderkooi, J. M., Coleman, R. G., Gryczynski, I., and Gryczynski, Z. (2006). Protonation of excited state pyrene-1-carboxylate by phosphate and organic acids in aqueous solution studied by fluorescence spectroscopy. *Biophys. J.* **91,** 3864.

Zernike, F., and Midwinter, J. E. (1973). Applied nonlinear optics. John Wiley & Sons, New York.

Zhang, L., Kao, Y., Qiu, W., Wang, L., and Zhong, D. (2006). Femtosecond studies of tryptophan fluorescence dynamics in proteins: Local solvation and electronic quenching. *J. Phys. Chem. Lett. B* **110,** 18097.

Zhao, L. J., Lustres, J. L. P., Farztdinov, V., and Ernsting, N. P. (2005). Femtosecond fluorescence spectroscopy by upconversion with tilted gate pulses. *Phys. Chem. Chem. Phys.* **7,** 1716.

Zwicker, H. R. (1981). *In* Photoemissive detectors in Optical and infrared detectors 2nd ed. p. 149. Springer-Verlag, New York.

CHAPTER NINE

USE OF FLUORESCENCE RESONANCE ENERGY TRANSFER (FRET) IN STUDYING PROTEIN-INDUCED DNA BENDING

Anatoly I. Dragan *and* Peter L. Privalov

Contents

1. Introduction	186
2. Preparation of Labeled DNA Duplexes	187
3. Fluorescence Resonance Energy Transfer	189
4. FRET in Studying Large Protein-Induced DNA Bends	191
5. Dependence of the Protein-Induced DNA Bend on the Forces Involved in Binding	194
6. FRET in Studying Small Protein-Induced DNA Bends	194
6.1. Construction of a U-shaped double bulged fluorophores labeled DNA	195
6.2. FRET titration of the U-shaped DNA	197
7. Conclusions	198
Acknowledgment	198
References	199

Abstract

The specific association of many DNA-binding proteins with DNA frequently results in significant deformation of the DNA. Protein-induced DNA bends depend on the protein, the DNA sequence, the environmental conditions, and in some cases are very substantial, implying that DNA bending has important functional significance. The precise determination of the DNA deformation caused by proteins under various conditions is therefore of importance for understanding the biological role of the association. This review considers methods for the investigation of protein-induced DNA bending by measuring the change in fluorescence resonance energy transfer (FRET) between fluorophores placed at the ends of the target DNA duplex. This FRET technique is particularly efficient when the protein-induced bend in the DNA is considerable and results in a significant decrease in the distance between the DNA ends

The Institute of Fluorescence, University of Maryland Biotechnology Institute, Columbus Center, Baltimore, Maryland bearing the fluorophores. However, in the case of small bends the change of distance between the ends of short DNA duplexes, as typically used in protein binding experiments (about 16–20 bp), is too small to be detected accurately by FRET. In such cases the change of the distance between the fluorophores can be increased by using levers attached to the binding site, that is, using two bulges to construct a U-shaped DNA in which the central part contains the protein-binding site and the fluorophores are attached to the ends of the perpendicularly directed arms.

1. INTRODUCTION

Formation of many specific protein/DNA complexes results in significant deformation of the DNA, typically a bend at the protein-binding site. The extent of DNA bending is different for different proteins and depends on the specificity of the DNA sequence and solvent conditions. It is particularly large for proteins interacting with the minor groove of DNA (e.g., HMG box-containing proteins bend their target DNA up to 120°); in contrast, the DNA bend induced by proteins interacting with the major groove is usually more moderate (Dragan *et al.*, 2003; Love *et al.*, 1995; Masse *et al.*, 2002; Murphy *et al.*, 2001; Read *et al.*, 1995). The biological role of protein-induced DNA bending must be to facilitate the assembly of multicomponent complexes at key sites such as promoters and enhancers, so proteins that induce large bends are sometimes referred to as "architectural"; however, the smaller bends induced by major groove binders may also be crucial to their role in combinatorial complex assembly. Thus, protein-induced DNA bending is a phenomena of considerable interest, however, its experimental investigation faces certain difficulties.

The most widely used method to study protein-induced DNA bending is the circular permutation assay based on the dependence of the electrophoretic mobility of a bent DNA/DBD complex on the position of the bend along the duplex. The interpretation of the variability in mobility in terms of bend angle is model-dependent and the method is predisposed to certain artifacts (Kerppola, 1996), so more direct methods are desirable. DNA bend angles can of course be determined accurately using crystallography, but this approach is subject to some uncertainty due to crystal packing forces exerted on the DNA ends that might enhance or reduce the DNA distortions (Masse *et al.*, 2002; Murphy *et al.*, 2001). Solution methods have obvious advantages, in particular NMR studies of protein–DNA complexes (Love *et al.*, 1995; Masse *et al.*, 2002; Murphy *et al.*, 2001), but the size of larger complexes represents an obstacle to the wide use of the method. It is in this context that the fluorescence resonance energy transfer (FRET) technique is of particular interest.

FRET analysis is based on measurements of the FRET between acceptor and donor fluorophores placed on the ends of a target DNA duplex, which provide information on the distance between the fluorophores; that is, the distance between the ends of the duplex. Changes in the FRET between these fluorophores on protein association with DNA thus characterize the protein-induced deformation of the DNA. The great advantage of this method is that it can be used under varying conditions of ionic strength and temperature. Furthermore, the method provides quantitative information not only on the bending of the DNA but also on the protein binding that gives rise to it; that is, the binding constant and the Gibbs energy of binding, factors important for correlating the extent of DNA bending with the forces involved in forming the complex.

The FRET technique is especially efficient for measuring large DNA bends due to the substantial changes of distance between the fluorophores; however, in the case of small DNA-bends the change in distance between the ends of the short DNA duplexes of the order of 16–20 bp is too small to be detected accurately by FRET. In such cases, the change of the distance between the fluorophores can be increased by using levers attached to the binding site; that is, using two bulges to construct a U-shaped DNA in which the central part contains the protein-binding site and the fluorophores are attached to the ends of the perpendicularly directed arms (Hillisch *et al.*, 2001; Stuhmeier *et al.*, 2000). The efficiency of such a construct for measuring small protein-induced DNA bends has been demonstrated for the example of GCN4-bZIP interaction with DNA (Dragan *et al.*, 2004b). Here we describe applications of the FRET technique for determination of both large and small protein-induced DNA bends.

2. Preparation of Labeled DNA Duplexes

Preparation of high quality double-stranded samples of fluorophore-labeled DNA is a crucial issue for analytical fluorescence measurements, especially for measurements of the FRET effect upon titration with proteins. This requires mixing complementary oligonucleotides in equimolar concentrations, heating the mixture up to 80 °C and cooling slowly to room temperature. Success in annealing the duplexes depends on the accuracy of 1:1 mixing of the strands and this requires a reliable method of determining oligonucleotide concentration. UV absorption spectroscopy of single-stranded oligonucleotides carries the complication of variable degrees of secondary structure but this can be avoided by determining the concentration of the DNA strands spectrophotometrically after their complete digestion with snake venom phosphodiesterase-I (PDE1) (Sigma) in 100 mM Tris–HCl, pH 8.0. Figure 9.1A shows the absorption spectra of

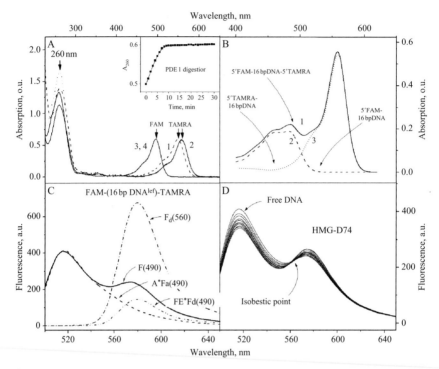

Figure 9.1 Spectral characteristics of labeled DNA samples. (A) Absorption spectra of 16 base single-stranded DNAs 5′-labeled with TAMRA and FAM (FAM-5′-ACTATAA CAATACAAG; TGATATTGTTATGTTC-5′-TAMRA) monitored before (2)–(4) and after (1)–(3) digestion with PDE1 in 100 mM Tris–HCl, pH 8.0, 20 °C. For digestion, 1 μl of PDE1 (0.01 units) was added to 1 ml of each sample with OD_{260} about 0.5 o.u. *Inset*: Dependence of A260 upon time for PDE1 digestion at 37 °C. (B) Absorption spectra of annealed double-labeled (1) and single-labeled DNA duplexes, (2)–(3), in 100 mM KCl, 10 mM K-phosphate, pH 6.0 at 20 °C. (C) Fluorescence spectra of double-labeled 16-bp DNA excited at 490 nm, F(490), and at 560 nm, F_d(560). Also shown the result of deconvolution of the spectra, F(490), into its FAM and TAMRA fluorescence components, A★F_a(490) and FE★F_d(490), respectively, according to Eq. (2). (D) FRET changes of florescence spectra of the double-labeled 16-bp DNA upon titration with the HMG–D74 DNA binding domain.

single-strands of 16-bp DNA labeled with FAM and TAMRA before and after PDE1 digestion and a typical kinetic curve. The extinction coefficients of the individual nucleotides at 260 nm at pH 8.0 are: 15,200 $M^{-1}cm^{-1}$, 8400 $M^{-1}cm^{-1}$, 7050 $M^{-1}cm^{-1}$, and 12,010 $M^{-1}cm^{-1}$ for dA, dT, dC, and dG, respectively. In the case of labeled single-strands the contribution of the fluorophores must also be taken into account. The most common are FAM and TAMRA, which have extinction coefficients at 260 nm of 28,000 $M^{-1}cm^{-1}$ and 29,000 $M^{-1}cm^{-1}$, respectively. The concentration of the annealed duplexes is finally checked by the same procedure after complete digestion by PDE1 using, however, 10 times higher concentration of the enzyme since hydrolysis of duplexes proceeds much slower than single strands. The absorption spectra of annealed double-labeled 16-bp DNA and the related single-labeled DNAs of equal concentrations in buffer (10 mM K-phosphate, 100 mM KCl, pH 6.0) are shown in Fig. 9.1B.

The completeness of hybridization has to be verified by sizing chromatography and gel-electrophoresis with visualization of the bands by excitation of the attached chromophores.

3. FLUORESCENCE RESONANCE ENERGY TRANSFER

The efficiency of FRET varies as the sixth power of the separation between the donor and acceptor, R_{da}, (Förster, 1946):

$$E = 1/[1 + (R_{da}/R_o)^6] \quad (9.1)$$

Here R_o is the characteristic Förster distance for 50% energy transfer efficiency which equals $9.79 \times 10^3 (k^2 \times n^{-4} \times \Phi_d \times J)^{1/6}$, where n is the refractive index of the medium (1.33 for water), Φ_d is the fluorescence quantum yield of the donor, J is the overlap integral between the emission spectrum of the donor and the excitation spectrum of the acceptor; k^2 is the orientation factor, which in the case of rapid randomization equals 2/3. For the well-known donor–acceptor pair, FAM–TAMRA, R_o is about 50 Å. As follows from the Eq. (1), using the FRET technique one can measure R_{da} values of the same order of magnitude as the Förster distance. For example, the R_{da} distance between FAM and TAMRA chromophores attached to the 5′-ends of a 16-bp DNA linear duplex is 61 Å (Dragan et al., 2003). Bending of the duplex lead to decrease of the end-to-end distance and, correspondingly, to increase of the FRET efficiency (FE).

FE can be determined from the quenching of donor fluorescence due to energy transfer or from sensitization of the acceptor fluorescence. The latter is the more accurate and reliable method because it normalizes the FRET signal for the quantum yield of the acceptor, for the concentration of the duplex molecule, and for any error in the effectiveness of acceptor labeling (Clegg, 1992). To obtain the FRET signal, the total fluorescence spectrum excited at 490 nm, $F(490)$, has to be deconvoluted into its donor and acceptor components: the FAM spectrum, $F_d(490)$, and the spectrum of TAMRA, $F_a(560)$, (Fig. 9.1C):

$$F(490) = A \times F_d(490) + \text{FE} \times F_a(560) \quad (9.2)$$

Here A and FE are the fitted weighting factors of the two spectral components. FE is linearly dependent on the efficiency of energy transfer, E:

$$\text{FE} = E[\varepsilon_d(490)/\varepsilon_a(560)] + \varepsilon_a(490)/\varepsilon_a(560) \qquad (9.3)$$

where $\varepsilon_d(490)$, $\varepsilon_a(560)$, and $\varepsilon_a(490)$ are the extinction coefficients of donor and acceptor at 490 and 560 nm. Their ratio can be determined from absorption/excitation spectra of the double- and single-labeled DNA samples (Fig. 9.1B).

It should be noted that in the case of protein-induced deformation of DNA, the measured FRET effect, FE, depends not only on the extent of DNA bending but also on the concentration of the complex in the reaction mixture as protein is added to the DNA (Figs. 9.1D and 9.2A). For the bimolecular reaction, Protein + DNA $\Leftrightarrow$ Complex, the concentration of the complex is: $[\text{DNA}]_{\text{comp}} = K_a [\text{DNA}]_{\text{free}} \cdot [\text{Protein}]_{\text{free}}$. The FE-parameter is proportional to the fraction of bound DNA, $v = \text{FE}/(\text{AFE} - \text{FE}_o)$, where FE_o is the FRET effect for free DNA and asymptotic FRET effect (AFE) is the asymptotic value of FE, that is, its value for 100% complex. The AFE can be obtained by fitting the data to a binding isotherm, that is, the dependence of FE upon concentration of added protein, using the following equation:

$$\text{FE} = \text{FE}_o - (\text{AFE} - \text{FE}_o)\{0.5(1 + P/D_o + 1)/(K_a D_o)) \\ - [0.25\left(1 + P/D_o + 1/(K_a D_o)\right)^2 - P/D_o]^{0.5}\}, \qquad (9.4)$$

Here P and D_o are the concentrations of protein and DNA, K_a is the association constant.

Using the measured parameter FE and a calculated R_o, the observed R_{da} distance between the donor and acceptor can be calculated from Eq. (1). Thus, to get R_{da}, the bending characteristic of a protein–DNA complex, one has to proceed with the following steps:

a. Perform a FRET titration to get a binding isotherm, that is, FE as a function of protein concentration.
b. By fitting the isotherm, determine the AFE value.
c. Calculate the R_{da} distance using Eqs. (1)–(3).

Transforming measured R_{da} distances into DNA bend angles can be done by setting up an empirical function using as calibrants the results obtained for complexes with well-known structure, or by using a appropriate geometric model of the DNA in the complex. Examples of the practical use of FRET to study the DNA binding and bending by proteins are given below.

4. FRET in Studying Large Protein-Induced DNA Bends

Among the DNA-binding proteins that induce large DNA bends, the most studied are representatives of the HMG box family, some of which bend their cognate DNAs up to 120° under physiological conditions (Dragan et al., 2003; Love et al., 1995; Masse et al., 2002; Murphy et al., 2001; Read et al., 1995).

The binding isotherms obtained by FRET titration of various 16-bp DNA duplexes (DNA^{Lef}: AGAGCTTAAAGGGTG; DNA^{Sox}: ACTATAACAA TACAAG; DNA^{Sry}: AGCTGCACAAACACCG; DNA^{AT}: AGAGCGA TATCGCGTG) with their specific HMG box DNA-binding domains (DBDs) are shown in Fig. 9.2A. In Fig. 9.2B, are shown binding isotherms obtained by the FRET titration of a 16-bp DNA^{Lef} with the HMG-D100 box DBD in the presence of different concentration of salt (KCl). The data show that the binding isotherms are very specific for a given DBD–DNA complex and for any given complex they depend on the KCl concentration. Fitting the obtained FE functions using Eq. (4) gives the values of AFE, and the association constant, K_a. The dependencies of the derived association constants upon concentration of salt are shown in Fig. 9.3A and B on a logarithmic scale.

The asymptotic value of a FRET effect (AFE) reflects the extent of deformation of the DNA upon 100% protein binding (Fig. 9.2). Binding of

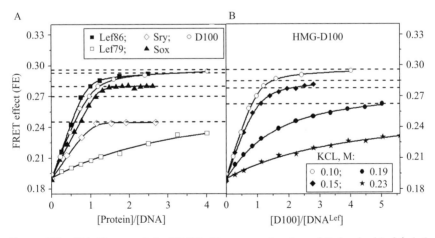

Figure 9.2 (A) Change of the FRET effect upon titration of 16-bp double-labeled specific DNAs with their cognate HMG DBDs at 20 °C in 10 mM K/phosphate pH 6.0, 100 mM KCl. The dashed lines correspond to the asymptotic values of the FRET effect (AFE). (B) Isotherms of DNA^{Lef} FRET titration by the HMG–D100 box at four different KCl concentrations in 10 mM K/phosphate pH 6.0. For 100 mM KCl, [DNA] = 50 nM, whilst for the three higher concentrations of KCl, [DNA] = 500 nM. The dashed lines correspond to the asymptotic values of the FRET effect (AFE).

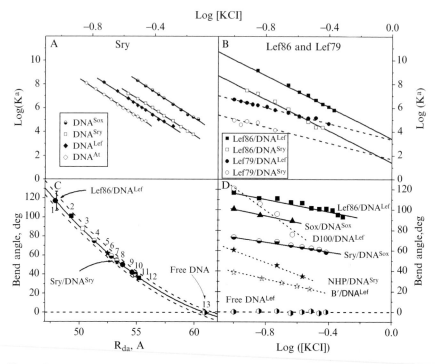

Figure 9.3 Dependence of the logarithm of the association constants of the Sry HMG box with four different DNAs (A) and of the Lef86 HMG box and its truncated form, Lef79, with DNA^{Lef} (B) upon logarithm of KCl salt concentration at 20 °C in 10 mM K/phosphate pH 6.0. (C) Dependence of the DNA bend angle on the distance between the ends of the 16-bp duplexes, R_{da}. Plotted data: Lef86 with DNA^{Lef} (1), DNA^{Sox} (8), DNA^{Sry} (9). Lef79 with DNA^{Lef} (3); Sox with DNA^{Sox} (2), DNA^{Sry} (10), DNA^{Lef} (12); Sry with DNA^{Sox} (4), DNA^{Lef} (6), DNA^{Sry} (7); NHP with DNA^{Sry} (5); Box-B' with DNA^{Lef} (11); free DNA (13). Arrows indicate the calibration data points. (D) Dependence of the protein-induced DNA bend angles on the logarithm of KCl salt. Note that the conformation of the free DNA detected by FRET does not change upon increasing KCl concentration, meaning that the FRET effect itself is not sensitive to the change of salt concentration.

HMG box DBDs to different DNAs of the same length (16 bp) results in different values of the AFE, showing that the induced bends in the DNA are not the same and depend on the particular protein. It is also seen that the AFE depends significantly on the KCl concentration (Figs. 9.2 and 9.3) and on the correspondence of the DNA sequence to the given DBD (Fig. 9.3). As the AFE directly depends on the distance between the donor and acceptor fluorophores, R_{da}, one can estimate the distance between the DNA ends in the complexes. The AFE data show that in solutions containing 100 mM KCl, the distance between the ends of the free 16-bp DNA (plus the length of the label connectors) is 60.9 ± 0.1 Å and in the DBD/DNA complexes this distance is significantly reduced.

As all the FRET experiments used DNA duplexes of the same 16-bp size, the observed change in the distance between their ends, R_{da}, upon binding of different proteins can be regarded as a measure of the protein-induced bending of the DNA. However, the change in R_{da} is a relative parameter of the DNA deformation that depends on the length of the DNA duplex used. A more useful parameter is the bend angle that does not depends on the length of the DNA duplex used in the experiment. In the case of the HMG box DBDs, which are all similar in structure and smoothly bend DNA over several base pairs, the deformation expressed as the change of R_{da} of duplexes of identical size can be transformed into a bend angle using as standards the known structures of the sequence-specific HMG box/DNA complexes, namely Lef86/DNA^{Lef} and Sry/DNA^{Sry} (Love et al., 1995; Murphy et al., 2001), and free DNA (Fig. 9.3C; Dragan et al., 2003). Table 9.1 shows that the maximal bending is seen for Lef86 binding to its cognate DNA^{Lef}; removal of the 8 residue C-terminal extension to generate Lef79 decreases the bend angle by 29°; DNA^{Sry} is bent to a lower extent than DNA^{Sox} by the HMG box of Sry.

Table 9.1 FRET and bending parameters of the 16 bp DNAs labeled at 5′-ends with FAM and TAMRA upon binding with HMG box DBDs, according to Dragan et al. (2004b)

HMG box DBD	Parameters	DNA^{Lef}	DNA^{Sry}	DNA^{Sox}
Sry	AFE	0.248	0.244	0.261
	R_{da}	52.8	53.3	51.3
	Bend angle	59 ± 4	54 ± 3^a	75 ± 6
Lef86	AFE	0.293	0.233	0.240
	R_{da}	47.9	54.6	53.8
	Bend angle	117 ± 10^a	39 ± 4	49 ± 5
Lef79	AFE	0.270	-	-
	R_d	50.4	-	-
	Bend angle	88 ± 8	-	-
Sox	AFE	0.228	0.231	0.280
	R_{da}	55.2	55.0	49.3
	Bend angle	36 ± 4	40 ± 4	101 ± 9
D100	AFE	0.296	-	0.285
	R_{da}	47.6	-	48.8
	Bend angle	121 ± 11	-	106 ± 9
D74	AFE	0.274	-	0.271
	R_{da}	47.6	-	50.2
	Bend angle	94 ± 8	-	90 ± 8

^{a} Bend angles determined from the NMR structures of the Sry/DNA and Lef86/DNA complexes (Murphy et al., 2001; Love et al., 1995); AFE is an asymptotic FRET effect for 100 % complex; R_{da}, the distance between the labeled ends of DNA, is given in angstroms and bend angles in degrees; the samples were in 10 mM potassium phosphate, pH 6.0, 100 mM KCl.

5. DEPENDENCE OF THE PROTEIN-INDUCED DNA BEND ON THE FORCES INVOLVED IN BINDING

The great advantage of the FRET titration experiment is that it provides the binding isotherm in terms of FE, from which one can determine not only the protein-induced DNA bend but also the energetics of this process, that is, the DNA–protein association constant, K_a, and thus the Gibbs energy of association, $\Delta G^a = -2.3RT \log K_a$. Moreover, by performing such experiments in solution with different salt concentrations one can separate out the electrostatic and nonelectrostatic components of the Gibbs energy of association, which is important for understanding the forces involved in DNA binding and bending (Dragan et al., 2004c). Indeed, according to salt titration experiments the logarithm of the DNA association constant is a linear function of the logarithm of the salt concentration (Fig. 9.3A and B):

$$\log K_a = \log K_{nel}^a - N \log [\text{salt}] \qquad (9.5)$$

The first term on the right hand side of this equation results from nonelectrostatic (nel) interactions between DNA and protein, the second results from electrostatic effects associated with release of counterions, and N is the number of ions released from the DNA phosphates upon protein binding (Manning, 1978; Record, et al., 1978). When the salt concentration approaches 1 M the electrostatic term in this equation vanishes and $\Delta G^a = \Delta G_{nel}^a = -2.3RT \log K_a$ represents the nonelectrostatic component of the Gibbs energy of the complex formation. The electrostatic component of the Gibbs energy of complex formation is then: $\Delta G_{el}^a = \Delta G^a - \Delta G_{nel}^a$.

Figure 9.4A and B illustrate the dependence of the DNA bend angle induced by sequence specific (SS) and nonsequence-specific (NSS) HMG box DBDs on the electrostatic forces involved. One can see that in both cases the DNA bend depends on the value of the electrostatic force between DNA and protein but this dependence is especially steep in the case of NSS protein/DNA complexes. The mechanisms of this effect are discussed in (Dragan et al., 2004c).

6. FRET IN STUDYING SMALL PROTEIN-INDUCED DNA BENDS

In the case of small protein-induced bending of DNA, the change in the distance between donor and acceptor fluorophores placed on the ends of the DNA is small and, correspondingly, the change in FRET effect upon

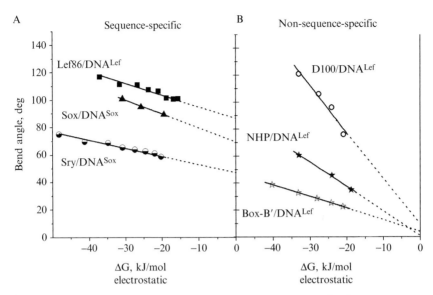

Figure 9.4 The dependence of the DNA bend angles induced by the (**A**) sequence specific (SS) DBDs and (**B**) by nonsequence specific (NSS) DBDs on the electrostatic component of Gibbs energy of binding. Bend angles measured from AFE values obtained in individual titrations of protein into DNA in 10 mM K-phosphate pH 6.0 at 20 °C at several concentrations of KCl.

protein binding is small. The efficiency FRET can be increased, however, by using the lever effect, realized by using U-shaped DNA constructs, as suggested by (Hillisch et al., 2001; Stuhmeier et al., 2000).

6.1. Construction of a U-shaped double bulged fluorophores labeled DNA

A U-shaped DNA construct containing a binding site at the center and "levers" attached to it through A5-bulges, which kink DNA almost perpendicularly, is shown in Fig. 9.5A. It is important to note that the U-shaped construct becomes planar when the length of the central target duplex (the distance between the two levers) is 9 bp (Stuhmeier et al., 2000). However, a central duplex of 9 bp is normally too small to accommodate a protein-binding site (usually 6–7 bp) and phase it in such a way to get in-plane DNA bending on binding protein. Extension of the central duplex can be done by insertion of 10-bp DNA, that is, a full turn of DNA, since that does not result in deviation of the U-shaped construct from planarity (Fig. 9.5A). Insertion of a 10-bp DNA segment opens out space to allow protein binding to the DNA but it also increases the distance between the ends of the arms by 34 Å, from 54 to 88 Å,

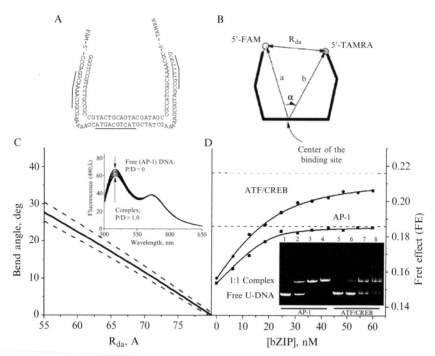

Figure 9.5 (A) Design and sequence of the double-labeled U-shaped DNA construct for FRET analysis of the small protein-induced DNA bends. Underlining shows insertions of the 10-bp DNA segments in the central part containing the protein-binding site and A4-tract in the arms of U-shaped DNA. (B) Geometric model of the DNA construct used for the quantitative calculation of the bend angle of the binding site from the measured R_{da} by Eq. (5). (C) The calibration function derived from the geometric consideration of the U-shaped DNA construct according to Eq. (5). Inset: Fluorescence spectra of the double-labeled U-DNA at the presence of different concentrations of GCN4 bZIP show the effect of energy migration from donor (FAM, spectrum at 520 nm) to acceptor (TAMRA, spectrum at 580 nm) (D) FRET isotherms of binding bZIP GCN4 to the AP-1 and ATF/CREB binding sites. The asymptotic values (AFE) of these isotherms (dotted lines), obtained by nonlinear least square fitting using Eq. (5), show that the bZIP dimer bends the ATF/CREB DNA more significantly than the AP-1 DNA. The concentration of DNA in both cases was 20 μM. Inset: Gel-shift experiment shows 1:1 binding of bZIP to U-DNA containing AP-1 (lanes 1–4) and ATF/CREB (lanes 5–8) sites. Free DNA is in lanes 1 and 5; lanes 2–4 and 4–8: the samples with increased bZIP/DNA ratio. For visualization of the bands intrinsic fluorescence of FAM and TAMRA chromophores was used (Dragan et al., 2004b).

thus decreasing the efficiency of energy transfer between the acceptor and donor fluorophores placed at the ends of the arms. To decrease the distance between the ends of the arms in the absence of bending, they were modified by inserting adenine-tracts (A4-tracts) that bend the arms inwards. According to the recent NMR analysis of a DNA duplex containing an A4-tract,

5′-GGCAAAACGG-3′, it is bent by 9° towards the minor groove (Barbic et al., 2003). This sequence was inserted into both arms in such a way as to bend the arms in the plane of the construct. Insertion of these A4-tracts should decrease the R_{da} distance by 8 Å. With these modifications, the R_{da} distance is expected to be 80 Å and FRET experiments have shown that this distance is indeed very close to the calculated 80 Å and decreases significantly upon GCN-bZIP binding to its recognition site in the base of the construct.

The geometric model of the U-shaped DNA construct shown in Fig. 9.5B permits transformation of the FRET measured dye-to-dye distance (R_{da}) into bend angle of the DNA binding site. According to this model the distances between the dyes, the center of the binding site (a and b), and the angle (α) between them is expressed by the equation:

$$(R_{da})^2 = a^2 + b^2 - 2ab \cos \alpha \tag{9.6}$$

Taking into account the known position of the dyes relative to the DNA (Hillisch et al., 2001), the length of the arms, the central duplex of DNA and the position of the binding site, we estimated that $a \cong b = 62 \pm 3$ Å. The protein-induced bend in the DNA, $\Delta\alpha$, can then be expressed by the equation:

$$\Delta\alpha = a \cos (1 - R_{da}^2)°/2a^2 - a \cos (1 - R_{da}^2/2a^2), \tag{9.7}$$

where $(R_{da}) = 80$ Å is a model distance between the dyes in the absence of protein and R_{da} is the measured dye-to-dye distance. This function, for the case when $a = 62$ Å, is plotted in Fig. 9.5C as a solid line. The dashed lines show the range of uncertainty in the calculated bend angle and it is notable that this range is quite small.

6.2. FRET titration of the U-shaped DNA

The U-shaped DNA is quite a long molecule and potentially has many nonspecific binding sites for protein, but only one specific target site. Nevertheless, provided the protein affinity for the target site is high, only 1:1 complexes are formed at low concentrations of protein (Fig. 9.5D). This was demonstrated for the example of GCN4 bZIP complex with the AP-1 and ATF/CREB binding sites (Dragan et al., 2004b). The binding isotherms for GCN4 bZIP with the U-DNA are shown in Fig. 9.5D. The FE parameter for free DNA is 0.1534 ± 0.0006. The fittings of the FRET titration data shown in Fig. 9.5D give two main parameters of binding and bending: the association constant (K_a) and the AFE (Table 9.2). The values of the association constants are in accordance with those determined by fluorescence anisotropy measurements (Dragan et al., 2004a) and this agreement confirms that binding of

Table 9.2 FRET and bending parameters of the target U-DNAs upon binding with GCN4-bZIP cross-linked dimer

U-shaped DNA	K^a, M^{-1}	FE	FRET Efficiency (E)	R_{da}, Å	Bend angle, deg
Free AP-1 DNA	–	0.1534 ± 0.0006	0.1163 ± 0.0005	70.1 ± 0.5	11 ± 1
Complex: bZIP:AP-1 DNA	6.7 × 10⁸	0.1857 ± 0.0007^a	0.1295 ± 0.0006	61.8 ± 1.5	20 ± 2
Free ATF/CREB DNA	–	0.1562 ± 0.0006	0.1252 ± 0.0006	69.1 ± 0.5	12 ± 1
Complex: bZIP:ATF/CREB DNA	15.6 × 10⁸	0.2140 ± 0.0007 ^{a}	0.3100 ± 0.0007	57.1 ± 1.5	25 ± 2

^{a} Asymptotic values of FRET Effect (AFE).
Buffer: 10 mM potassium-phosphate, 100 mM KCl, (pH6.0); 20 °C.

GCN4 bZIP to the AP-1 and ATF/CREB sites of the U-DNAs is the same as to short linear duplexes, that is, there are no additional constraints on binding in consequence of using the U-shaped construct. Increase of the FE upon titration demonstrates a decrease of R_{da} (the distance between the fluorophores on the ends of the DNA), that is, the binding site bends towards the protein. Calculations performed using obtained values of AFE (Table 9.2) and Eq. (6) give bend angles of $20 \pm 2°$ and $25 \pm 2°$ for the GCN4 bZIP complexes with the AP-1 and ATF/CREB sites, respectively.

7. Conclusions

The advantage of the FRET technique in studying protein–DNA complexes is that it is a solution technique that provides information both on the energetics of formation of these complexes in solution and the protein-induced DNA deformation. Furthermore, this can be done under a variety of conditions (e.g., temperatures and solvents). Using a U-shaped DNA construct with fluorophores on each end, one can also use FRET to determine small DNA bend angles, the functional meaning of which may be just as important as of large bends.

ACKNOWLEDGMENT

The financial support of NIH GM48036–12 and NSF 0519381 are gratefully acknowledged.

REFERENCES

Barbic, A., Zimmer, D. P., and Crothers, D. M. (2003). Structural origins of adenine-tract bending. *Proc. Natl. Acad. Sci. USA* **100,** 2369–2373.

Clegg, R. M. (1992). Fluorescence resonance energy-transfer and nucleic-acids. *Methods Enzymol.* **211,** 353–388.

Dragan, A. I., Klass, J., Read, C., Churchill, M. E. A., Crane-Robinson, C., and Privalov, P. L. (2003). DNA binding of a non-sequence-specific HMG-D protein is entropy driven with a substantial non-electrostatic contribution. *J. Mol. Biol.* **331,** 795–813.

Dragan, A. I., Frank, L., Liu, Y., Makeyeva, E. N., Crane-Robinson, C., and Privalov, P. L. (2004a). Thermodynamic signature of GCN4-bZIP binding to DNA indicates the role of water in discriminating between the AP-1 and ATF/CREB sites. *J. Mol. Biol.* **343,** 865–878.

Dragan, A. I., Liu, Y., Makeyeva, E. N., and Privalov, P. L. (2004b). DNA-binding domain of GCN4 induces bending of both the ATF/CREB and AP-1 binding sites of DNA. *Nucleic Acids Res.* **32,** 5192–5197.

Dragan, A. I., Read, C. M., Makeyeva, E. N., Milgotina, E. I., Churchill, M. E., Crane-Robinson, C., and Privalov, P. L. (2004c). DNA binding and bending by HMG boxes: Energetic determinants of specificity. *J. Mol. Biol.* **343,** 371–393.

Förster, T. (1946). Energiewanderung und fluoreszenz. *Naturwissenschaften* 166–175.

Hillisch, A., Lorenz, M., and Diekmann, S. (2001). Recent advances in FRET: Distance determination in protein-DNA complexes. *Curr. Opin. Struct. Biol.* **11,** 201–207.

Kerppola, T. K. (1996). Fos and Jun bend the AP-1 site: Effects of probe geometry on the detection of protein-induced DNA bending. *Proc. Natl. Acad. Sci. USA* **93,** 10117–10122.

Love, J. J., Li, X., Case, D. A., Giese, K., Grosschedl, R., and Wright, P. E. (1995). Structural basis for DNA bending by the architectural transcription factor LEF-1. *Nature* **376,** 791–795.

Manning, G. S. (1978). The molecular theory of polyelectrolyte solutions with applications to the electrostatic properties of polynucleotides. *Q. Rev. Biophys.* **11,** 179–246.

Masse, J. E., Wong, B., Yen, Y. M., Allain, F. H. T., Johnson, R. C., and Feigon, J. (2002). The *S. cerevisiae* architectural HMGB protein NHP6A complexed with DNA: DNA and protein conformational changes upon binding. *J. Mol. Biol.* **323,** 263–284.

Murphy, E. C., Zhurkin, V. B., Louis, J. M., Cornilescu, G., and Clore, G. M. (2001). Structural basis for SRY-dependent 46-X,Y sex reversal: Modulation of DNA bending by a naturally occurring point mutation. *J. Mol. Biol.* **312,** 481–499.

Read, C., Cary, P. D., Crane-Robinson, C., Driscoll, P. C., Carillo, M. O. M., and Norman, D. G. (1995). The structure of the HMG box and its interaction with DNA. *In* "Nucleic Acids and Molecular Biology." (F. Eckstein and D. M. J. Lilley, eds.), pp. 222–249. Springer-Verlag, Berlin, Heidelberg.

Record, M. T., Jr., Anderson, C. F., and Lohman, T. M. (1978). Thermodynamic analysis of ion effects on the binding and conformational equilibria of proteins and nucleic acids: The roles of ion association or release, screening, and ion effects on water activity. *Q. Rev. Biophys.* **11,** 103–178.

Stuhmeier, F., Hillisch, A., Clegg, R. M., and Diekman, S. (2000). Fluorescence energy transfer analysis of DNA structures containing several bulges and their interaction with CAP. *J. Mol. Biol.* **302,** 1081–1100.

CHAPTER TEN

Fluorescent Pteridine Probes for Nucleic Acid Analysis

Mary E. Hawkins

Contents

1. Introduction	202
2. Pteridine Analog Characteristics	205
2.1. Intensity and spectral shifts	205
2.2. Stability	206
2.3. Participation in base pairing	207
3. Procedures for Oligonucleotide Synthesis with Pteridine Analogs	208
3.1. Conservation of phosphoramidite	208
3.2. Deprotection	209
3.3. Purification procedures	210
4. Characterization of Pteridine-Containing Sequences	210
5. Applications	211
5.1. Using fluorescence intensity changes	211
5.2. Fluorescence characterization of A-tracts using 6MAP	212
5.3. Temperature-dependent behavior of A-tract duplexes	214
5.4. Bulge hybridization	216
5.5. Anisotropies of pteridine-containing sequences to examine protein binding	218
5.6. Lifetimes, steady-state and time-resolved anisotropies of 3MI- and 6MI-containing sequences	219
5.7. Probing the hairpin structure of an aptamer using 6MI	221
5.8. Two-photon excitation of 6MAP	223
5.9. Single molecule detection of 3MI	226
6. Summary	227
Acknowledgments	228
References	228

Abstract

This chapter is focused on the fluorescent pteridine guanine analogs, 3MI and 6MI and on the pteridine adenine analog, 6MAP. A brief overview of commonly

National Institutes of Health, National Cancer Institute, Bethesda, MD 20892 used methods to fluorescently label oligonucleotides reveals the role the pteridines play in the extensive variety of available probes. We describe the fluorescence characteristics of the pteridine probes as monomers and incorporated into DNA and review a variety of applications including changes in fluorescence intensity, anisotropies, time resolved studies, two photon excitation and single molecule detection.

1. INTRODUCTION

As we begin to tap into the vast potential for biochemical applications using sequences of DNA as probes and binding ligands, it will be very useful to understand the structure of these molecules and how a specific sequence of DNA can affect its function. It is clear that each specific sequence contributes to the features of a given locus in terms of flexibility or rigidity and consequently its association with other molecules. The overall function of DNA is dependent on these characteristics as it conveys the information needed for many vital processes in the cell. Subtle variations contribute to interactions between DNA and other factors, made even more complex by constantly changing conditions in response to alterations in the environment. They are all keys to the function of the DNA.

Fluorescence techniques are a natural choice to study these subtle variations because they provide information on the status of the DNA through measurement of fluorescence intensity, spectral shifts, lifetimes, steady-state anisotropies, and time-resolved anisotropies. It would be ideal to use the inherent fluorescence of DNA to study its structure and flexibility in a totally unaltered state, without introducing external probes of any kind. Although the very low fluorescence quantum yields of the native bases make it very challenging to do this, Georghiou et al. (1996) have used thymine fluorescence in poly(dA)–poly(dT) and in a dA–dT 20-mer in a structural study. This work reveals some features of the natural status of these bases in solution, the first to examine the movement of sequences using time-resolved intrinsic fluorescence anisotropy. It also provides us with an unadulterated look at the native features of the DNA for the sequences they studied. For most studies, however, the limitations that result from the extremely low yields of the native bases require that we find other ways to analyze these systems.

One way to achieve enhanced fluorescence labeling of DNA is to add an extrinsic fluorescent molecule to the structure. The introduction of any non-native molecule will have an impact on the structure through changes in the electronic field in its immediate environment. Some probes will have less negative impact than others simply because of their chemical structures. On the grosser level, we can assess the impact of a probe's structure on a sequence by measuring the melting temperature (T_m) of the duplex with and without the probe present. Similar T_m measurements lead us to assume that at least the tertiary structure likely is intact. On a more subtle level, however, any change in the chemical nature of a sequence changes its character enough to potentially modify recognition by interacting molecules such as a highly selective enzyme. There are many systems that will tolerate addition of a fluorescent probe through incorporation, tagging, or intercalation, however, care should be used in designing the study and the proper controls must be carefully considered.

There are several ways to provide DNA with fluorescence. One is through saturation of the DNA with an intercalating agent. This method allows a view of the global status of the entire molecule as the probe distributes itself throughout the entire sequence. One example of the intercalation method uses ethidium bromide as a probe which naturally stacks between the bases of the sequence. Global dynamics of sequences have been studied using this technique. For a review, see Schurr and Fujimoto (1988). As with any extrinsic labeling system, there can be unwanted effects introduced by these fluorophores. Intercalating agents have been found to cause distinct changes in molecular elasticity and can extend and frequently partially unwind DNA double strands, so this must be considered when interpreting results (Sischka et al., 2005). For many studies, however, this subtlety does not pose a problem and a great deal of valuable information can be obtained using this approach.

Another technique employs specific labeling with a linker-attached fluorescent molecule, usually at the terminal position of a sequence. This method has been used extensively and very successfully for labeling molecules. There are many studies using fluorophores which are attached to a sequence using carbon chain linkers of various lengths. Some of these linker-attached probes have been found to naturally have individual character that results in variations on the data (Unruh et al., 2005b; Vamosi et al., 1996). Fluorescein, a very commonly used probe, is often linked to the terminus of a sequence through a carbon linker. Hill and Royer (1997) have found that, in some cases, only about 15% of the linker-attached fluorescein undergoes rotation coupled to global motion of the DNA. In another study by Unruh et al. (2005a) the fluorescence properties of fluorescein, Texas Red, and Tetramethylrhodamine (Tamra) were compared in identical sequences. They determined that, of this group, Texas Red was the most reflective of global DNA motion. Experiments using fluorescein confirmed the findings of Hill and Royer (1997), and Tamra was found to be sensitive to the environmental which was seen to be a complicating factor for this type of measurement. Texas Red behaves uniquely because it appears to form an association with the bases of the sequence in the major groove. However, while this does much to stabilize it and permit measurement of global rotation of the DNA, it may actually interfere with the natural structure of the sequence and prevent native interactions in its vicinity.

The introduction of carbon linkers of varying lengths permits placement of the probe in areas where it might otherwise be structurally obtrusive.

A third method is through the incorporation of a nucleoside analog which is attached to the sequence through the same deoxyribose linkage as native DNA, preferably using automated DNA synthesis. This method has several advantages beyond the potential for visualization of the sequence. One is that the process of incorporation using automated DNA synthesis insures that the probe is covalently attached in every molecule of DNA. Also, there is no postsynthesis step for incorporation of the label, which can be very costly in terms of time and materials. The placement of the probe in the sequence permits observance of local changes that occur as it reacts with other molecules. A well known probe of this type, the adenine analog, 2-amino purine (2-AP), has been used extensively in studies of DNA base stacking and base pairing. One study examines properties of the TATA sequence, important because it is a part of a sequence involved in transcription initiation (Rai et al., 2003). Xu and others have focused heavily on defining the most fundamental interactions between 2-AP and its neighboring bases (Davis et al., 2003; Xu and Sugiyama, 2006). The dynamics of DNA have been studied through fluorescence lifetime and anisotropy decay kinetic measurements in 2-AP-containing sequences (Larsen et al., 2001; Lee et al., 2007; Ramreddy et al., 2007). There are also a number of studies using pyrrolocytosine as a cytosine analog (Berry et al., 2004; Dash et al., 2004; Hardman and Thompson, 2006; Liu and Martin, 2001). New probes are constantly being developed (for a review of fluorescent nucleotide analog probes, see Rist and Marino, 2002). A survey of these probes shows that in some instances, probes that are insensitive to the sequence environment are best, while in others, a probe that reports on the immediate environment is preferred. The wide selection that is becoming available makes it an interesting task to select one that best fits the needs of a given system.

In this chapter, we will describe the features of the pteridine nucleoside analogs. In terms of size and structure, the highly fluorescent pteridine analogs, 3MI, 6MI, and 6MAP, are similar to the native purines, guanine, and adenine. This allows the pteridine molecule to be substituted for a purine and incorporated into an oligonucleotide through a deoxyribose linkage (Fig. 10.1).

A resulting feature of this incorporation is that, within an oligonucleotide, the pteridine ring system is intimately associated with neighboring bases and reflects changes in these associations through changes in fluorescence (Hawkins, 2001, 2003). The pteridines are significantly quenched primarily through base stacking interactions (and to a lesser degree through base pairing) and the degree of quench is dependent on the sequence of the DNA in the vicinity of the probe. Because of the high quantum yields of the monomer forms (listed below) nanomolar concentrations of pteridine-containing oligonucleotides can be detected easily (Hawkins et al., 1995).

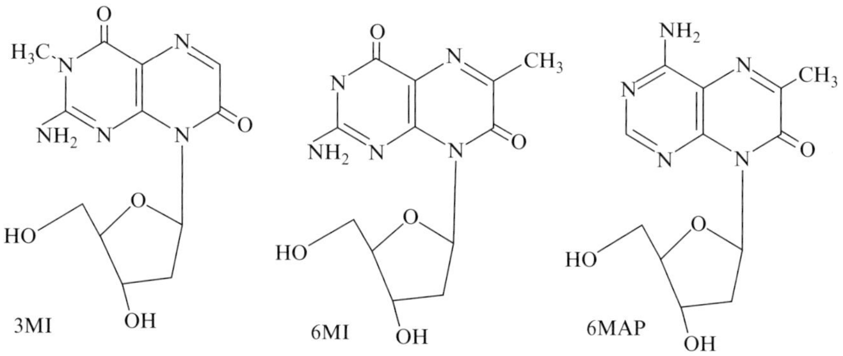

Figure 10.1 Chemical structures of 3MI, 6MI, and 6MAP.

One can engineer a pteridine-containing sequence that is surrounded with purines for a quenched signal or surrounded with pyrimidines for a brighter signal. The absorption and emission maxima of these probes are red shifted from the maxima of native DNA making them useful for DNA studies (Hawkins et al., 1997).

Although it is clear that there are differences between the pteridine analogs and the corresponding native purines, the ability to position them as an integrated part of the DNA can provide a view of the subtle changes that accompany distortions in tertiary structure brought about by environmental changes. The associations between the incorporated pteridine analog and neighboring bases in a sequence provide information that is directly linked to structural changes in the DNA.

In this chapter, the properties of these pteridine nucleoside analogs are discussed to assist the researcher with the use of these probes. Procedures for synthesis and purification of pteridine-containing oligonucleotides are described. Examples are given of applications using fluorescence intensity changes, anisotropy, time-resolved studies, single molecule detection, and two-photon counting.

Note: The probes 3MI, 6MI, and 6MAP are commercially available through Fidelity Systems, Inc., Gaithersburg, MD (301) 527-0804; fsi1@fidelitysystems.com.

2. Pteridine Analog Characteristics

2.1. Intensity and spectral shifts

In the monomer form (unincorporated), the quantum yields for 3MI, 6MI, and 6MAP are 0.88, 0.70, and 0.39, respectively. In general, the adenine analog, 6MAP, is quenched more severely when incorporated into an oligonucleotide than 3MI or 6MI, with quantum yields of 6MAP-containing single strands ranging from <0.01 to 0.04 depending on the neighboring sequence (Hawkins et al., 2001). The single-stranded sequence-dependent variation in quantum yield for the guanine analog, 3MI, ranges from 0.03 to 0.29 (Hawkins et al., 1995). Although 6MI seems to follow the same quenching patterns as 3MI, several notable exceptions have been found to show strong increases in fluorescence intensity upon duplex formation. The sequences involved in this anomaly are currently under investigation. In a recent study (described below) time-resolved anisotropy data suggests that 6MI may be mechanically coupled with neighboring bases to a larger extent than 3MI.

Variations in quenching patterns seen in pteridine probes incorporated into oligonucleotides have underscored the complexity of nucleic acid sequences. Driscoll et al. (1997) have determined that within an oligonucleotide, the distance between a 3MI molecule and the 5′-terminus does not seem to impact the fluorescence. A 3′-purine neighbor (especially adenine) appears to induce more quenching than a 5′-purine neighbor. The greatest quenching for 3MI is seen with adenines surrounding the probe and the lowest quench appears to occur when a probe is neighbored by thymine residues. The source of these effects was primarily attributed to *static* quenching mechanisms (or perhaps quasistatic, as studies of the first 50 ps have not been done).

Changes in pH have been shown to cause up to 10 nm shifts in the emission spectra of 6MI but this is seen to a lesser degree (~1–3 nm) in emission spectra of 3MI. Similarly spectral shifts are also seen when comparing 6MI monomer form to those of 6MI incorporated into single and double strands. Seibert et al. (2003) have compared the effects of pH on the ground state and lowest energy excited equilibria for 6,8-dimethylisoxanthopterin (6,8-DMI) an analog to 6MI in which the deoxyribose moiety is replaced with a methyl group (ruling out the possibility of interactions with the sugar). They found that the absorbance and fluorescence emission spectra are shifted to lower energies as the pH is increased and determined pKa values of 8.3 for absorbance and 8.5 for fluorescence. This group also demonstrated the involvement of the 3-position of the pteridine in this process by comparison of pH-dependent iodide ion quenching of 6,8,-DMI fluorescence with pH-independent absorption and fluorescence of 3,6,8-trimethylisoxanthopterin (an analog in which the 3-protonation site is blocked by a methyl group).

2.2. Stability

We have seen no evidence of degradation of the pteridine probes when exposed to ambient light. As one might expect, however, the phosphoramidite forms of the pteridines are vulnerable to hydrolysis in the same way that the phosphoramidite forms of native bases are. To maintain pteridine phosphoramidites for long periods (years), we store them with desiccant in a −80 °C freezer.

Of these three probes, 3MI displays the greatest stability during fluorescence analysis. Using time-based acquisition (cuvette exposed to a spot-containing UV power <~200 uW), we have monitored the fluorescence emission of a 3MI-containing oligonucleotide exposed at the excitation maximum for 2 h at 37 °C and observed no detectable change (loss) in the fluorescence intensity that could be an indication of photolysis (Hawkins et al., 1995). Slight degradation of the fluorescence emission signals for 6MI- and 6MAP-containing oligonucleotides have been observed during time-based acquisition exposed at the excitation maximum at 37 °C. Three successive scans of oligonucleotides containing 6MI or 6MAP revealed no detectable loss in fluorescence intensity, however, which is an indication that short term exposure is less damaging (M. E. Hawkins, unpublished results).

All three of these pteridine probes have shown remarkable stability through the caustic conditions generated during automated DNA synthesis. As a quality control experiment, oligonucleotides containing each of the three probes were completely digested using P1 nuclease (*Penicillium citrinum*, Boehringer Mannheim Biochemica). The expectation for a stable probe would be that the fluorescence of the probe would be recovered by removing it from the quenched environment of the surrounding bases. For each probe, the ratio of the integral of the fluorescence emission spectrum of the oligonucleotide prior to the digestion to that of the corresponding digested product was equivalent to the ratio of the quantum yield of the probe-containing oligonucleotide to the quantum yield of the monomer form of the corresponding probe. This procedure is described in more detail in Section 4. The findings demonstrate that each probe remains intact and that the quench in signal originates from interactions with neighboring bases and not from degradation caused by exposure to chemicals during synthesis (for 6MAP, Hawkins et al., 2001; for 3MI and 6MI, M. E. Hawkins, unpublished results).

2.3. Participation in base pairing

For 3MI, it is apparent that the methyl group in the 3-position should block formation of a hydrogen bond between the incorporated probe and cytosine. Melting temperatures (T_ms) of 3MI-containing oligonucleotides paired with a complementary strand (where 3MI is paired with cytosine) are similar to T_ms of a *single base pair mismatches* in the identical positions, and the degree of T_m depression is sequence dependent (Hawkins et al., 1995).

6MI, however, displays evidence of base pairing with cytosine. The melting temperatures of 6MI-containing oligonucleotides paired to complementary strands are almost identical to those of control oligonucleotides of identical sequences. The emission maximum of 6MI also undergoes a substantial shift (~10 nm to the red) at pH 8.0 as compared to that measured at pH 6.0. A comparable shift to the blue is seen when 6MI transitions from monomer to double strand at pH 7.2. These spectral shifts are suggestive of a change in status of the 3-position proton which is related to duplex formation.

6MAP also displays evidence of hydrogen bonding (with thymine) as determined by T_m measurements. To further investigate this, we measured the T_ms of 6MAP-containing strands paired with each of the four native bases: adenine, thymine, cytosine, and guanine, as a pairing partner in the complementary strand. The results clearly show that 6MAP pairing with thymine is the most stable (Hawkins et al., 2001).

3. Procedures for Oligonucleotide Synthesis with Pteridine Analogs

3.1. Conservation of phosphoramidite

Automated DNA synthesis is a fairly routine procedure and because most of the reagents are relatively inexpensive, the standard synthesis protocols are quite liberal in use of reagents. There are some measures that can conserve the amount of pteridine phosphoramidite that is consumed during a synthesis. Most of our oligonucleotides are made using an ABI (Applied Biosystems, Foster City, CA) DNA synthesizer and the first point where the phosphoramidite may be unnecessarily lost is in the standard "bottle change" procedure. We dissolve the probe by hand using a freshly opened bottle of low water acetonitrile (ACN) at a ratio of 10 μl ACN per mg of probe phosphoramidite. For volumes of phosphoramidite solution under 1 ml, we place a smaller vial (12 × 32 mm P/N 5185-5821, Agilent Technologies, Wilmington, DE) containing the pteridine phosphoramidite inside one of the standard phosphoramidite bottles. The tubing must be trimmed for this application so that it just fits into the bottom of the inserted vial. After placing the phosphoramidite in the bottle, we use a modified "user program" to install the probe-containing vial on the synthesizer as advised by the company (ABI, Foster City, CA) which has a reduced phosphoramidite "flush to waste" time from "2" to "1 s."

We use 200 nM low volume columns (LV200, ABI, Foster City, CA) for any synthesis involving the probes. The low volume column has a frit that reduces the bed volume of the column thus requiring lower volumes of each reagent to accomplish each wash. The appropriate procedure to accommodate these columns on the synthesizer can be obtained from the company. The combination of these two modifications has increased the number of incorporations we can obtain from a given volume of phosphoramidite. We also avoid using the "Begin Procedure" when the pteridine phosphoramidite is on the machine because this procedure involves flushing all lines with fresh reagents. Once the probe is on the machine sequences are made without delay so that the begin procedure is not needed. Other manufacturers of synthesizers may have suggestions of their own on how to reduce unnecessary losses of probe phosphoramidites.

3.2. Deprotection

Both 3MI and 6MAP are deprotected in the standard manner through incubation at 55 °C for 15 h in concentrated ammonium hydroxide.

The 6MI phosphoramidite has a paranitrophenyl group to block unwanted chemistry at the 3-position during DNA synthesis and because this group is not removed during standard deprotection using ammonium hydroxide (W. Pfleiderer, personal communication), additional steps must be taken to remove it. Oligonucleotides are not very soluble in 10% 1,8-diazabicyclo(5.4.0)undec-7-ene (DBU) in acetonitrile, the reagent required for this procedure, therefore, 6MI must be deblocked while the oligonucleotides are still attached to the column. When synthesizing 6MI-containing oligonucleotides select the "manual end procedure" and mark the identity of each sequence on the appropriate column. The oligonucleotides can be manually deprotected as described below.

After the synthesis is finished (and ready for "manual end procedure") set up each CPG column with an empty 3 ml luer lock syringe on each side removing the plunger from one syringe. Add 425 μl ACN to the plunger-less syringe (held upright) and pull it through to the opposing side using the plunger. Flush it through to the plunger-less side, add 75 μl DBU to the ACN and pull it through again. Insert the other plunger and use it to flush the samples back and forth at least three times. Put in a dark place at room temperature for 5 h. Flush the solution back and forth about once an hour.

After the incubation, pull the solution through and save it (eluate-1). Add 500 μl ACN to the column and flush it through several times. Add this wash to eluate-1 of each corresponding sample. Save the CPG columns for the next step. Monitoring the UV absorption of samples from each step has revealed that a substantial amount of oligonucleotide is released from the column during the deprotection steps. The eluate-1 samples may be evaluated to determine whether it is necessary to retrieve the oligonucleotide from them before continuing. If the optical density indicates the presence of oligonucleotide in these samples, evaporate the DBU/ACN mixture (eluate-1) using a Speed-vac. Perform an ethanol precipitation on the resulting DBU/ACN oily residue saving the pellet for the next step. Add 1 ml aliquots of NH_4OH to each sample (CPG column) using two syringes as described above, and allow them to stand for 1 h. Add the NH_4OH eluate (eluate-2) from each column to the corresponding DBU/ACN eluate-1 pellet, mix thoroughly and heat for 15 h at 55 °C. Speed vac and purify as usual.

3.3. Purification procedures

Oligonucleotides are purified by 20% denaturing polyacrylamide gel electrophoresis (19:1 acrylamide:bis). The oligonucleotide band, visualized using UV shadowing, is excised and extracted from the gel slice using an electroelution device (Schleicher & Schuell, Keene, NH). HPLC purification techniques may also be used. Following ethanol precipitation, the oligonucleotides are stored in a −80 °C freezer.

4. Characterization of Pteridine-Containing Sequences

The absorption of the pteridine at 330–350 nm should be detectable using a UV–VIS spectrophotometer and in a 20-mer should be about ten times less than the absorbance at 260 nm (representing mostly native bases). Naturally, this ratio will vary depending on the length and the sequence of the strand.

The quantum yield of the probe-containing oligonucleotide can be measured using quinine sulfate (QS) as a standard as previously described (Hawkins *et al.*, 1997).

When calculating the extinction coefficient of a probe-containing sequence, we use a method described by Eckstein (1991) with the extinction coefficients of 3MI as measured in methanol as log ϵ at 216 (254) 292 and 350 nm equal to 4.54 (3.69) 3.93 and 4.13, respectively (numbers in parentheses are for a shoulder) (W. Pfleiderer, personal communication).

The degree to which the presence of a probe may disrupt double-strand formation may be assessed from the melting temperature (T_m). We typically measure T_ms in 10 mM Tris at pH 7.5 in the presence of 10 mM NaCl.

One way to verify that the probe has not been degraded during the exposure to caustic chemicals during DNA synthesis is to totally digest the product strand and compare the fluorescence with that which would be expected for an equal amount of probe monomer. After determining the relative quantum yield of the sequence, set up a sample (100 μl volume) to digest with 3 Units of P1 Nuclease (*P. citrinum*, Boehringer Mannheim Biochemica, Germany). Scan the sample before digestion and then again after total digestion (overnight incubation). The ratio between the scans of the pre- and postdigest should approximately equal the ratio of the relative quantum yields of the sequence and the monomer. When this test was performed on oligonucleotides containing each of the three probes, the ratios obtained were almost identical to the values obtained from the starting sequence and the monomer form of the corresponding probe.

5. Applications

5.1. Using fluorescence intensity changes

Because the fluorescence properties of the incorporated pteridine nucleoside analogs are so heavily impacted by interactions with neighboring bases, any event that changes these associations can directly influence the fluorescence intensity of the probe. One of the simplest ways to monitor these interactions is through changes in fluorescence intensity.

The first experiment to take advantage of this effect using the pteridine probes was designed to measure a cleavage activity of the retrovirally coded protein, human immunodeficiency virus-1 (HIV-1) integrase (IN) (Hawkins et al., 1995). This protein functions in a stepwise manner resulting in the integration of the HIV-1 genome into the host cell DNA with the first step being the cleavage of a specific dinucleotide from each end of the HIV genome. The pteridine probe, 3MI, was incorporated into a model substrate at a specific cleavage site known to be processed by the integrase. As the protein cleaves at the specific site, the probe is removed from the base stacked environment and the increases in fluorescence intensity can be monitored in real time. This is a distinct advantage over the previously used P-32-based assay using polyacrylamide gels and autoradiography for analysis.

Another example of using fluorescence intensity changes is the use of 3MI to assess the DNA-binding characteristics of the nonspecific multifunctional histone-like DNA-binding protein, HU, by Wojtuszewski et al. (2001) In this study, a change in fluorescence intensity was related to a change in base stacking due to local unwinding or bending of the DNA helix that occurs in the presence of HU concentrations below 4 μM. When measured as a function of HU concentration, the fluorescence intensity increase demonstrates saturable binding. Of the 3MI-containing strands examined, two 34-mers displayed a fluorescence intensity change while a 13-mer showed no change. This result was interpreted by the authors to indicate an absence of induced bending in the HU-binding interaction with the 13-mer. Binding stoichiometry, as determined by fluorescence intensity and confirmed by analytical ultracentrifugation, indicated that either the binding site size or the mode of binding is altered between the longer and shorter strands. The fluorescence intensity-based data was combined with changes in the anisotropy of the 3MI-containing oligonucleotides (discussed briefly below) to assess the binding characteristics.

Myers et al. (2003) have used 6MI-containing sequences to map specific binding between single-stranded DNA and unwinding protein (UP1), a subunit of heterogeneous ribonucleoprotein A1, a protein strongly involved in RNA processing. These researchers designed the 6MI-containing substrate sequences after examining the structure of UP1 bound to the heterogeneous telomeric repeat sequence d(TTAGGG)$_n$ as determined by X-ray diffraction. They substituted 6MI for guanine in two different positions, TR2-6F where 6MI is expected to become unstacked when bound to protein, and TR2-11F, where it is expected to remain stacked when bound. X-ray diffraction studies of the 6MI-containing sequences complexed to UP1 demonstrated that the presence of 6MI in the sequences did not significantly change the tertiary structure of the complex. The changes in fluorescence intensity seen upon binding these sequences to the UP1, three- to fourfold for TR2-6F and 1.2-fold for TR2-11F, confirmed structural predictions based on the crystallographic data. The magnitude of change in fluorescence intensity seen with 6MI in the unstacking position, TR2-6F, permitted the determination of binding isotherms under tight-binding conditions with protein at 10 nM. These authors demonstrate the usefulness of 6MI as a probe for determination of equilibrium binding between protein and nucleic acids.

5.2. Fluorescence characterization of A-tracts using 6MAP

In a study by Augustyn et al. (2006), the fluorescence properties of the adenine nucleotide analog, 6MAP, have been investigated within the context of DNA "A-tracts," DNA sequences consisting of two or more adjacent adenine residues. When A-tracts are repeated in phase with the DNA helix, these sequences exhibit curvature or bending. Gel mobility measurements showed previously that A-tracts of six residues exhibit the maximum amount of curvature (Crothers and Shakked, 1999). The fluorescent analog, 6MAP, was substituted for adenine residues in five oligonucleotides both to asses the fluorescence properties of 6MAP in the context of these sequences and to probe the relative differences in structure of these sequences. The sequences were (see Table 10.1) characterized in both single- and double-stranded forms.

The fluorescence signal of 6MAP is quenched upon incorporation into a single strand by over 95% in all five oligonucleotides, relative to free monomer.

Table 10.1 Oligonucleotides containing 6MAP

Name	Sequence[a]	Φ_{rel} (SS)[b]	Φ_{rel} (duplex)[b]
A3-1	5'-CGCAFATTTCGC-3'	0.017	0.013
A3-2	5'-CGCAAFTTTCGC-3'	0.036	0.019
T3-1	5'-CGCTTTAFACGC-3'	0.022	0.012
T3-2	5'-CGCTTTFAACGC-3'	0.033	0.027
AT-1	5'-CGCATFTATCGC-3'	0.039	0.015

[a] F denotes the position of the fluorophore in the oligonucleotide strand. For control sequences, F = A.
[b] Φ_{rel} represents quantum yields relative to that of the 6MAP monomer (0.39) as measured relative to quinine sulfate (Hawkins et al., 2001).

The adjacent nucleotide neighbors strongly influence the degree of quenching, with the lowest quantum yields observed for the A3-1 and T3-1 oligonucleotides where 6MAP is situated between two adenine residues (Table 10.1). For the case of A3-2, T3-2, and AT-1 oligonucleotides, with at least one pyrimidine neighbor adjacent to 6MAP, the fluorescence intensity is not as quenched. In general, the emission maximum shifts seen upon incorporation into the single strand are relatively small.

Duplex formation further reduces 6MAP fluorescence intensity. The AT-1 oligonucleotide containing 6MAP has the largest relative quantum yield as a single-stranded molecule and experiences the greatest fluorescence quench upon duplex formation (62%). The A3-1 and T3-1 duplexes, in which 6MAP is situated between two adenines, have the lowest quantum yields in the single strand and the A3-1 duplex displays the least fluorescence quenching upon duplex formation. Nearest neighbor interactions are not sufficient to explain the relative quantum yields of the duplexes and undoubtedly, other factors, such as local structure of the duplex, have an impact. An interesting example of the effect of local structure can be seen in a comparison of T3-2 and A3-2 duplexes, where 6MAP is located in between adenine and thymine residues. In the single-stranded forms, the two dodecamers have similar fluorescence quantum yields; however, upon duplex formation the quantum yield of the A3-2 duplex decreases substantively compared to the T3-2 duplex. Since 6MAP has the same neighboring bases in both duplexes, the difference in quenching upon duplex formation most likely arises because of the change in orientation of the neighboring bases, AFT in A3-2 versus TFA in T3-2 (Table 10.1).

Duplex formation also results in a shift of peak emission to a shorter wavelength for each of the sequences studied. The A3-1 and AT-1 duplexes exhibit the largest shifts, -14 and -13 nm, respectively. Emission shifts appear to be more reflective of local nucleic acid duplex structure rather than local environment caused by neighboring residues (base stacking), since the A3-1 and T3-1 duplexes have the same neighboring residues and display quite different shifts; A3-1 (-14 nm) and T3-1 (-8 nm). Similarly, the A3-2 duplex, with 6MAP located at the $3'$-end of the A-tract, exhibits the smallest shift in emission maximum, possibly reflecting local structure close to the probe.

Nearest neighbor interactions have a greater influence on the stability of duplexes containing 6MAP. Thermal melting temperatures (T_m) as measured by UV–VIS absorption spectroscopy reveal that 6MAP-containing sequences are destabilized by 2–5 °C as reported previously (Hawkins et al., 2001). The greatest perturbations in stability ($\Delta\Delta H = 5.5$ kcal/mol as determined from the UV melting curves) were observed for the A3-1 and T3-1 sequences, in which 6MAP was situated between two adenines. The smallest perturbation ($\Delta\Delta H = 2.5$ kcal/mol) was detected for the AT-1 duplex, in which 6MAP is between two thymine residues.

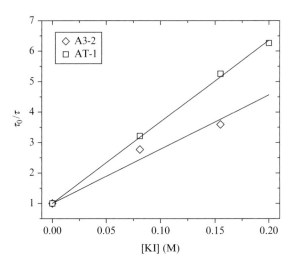

Figure 10.2 Stern–Volmer plot of the longest lived component for the A3-2 and AT-1 duplexes. Ksv = 18 mol per KI for the A3-2 duplex and 27 mol per KI for the AT-1 duplex. Measurements were performed in 0.0, 0.08, 0.15, and 0.20 M KI solutions with 5 mM Na$_2$S$_2$O$_4$ in a 10 mM Tris buffer at pH 7.4. KCl was added to maintain a constant ionic strength of 0.2 M.

The environment of the probe within different duplexes was investigated using KI quenching. A Lehrer analysis was used to analyze the steady-state fluorescence quenching behavior of KI on the dodecamers (Lakowicz, 1999). These data reveal that the fractional accessibility of 6MAP in all five duplexes is at least 83%. Time-resolved experiments indicate that the fluorescence behavior of 6MAP in the duplexes is relatively complex exhibiting at least two lifetime components; the longer component, 4.6 ns, similar to free 6MAP monomer (3.8 ns) (Hawkins et al., 2001), and a shorter component of 0.4 ns. The 4.6 ns component is strongly affected by KI, while the 0.4 ns lifetime component remains relatively constant in the presence of increasing KI quencher (Fig. 10.2).

These studies indicate that 6MAP situated between two pyrimidines is the most accessible to quencher. Comparable quenching constants to those found for A3-2 were obtained for the other duplexes, in which 6MAP is adjacent to at least one purine residue, an indication that accessibility is reduced significantly even with one purine neighbor.

5.3. Temperature-dependent behavior of A-tract duplexes

6MAP fluorescence has also proven to be an effective means to site specifically probe A-tract structure and stability. Considerable spectroscopic and calorimetric evidence exists which suggests that A-tract duplexes undergo a transition from a bent to a straight conformation with a midpoint of ~35 °C (Chan *et al.*, 1990, 1993; Mukerji and Williams, 2002; Park and Breslauer, 1991). This premelting transition is characteristic of duplexes containing A-tracts in a 5'- to 3'-orientation, as in A3-1 and A3-2, but not a 3'- to 5'-orientation, as in T3-1 and T3-2 (Park and Breslauer, 1991). We have monitored the fluorescence behavior of the single- and double-stranded oligonucleotides (shown in Table 10.1) as a function of increasing temperature. We observe that the fluorescence intensity of the monomer probes and the single-stranded oligonucleotides decrease relatively linearly as a function of temperature which is attributed to an increase in the relative efficiency of other processes such as internal conversion and vibrational relaxation.

The 6MAP-containing duplexes also exhibit emission maxima shifts as a function of increasing temperature. For each of the four duplexes at 80 °C, the emission maximum shifts to the same value (±1 nm) as that observed for the corresponding single-stranded oligonucleotide at 80 °C. This behavior is consistent with exposure of the 6MAP probe to solvent and an absence of stacking interactions at high temperature. The fluorescence intensity of the duplexes can also be monitored as a function of temperature. Since the fluorescence yield of the single-stranded oligonucleotides is greater than that of the duplexes (Table 10.1), melting of the duplexes leads to an initial decrease in fluorescence intensity followed by an increase in intensity. A ratio of the fluorescence intensity of the duplex and the single-stranded oligonucleotides yields a sigmoidal curve that is characteristic of the cooperative DNA melting process (Fig. 10.3).

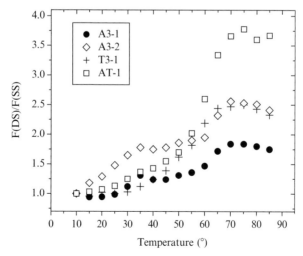

Figure 10.3 Fluorescence of the A3-1, A3-2, T3-1, and AT-1 duplexes as a function of temperature. The fluorescence of the duplex is ratioed against that of the single strand. Measurements were performed in a 0.5 M NaCl, 10 mM Tris buffer at pH 7.4.

By using a ratio of the data, the decrease in fluorescence intensity as a consequence of increased temperature is suppressed. Similar profiles are obtained if the ratio is done with the monomer instead of the single-stranded oligonucleotide. The maximum shift in peak emission occurs at the same temperature as the T_m as measured from the fluorescence intensity change.

A premelting transition is observed in the melting profiles of the A3-1 (•) and A3-2 (◊), but not the T3-1 (+) duplexes (Fig. 10.3). The influence of 6MAP on melting behavior cannot be ignored, however, this influence should also be seen in the control duplexes T3-1 and AT-1, which show no evidence of a premelting transition. In this study, 6MAP is used to investigate specific sites in DNA duplexes of identical base composition, but different sequence, and the fluorescence signals observed clearly reveal subtle differences in DNA structure and that the locus of the premelting transition is in the A-tract.

5.4. Bulge hybridization

In many cases, a quench in fluorescence intensity is seen when a pteridine-containing oligonucleotide is annealed to a complementary strand, however, the changes are usually small and reported through a loss in fluorescence intensity (Hawkins et al., 1997). We have developed a technique that reports on hybridization of sequences in solution by displaying substantial increases in fluorescence intensity (Hawkins and Balis, 2004).

The quench in fluorescence intensity seen with pteridines incorporated into a single strand has been attributed mostly to base stacking interactions. When a 3MI-containing oligonucleotide is annealed to a complementary strand that does not contain a base pairing partner for the fluorophore, the probe can be pushed out of base stacking interactions resulting in increased fluorescence intensity. This sequence-specific technique can result in increases in fluorescence intensity of up to 27-fold (Fig. 10.4).

Almost all sequences we have tested display some increase in fluorescence intensity upon hybridization, however, the largest increases are seen with two adenines on each side of 3MI. A specific sequence containing one, two, or three 3MI probes per strand was tested to determine the optimum condition for that sequence. Each was also tested for potential disruption of annealing from the probes by analyzing T_ms as compared to the control (71.4 °C). It was found that with one probe per strand there was an eightfold increase and a T_m of 70.4 °C, for two it was eightfold and 68.6 °C, for three it was 10-fold and 66.6 °C and for four it was 10-fold and 63.6 °C. In another experiment we examined the optimum length of strand for bulge formation. The results are strongly dependent on annealing conditions including buffer, salt, and overall sequence. In the one sequence that was tested for this, using 10 mM Tris, pH 7.5 in 10 mM NaCl, we found that the greatest increases were in strands of at least 21 bases in length.

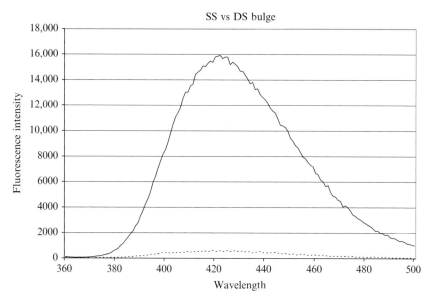

Figure 10.4 Fluorescence emission scans of the single-stranded 3MI-containing oligonucleotide, 5′-cct cta aga ggt gta aFa atg tgg aga atc tcc-3′, as a single strand (dashed line) and annealed to its complementary strand (5′-gga gat tct cca cat ttt aca cct ctt aga gg-3′) which does not contain a base pairing partner for 3MI (solid line). Samples were measured at 25 °C in the presence of 10 mM NaCl.

Results will most likely differ depending on the neighbors for the probe, overall sequence, and the distance of the probe from a 3′- or 5′-terminus and should be determined for each sequence experimentally. While this study was done with 3MI as the probe, selected sequences containing 6MI or 6MAP have been found to behave in a similar manner (M. E. Hawkins, unpublished results).

The usefulness of this application has been demonstrated by measuring the presence of positive polymerase chain reaction (PCR) product for HIV-1. The probe in this study (5′-taa ata aFaa tag taaF gaa tgt ata gcc cta cc-3′ with F = 3MI) is complementary to a region in the 115 bp PCR product defined by the HIV-1 *gag* primers SK38/SK39 (Ou *et al.*, 1988). Reactions contained varying amounts of template from the plasmid, pSum9, generated from a full-length HIV molecular clone (Tanaka *et al.*, 1997). The 3MI-containing sequence was present in the reaction mixture during the amplification process and results were compared to a sample identical in all components and handling except that it contained no template. Experiments were performed in duplicate and repeated at least three times. Increases in fluorescence intensity were proportional to the amount of template added and resulted in signal changes of up to three- to fourfold (Fig. 10.5).

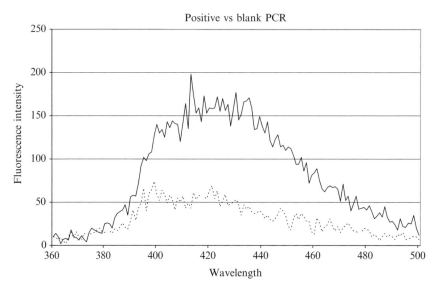

Figure 10.5 Fluorescence emission scans from positive (solid line) and blank (dashed line). The positive sample contained 2.5×10^6 copies of the template. Both samples contained identical concentrations of all components for the amplification including 25.6 nM of the 3MI-containing PCR probe and were subjected to identical amplification procedures with the negative control containing no template. All samples were measured at 25 °C.

A 3MI-containing strand that was not complementary to the PCR product was used as a negative control and showed no increase in fluorescence intensity after amplification. Blocking the probe to prevent it from functioning as a primer would permit the use of higher concentrations of the probe without interfering with the primers. Monitoring the results in real time would also improve this technique.

The results of the bulge hybridization study revealed that the optimum sequence for obtaining a strong increase in fluorescence intensity upon binding would include the probe having two adenines on each side, containing one or two probes in a sequence at least 21 bases long. The design of probe-containing sequences for this application should include a test of the probe's performance under conditions of the system under study.

5.5. Anisotropies of pteridine-containing sequences to examine protein binding

Anisotropies of 3MI-containing sequences were first used to probe protein/DNA binding by Wojtuszewski *et al.* (2001) in a study of HU protein (described in Section 5.1). In this study, the authors found that the anisotropy of a 3MI-containing 13-mer ranged from 0.04 to 0.076 in the transition from unbound DNA to HU-bound DNA while the corresponding anisotropy of a 3MI-containing 34-mer sequence went from 0.12 to 0.22 during the same transition. Based on these results combined with results from observation of changes in fluorescence intensity and data from analytical ultracentrifugation, the authors propose a model for binding stoichiometry that is supportive of three HU molecules binding to the 34 bp sequence (accompanied by bending of the DNA) and two HU molecules binding to the 13 bp duplex. When changes in fluorescence intensity are observed upon binding, it is likely that changes in lifetime that accompany protein binding will result in the observation of a lesser steady-state anisotropy change. Fluorescence intensity changes are likely to occur when pteridines are used in binding studies because of the probe's association with its environment and this must be considered when interpreting results as these authors have done.

Nucleoside analog probes may provide some advantages over linker-attached probes in the analysis of anisotropy data. The presence of a linker arm can potentially complicate the interpretation of results. An example of this is stated in a study by Hill and Royer (1997), where it was determined that 85% of the signal change seen upon binding a fluorescein probe attached through linker chemistry was due to rotational motions associated with the linker, *not* due to the global motions of the DNA.

5.6. Lifetimes, steady-state and time-resolved anisotropies of 3MI- and 6MI-containing sequences

In an ongoing study by Wojtuszewski *et al.*, we have examined the fluorescence characteristics of oligonucleotides containing each of the two probes, 3MI and 6MI. The sequence 5′-act aFa gat ccc tca gac cct ttt agt cag tFt gga-3′ was used as a model for this analysis. 3MI or 6MI was incorporated into one of the positions shown for each sequence and analyzed in a single or double-stranded environment. In this study the sequence containing the aFa (near the 5′-terminus) environment is expected to be more quenched and the sequence containing the tFt environment is expected to be less quenched.

Measured steady-state anisotropies of 3MI-containing single strands were 0.088 and 0.093 in the less quenched and more quenched environments, respectively. The variation between measurements (averaged from at least three samples each) for duplex environment with 3MI were 0.072 and 0.054, respectively. For 6MI the variation between less quenched and more quenched was 0.103 and 0.154 for the single and 0.154 and 0.206 for the duplex, respectively. Steady-state anisotropy measurements are the result of an average of stacked and partially stacked states within each of these structures. An interesting finding in the 6MI-containing set is that in the

Table 10.2 Lifetimes

	α_1	τ_1 (ns)	α_2	τ_2 (ns)	α_3	τ_3 (ns)	$\langle\tau\rangle$ (ns)	τ_m (ns)
3MI ss	21	1.03	28	4.28	51	0.15	1.49	3.60
3MI ds	10	1.25	14	5.72	75	0.14	1.04	4.61
3MI	2	3.54	98	6.58			6.55	6.52
6MI ss	19	1.00	10	4.7	71	0.21	0.81	3.00
6MI ds	17	0.93	5	5.59	78	0.34	0.70	2.56
6MI	20	5.45	80	6.58			6.39	6.35

Abbreviations: τ_i, lifetime for each component of a multiexponential model; α_i, pre-exponential for each component of a multiexponential model; $\langle\tau\rangle$, species-concentration-weighted lifetime; τ_m, intensity-weighted lifetime.

duplex form of the less quenched sequence, the quantum yield of the probe almost doubles (compared to the single strand). This is different from what we would expect and very different from the behavior of 3MI-containing sequences.

The lifetimes, mean lifetimes, and amplitudes for the identical more quenched 36-mer sequences (aFa) defined above are shown in Table 10.2. Changes in intensity-weighted lifetime (τ_m) between the single and double strands of these two probes in their single- and double-stranded forms may provide some insight into the type of quench the probes are experiencing within these structures. The amplitude of very short lifetimes (0.14–0.34 ns) is combined with the quantum yield defect; together these are taken to represent the portion of total fluorescence that is quasistatically quenched by the neighboring bases through base stacking interactions. Thus, the 3MI-containing sequences are 51–75% quasistatically quenched. The τ_1 and τ_2 values, in contrast, represent the monomer-like behavior that is representative of the probe in a more solvent exposed environment. These values are, of course, an average of multiple unstacked and partially unstacked populations subject to some forms of dynamic quenching. The fact that the changes in lifetime are disproportionate to changes in fluorescence intensity seen during these transitions (not shown) is an indication that the factors contributing to the changes are comprised of a combination of dynamic and static quenching interactions. We have not tested associative models for these data, however, the fact that the changes in lifetime are modest indicates that the nonassociative models will yield similar results. It may be difficult to do associative analysis when the lifetimes of the two different states (in this case, single- and double-stranded) are so similar (Brand et al., 1984; Davenport et al., 1986). In general, it might be expected that a double-stranded sequence (as compared to a single strand) would afford fewer degrees of rotational freedom for an incorporated probe. The results for 6MI (requiring fewer components to obtain a good fit) appear to

Table 10.3 Time-resolved anisotropy measurements

	β_1	φ_1 (ns)	β_2	φ_2 (ns)	β_3	φ_3 (ns)	χ^2
ss3MI	0.167	0.637	0.057	2.43			1.95
ds3MI	0.10	1.29	0.35	0.11	(−0.15)	(0.003)	2.04
ss6MI	0.071	6.66	0.23	0.857			1.12
ds6MI	0.32	2.09	(0.214)	(0.032)	(0.05)	(0.004)	1.17

Abbreviations: β_i, pre-exponentials for each component of a multiexponential model; φ_i, rotational correlation times for each component of a multiexponential model; χ^2 is derived from a model to assess goodness of fit.

confirm this. For 3MI, lifetime results agree with results from the steady-state anisotropies revealing similarity between single- and double-strand configurations. This suggests that 3MI does not participate in hydrogen bonding with cytosine, as expected from the steric hindrance from the 3-methyl moiety.

Table 10.3 shows the results from the time-resolved anisotropy studies. The 3MI-containing single strand displays much shorter rotational correlation times (0.64 and 2.43 ns) compared to those of the 6MI-containing single strand (6.66 and 0.86 ns). Based on the lack of a long correlation time component, the motion of 3MI appears to be poorly coupled to the overall motion of the DNA. Although we are quite certain that these results are sequence dependent, it seems that even in a single strand, 3MI's association with neighboring bases may be significantly weaker than those of 6MI.

Until now, we have based the assessment of probe positioning in a double strand totally on measurements of melting temperatures of probe-containing sequences (Hawkins et al., 1997). The current results suggest that 6MI may be more stably associated with, and consequently more accurately reflect the global motion of DNA than 3MI. Further studies are now in progress to explore the dependence of these results on the identity of neighboring bases.

5.7. Probing the hairpin structure of an aptamer using 6MI

In an ongoing study, Hawkins et al. are using the 37-base DNA aptamer, D17.4, as a model system for investigations of structure using 6MI. This aptamer which has the sequence of (5′-ggggcacgtttatccgtccctcc-tagtggcgtgcccc-3′) is known to bind with nM affinity to human immunoglobulin E, IgE, in a highly specific manner and is predicted to have the structural elements of a stem, a single base bulge and a hairpin loop (Gokulrangan et al., 2005; Wiegand et al., 1996). A comparison of steady-state anisotropies of 3MI and 6MI, each of which is directly incorporated into the 5′-end through a deoxyribose linkage, with that of Texas red, attached to the 5′-terminus of the sequence through a 21-carbon linker has

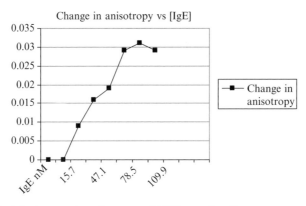

Figure 10.6 The increase in anisotropy of AP33 reveals a direct response to increasing concentrations of IgE protein. AP9 (6MI substituted for the 9th guanine) and AP16 (6MI substituted for the 16th guanine) showed no anisotropy changes in response to added IgE most likely because of the selectivity of binding in the area of the loop where they are located.

revealed large differences dependent on the nature of the probe. Steady-state anisotropies of sequences labeled at the 5′-terminal position with 3MI and 6MI were 0.013 and 0.018, respectively, while the sequence labeled at the same site with Texas Red was 0.16. Results for the pteridines suggest rapid segmental motion, in this case not attributable to a linker and so therefore most likely a feature of this site. For Texas Red, the linker attachment is thought to permit most of the probe to "fold back" and associate with the rigid annealed sequence, providing anisotropy values determined mostly by the molecule's overall motion (Unruh et al., 2005a). Others have explained the rapid motion seen in linker-attached probes at the terminal position of a sequence to the motion of the linker. 6MI and 3MI are attached to the sequence through the native deoxyribose linkage, therefore, this rapid motion is likely associated with fraying of the duplex ends. 6MI was also incorporated into sites within the aptamer in either the stem or loop environments of the molecule. Steady-state anisotropies of 6MI at these positions range from 0.14 to 0.03 depending on the location. The position most reflective of "global motion" appears to be in the stem of the sequence where the 6MI probe is substituted for the guanine located 33 bases from the 5′-end (AP33). The same sequence was also used in a binding study with IgE (results shown in Fig. 10.6).

Time-resolved studies should reveal whether these steady-state differences represent actual changes in the motion of the DNA aptamer, or if they are the result of concomitant changes in lifetimes and fluorescence intensities.

5.8. Two-photon excitation of 6MAP

The utility of fluorescent base analogs is their ability to act as dynamic reporters of changes in DNA conformation. 6MAP is similar to 2-Ap (Nordlund et al., 1989) in that it is an effective reporter of base stacking. It has a significantly larger Stokes shifted emission compared to 2-Ap (100 vs 50 nm) making it useful for certain applications involving microscopy and fluorescence resonance energy transfer. A problem with both analogs is that the lowest lying electronic transition of significant oscillator strength is in the near-UV (~317 nm for 2-Ap and 326 nm for 6MAP). Since many molecules absorb in this region it will be difficult to utilize single molecule techniques due to autofluorescence from buffer components, etc.

Two-photon excitation (TPE) has been used to ameliorate the problem of out of focus autofluorescence and is now widely implemented for biological imaging. TPE is a nonlinear process where two photons must be available simultaneously for efficient excitation. This implies high light intensities, which can be achieved by focusing short pulse mode-locked laser light. TPE is, therefore, only efficient in the focal volume of the laser. If a sufficiently short focal length optical system is used then excitation can be isolated to a very small volume, typically 1 μm^3. Since the exciting light has double the wavelength of the transition (i.e., 652 vs 326 nm) there will be no extraneous absorption outside of this focal volume and therefore a large reduction in the autofluorescence due to solvent components or other endogenous fluorophores. Secondly, due to the small focal volume and relatively large penetration depth of long wavelength light compared to near-UV excitation it is possible to selectively excite and image structures with a resolution on the order of the focal volume as well as perform studies on single molecules. Drawbacks to this approach have to do with the inefficiency of TPE compared to one-photon excitation and the difficulties associated with quantitation of the signal intensity. Stanley et al. (2005) have used TPE to assess whether 6MAP emits two-photon-induced fluorescence (TPIF) and their preliminary data suggests that it will make a useful TPE probe.

TPIF spectra were obtained from 6MAP monomer solutions using a nonlinear optical parametric amplifier (NOPA) as an excitation source. A full description of this work will be presented elsewhere. Samples of 6MAP (50–240 μM) were dissolved in Tris buffer, pH 7.4, 5% glycerol, and placed in a cuvette having a 1 cm excitation path length and a 0.2 cm emission path length. Solutions were stirred to avoid saturation and photobleaching effects. It should be noted that 6MAP has no one-photon absorption above about 380 nm. The energy of the ca. 300 fs pulse centered around 690 nm (1 kHz repetition rate) was controlled using a variable neutral density filter and measured with a sensitive joulemeter. This wavelength was chosen because of the availability of two-photon cross-section values for standard compounds such as fluorescein and rhodamine dyes (Xu and Webb, 1996). The maximum useable energy was limited by continuum generation in the solvent (about 2.2 µJ/pulse) and depended on the choice of focusing optics (5 cm focal length) as well as the pulse energy. Fluorescence from the focal volume was collected using a 5 cm lens, filtered to remove residual excitation light, and focused into a spectrograph with a cooled low noise CCD detector.

TPE of solutions of 6MAP with 690 nm light generated blue emission visible to the naked eye, which emanated from the focus of the excitation beam. TPIF spectra were obtained for both 6MAP and QS (Heller et al., 1974). QS was used as a standard for correcting the spectral responsivity of the measuring apparatus (for both one- and two-photon emission data).

When solutions of 6MAP are excited with focused ultrafast laser pulses from 614 to 700 nm blue fluorescence can be seen emanating from the focal point in the sample cuvette. For reference the one-photon absorption spectrum of 6MAP in buffer (δ) is shown in Fig. 10.6.

The absorption spectrum has structure with the largest extinction at about 330 nm and a prominent shoulder at 294 nm. The two-photon-induced fluorescence spectrum (TPIF) of a 240 µM solution of 6MAP in 50 mM Tris (pH 7.4) with 5% glycerol with excitation at 690 nm is shown in Fig. 10.6. The spectrum was corrected for the spectral responsivity of the instrument using QS. For comparison a one-photon-induced fluorescence (OPIF) emission spectrum of 6MAP at 20 µM concentration in the same buffer using 345 nm excitation is also shown, taken on a conventional fluorimeter with the same resolution whose spectral responsivity has been corrected using QS as indicated above.

Stanley and Yang performed a power dependence study to verify that the emission was due to TPE. The results of this study are shown in Fig. 10.7, where the logarithm of the TPIF is plotted versus the logarithm of the excitation power. A linear fit gave a slope of 2.0 ± 0.1 which is the value expected for a two-photon process. This value was obtained over the concentration range given above, indicating that dimerization does not affect the measured two-photon cross section:

$$\delta_{\text{TPE}}^{\text{6MAP}}(\lambda) = \frac{\delta_{\text{TPE}}^{S} \Phi_{f}^{S}[S]\langle F(t)\rangle_{\text{6MAP}}}{\Phi_{f}^{\text{6MAP}}[\text{6MAP}]\langle F(t)\rangle_{S}}. \tag{10.1}$$

To measure the two-photon cross section, $\delta_{\text{TPE}}^{\text{6MAP}}$, for 6MAP it is convenient to ratio the emission of the probe under constant measurement conditions to that for which δ_{TPE}^{S} have been determined (for the standard S). Here, δ_{TPE}^{S}, Φ_{f}^{S}, and Φ_{f}^{6MAP} are the two-photon cross section for the standard S, the fluorescence quantum yield for S, and the fluorescence

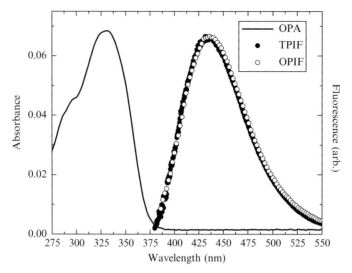

Figure 10.7 One (line)- and two (dots)-photon-induced fluorescence emission spectra of 6MAP in Tris buffer. The one-photon spectrum was taken at a concentration of 20 μM in 6MAP while the two-photon spectrum was obtained at 240 μM. Both spectra have been corrected for the spectral responsivity of the respective spectrometer using quinine sulfate.

quantum yield for 6MAP, respectively. The time-averaged excitation power is assumed to be the same for both measurements. Since the observable is the time-averaged emission intensity $\langle F(t) \rangle$ for the two species, knowledge of the fluorescence quantum yield is necessary to make a quantitative comparison. Xu and Webb (1996) have compiled a table of cross sections for a variety of commonly available laser dyes. Fluorescein was chosen as a standard because its TPE spectrum has been measured with 690 nm excitation.

TPIF emission spectra for both 6MAP and fluorescein (pH 13) were obtained under identical conditions as a function of laser intensity (data not shown). The cross section for TPIF of fluorescein from 690–900 nm has been measured previously and serves as a calibration to determine the cross section of 6MAP (Xu and Webb, 1996). The integrated areas of the emission spectra are compared as described above to obtain $\delta_{\text{TPE}}^{\text{6MAP}} = 0.4(\pm 0.1) \times 10^{-50}\,\text{cm}^4\text{s/photon}$ (0.4 GM), based on a fluorescence quantum yield of $\Phi_{\text{f}}^{\text{6MAP}} = 0.39$ as measured by Hawkins et al. (2001) The cross section was invariant over the range of 50–240 μM. This two-photon cross section is much lower than for most laser dyes, which is consistent with the much more extended π systems of large dye molecules compared to 6MAP. However, the value of $\delta_{\text{TPE}}^{\text{6MAP}}$ for 6MAP is comparable to other fluorophores currently used in two-photon microscopy, such as dansyl and DAPI (Xu *et al.*, 1996). In addition, the 6MAP two-photon cross section compares favorably with endogenous fluorophores of biological interest such as flavins and nicotinamide (Huang *et al.*, 2002). We therefore expect 6MAP to become a valuable two-photon probe for single molecule studies of DNA dynamics and conformation.

5.9. Single molecule detection of 3MI

Using fluorescence correlation spectroscopy (FCS), it has been found that 3MI is sufficiently bright and stable to be detected at the single molecule level (Sanabia *et al.*, 2004). The use of FCS or single molecule techniques to study the binding and dynamics of DNA is hampered in the UV range by high background fluorescence from many sources of the system including the optics, DNA, coverslip, lens, and other components. FCS measurements yield details such as fluorophore diffusivity, number of fluorophores in the focal volume, intensity, rate of photobleaching, triplet correlation time, triplet fraction, and triplet lifetime (Brand *et al.*, 1997; Dittrich and Schwille, 2001; Eggeling *et al.*, 1998; Thompson, 1991; Zander and Enderlein, 2002). Two parameters that are important indicators for a useful single molecule probe include the count rate per molecule (which must be well above the background count rate) and the photobleaching quantum yield, which when inverted gives the number of photons emitted, on average, before photobleaching occurs. The UV FCS system, based on a homebuilt inverted confocal fluorescence microscope, was developed and used to measure the count rate per molecule, photobleaching quantum yield, and triplet population of 3MI. This system uses an argon laser to obtain continuous wave excitation at 351.1 nm (selected with a BK7 prism) controlled with a half-wave plate followed by a polarizing cube. A telescope and a 50 μm pinhole allowed spatial filtering of the beam with a second telescope to adjust the diameter and convergence entering the objective. To minimize artifacts in analysis the diameter of the laser beam entering the objective was roughly one-half the back-focal plane aperture diameter, underfilling the back aperture (Hess and Webb, 2002). The beam was directed into an inverted oil immersion objective lens (Zeiss Fluar, 1.3 NA, 100×, f = 1.65, infinity corrected) using a dichroic filter (Omega 400DCLP) and measured by a silicon photodiode. Excitation intensities (calculated using power measurements) of samples held in a polydimethylsiloxane (PDMS, Sylgard 184) well on a glass coverslip (Fisher Premium #1) were monitored through a CCD video camera. With 351.1 nm excitation and the 400 nm Raman line blocked using dichroic and band-pass filters, signals passed through a 100 mm focal length achromatic doublet lens and a 50 μm diameter confocal pinhole to a Hamamatsu H7421-40 GaAsP photomultiplier photon-counting detector. Fluorescence intensity of solutions contained in the PDMS well was measured as a function of time with

5 μs resolution. A more detailed account of the theory and data interpretation is presented in the chapter (Sanabia *et al.*, 2004).

In summary, these researchers have shown that single 3MI molecules are detectable using single photon excitation with a signal to background ratio as high as 5, a count rate per molecule above 4 kHz, and a photobleaching quantum yield of 2.4×10^{-4}. The signal from a 3MI-containing oligonucleotide is also detectable on a single molecule level with a signal to background ratio greater than 1. Substantial improvement in both signal to background and photostability are possible with the use of two-photon excitation.

6. SUMMARY

Ideally, a fluorescent nucleoside analog incorporated into an oligonucleotide must be similar enough to native bases to permit native-like behavior of the DNA, but structurally different enough to be highly fluorescent. The electronic configuration within a molecule is the source of its fluorescence. The effect of neighboring bases on the electronic structure of an incorporated pteridine probe and consequently on its fluorescence properties is what allows tracking of structural characteristics of pteridine-containing DNA as it encounters and reacts with other molecules.

We have been unsuccessful at finding a pteridine probe that is more similar to guanine or adenine than those described here. In the initial studies of pteridine probes we found that the two most attractive analogs (probes 5 and 25; Hawkins *et al.*, 1997), essentially 6MI and 6MAP without the 6-position methyl moieties, were extremely unstable. The presence of at least one methyl group in either the 3- or 6-position appears to have a very strong stabilizing effect on these structures. Even with the structural similarities, however, pteridines and purines are electronically quite different. When a fluorescent pteridine molecule is substituted for one of the native bases in the DNA, the overall electronic environment of the DNA molecule is changed. Whether or not this substitution is acceptable to the system under study must be determined experimentally.

The most important features of the pteridine analog probes are that they are unobtrusive enough to be positioned within an oligonucleotide site specifically at or near a location of interest and that they are so intensely fluorescent that they can still be detected within the quenched environment of the oligonucleotide. Early attempts to characterize the properties of pteridine-containing oligonucleotides were focused on the immediate neighbors (two on either side) of the probe. Studies on bulge-forming 3MI sequences (Hawkins, 2003) have revealed that the immediate neighbors to the probe are only a part of the total picture. Strands with the identical neighboring sequence (AAFAA) but different overall sequences were found to respond quite differently when in a bulged configuration. The entire sequence of bases as well as the overall length and the distance of the probe from each end of the sequence most likely contribute to these variations. Even though subtle differences between sequences may be difficult for researchers to measure, these small variations may be very important to the overall recognition of a strand for physiological function. The incorporated pteridine probes may provide the sensitivity to electronic environment required to detect and understand these subtle variations.

ACKNOWLEDGMENTS

I would like to express my appreciation to the following for generously sharing results of their recent research: Robert J. Stanley and Aiping Yang at Temple University, Philadelphia, PA, for work on the two-photon excitation of 6MAP; Jason Sanabia, Lori Goldner, and Pierre-Antoine Lacaze at National Institutes of Standards and Technology, Gaithersburg, MD for their work on single molecule detection of 3MI; K. Augustyn, K. Wojtuszewski, J. Knutson, and I. Mukerji for the A-tract experiments using 6MAP; and K. Wojtuszewski, Aleksandr Smirnov, and J. Knutsen for the work on time-resolved anisotropies and lifetimes of 36-mers. I especially want to thank Drs. Ishita Mukerji and Jay R. Knutson for helpful discussions.

REFERENCES

Augustyn, K. E., Wojtuszewski, K., Hawkins, M. E., Knutson, J. R., and Mukerji, I. (2006). Examination of the premelting transition of DNA A-tracts using a fluorescent adenosine analogue. *Biochemistry* **45,** 5039–5047.
Berry, D. A., Jung, K.-Y., Wise, D. S., Sercel, A. D., Pearson, W. H., Mackie, H., Randolph, J. B., and Somers, R. L. (2004). Pyrrolo-dC and pyrrolo-C: Fluorescent analogs of cytidine and 2′-deoxycytidine for the study of oligonucleotides. *Tetrahedron Lett.* **45,** 2457–2461.
Brand, L. C. E. O., Zander, C. K. H. D., and Seidel, C. (1997). Single-molecule identification of coumarin-120 by time-resolved fluorescence detection: Comparison of one- and two-photon excitation in solution. *J. Phys. Chem. A* **101,** 4313–4321.
Brand, L., Knutson, J. R., Davenport, L., Beechem, J. B., Dale, R. E., and Kowalczyk, A. A. (1984). Time-resolved fluorescence spectroscopy: Some applications of associative behavior to studies of proteins and membranes. *In* "Dynamics of Molecular Biological Systems" (P. Bayley and R. E. Dale, eds.), 259-305. Academic Press.
Chan, S. S., Breslauer, K. J., Austin, R. H., and Hogan, M. E. (1993). Thermodynamics and premelting conformational changes of phased (dA)5 tracts. *Biochemistry* **29,** 6161–6171.
Chan, S. S., Breslauer, K. J., Hogan, M. E., Kessler, D. J., Austin, R. H., Ojemann, J., Passner, J. M., and Wiles, N. C. (1990). Physical studies of DNA premelting equilibria in duplexes with and without Homo dA*dT tracts: Correlations with DNA bending. *Biochemistry* **29,** 6161–6171.

Crothers, D. M., and Shakked, Z. (1999). DNA bending by adenine-thymine tracts. In "Oxford Handbook of Nucleic Acid Structure" (S. Neidle, ed.), pp. 455–470. Oxford University Press, Oxford.

Dash, C., Rausch, J. W., and Le Grice, S. F. (2004). Using pyrrolo-deoxycytosine to probe RNA/DNA hybrids containing the human immunodeficiency virus type-1 3' polypurine tract. *Nucleic Acids Res.* **32,** 1539–1547.

Davenport, L., Knutson, J. R., and Brand, L. (1986). Faraday Discuss. *Chem. Soc.* **81,** 81–94.

Davis, S. P., Matsumura, M., Williams, A., and Nordlund, T. M. (2003). Position dependence of 2-aminopurine spectra in adenosine pentadeoxynucleotides. *J. Fluorescence* **13,** 249–259.

Dittrich, P. S., and Schwille, P. (2001). Photobleaching and stabilization of fluorophores used for single-molecule analysis with one- and two-photon excitation. *Appl. Phys. B: Lasers Opt.* **73,** 829–837.

Driscoll, S. L., Hawkins, M. E., Balis, F. M., Pfleiderer, W., and Laws, W. R. (1997). Fluorescence properties of a new guanosine analog incorporated into small oligonucleotides. *Biophys. J.* **73,** 3277–3286.

Eckstein, F. (1991). "Oligonucleotides and analogues: A practical approach." In "The Practical Approach Series" (D. Rickwood and B. D. Hames, eds.). Oxford University Press, New York.

Eggeling, C. J. W., Rigler, R., and Seidel, C. (1998). Photobleaching of fluorescent dyes under conditions used for single-molecule detection: Evidence of two-step photolysis. *Anal. Chem.* **98,** 10090–10095.

Georghiou, S., Bradrick, T. D., Philippetis, A., and Beechem, J. M. (1996). Large amplitude picosecond anisotropy decay of the intrinsic fluorescence of double stranded DNA. *Biophys. J.* **70,** 1909–1922.

Gokulrangan, G., Unruh, J. R., Holub, D. F., Ingram, B., Johnson, C. K., and Wilson, G. S. (2005). DNA aptamer-based bioanalysis of IgE by fluorescence anisotropy. *Anal. Chem.* **77,** 1963–1970.

Hardman, S. J., and Thompson, K. C. (2006). Influence of base stacking and hydrogen bonding on the fluorescence of 2-aminopurine and pyrrolocytosine in nucleic acids. *Biochemistry* **45,** 9145–9155.

Hawkins, M. E. (2001). Fluorescent pteridine nucleoside analogs: A window on DNA interactions. *Cell Biochem. Biophys.* **34,** 257–281.

Hawkins, M. E. (2003). Fluorescent nucleoside analogues as DNA probes. In "DNA Technology" (J. R. Lakowicz, ed.). Vol. 7, pp. 151–175. Kluwer Academic/Plenum Publishers, New York.

Hawkins, M. E., Pfleiderer, W., Balis, F. M., Porter, D., and Knutson, J. R. (1997). Fluorescence properties of pteridine nucleoside analogs as monomers and incorporated into Oligonucleotides. *Anal. Biochem.* **244,** 86–95.

Hawkins, M. E., Pfleiderer, W., Jungmann, O., and Balis, F. M. (2001). Synthesis and fluorescence characterization of pteridine adenosine nucleoside analogs for DNA incorporation. *Analy. Biochem.* **298,** 231–240.

Hawkins, M. E., Pfleiderer, W., Mazumder, A., Pommier, Y. G., and Balis, F. M. (1995). Incorporation of a fluorescent guanosine analog into oligonucleotides and its application to a real time assay for the HIV-1 integrase 3'-processing reaction. *Nucleic Acids Res.* **23,** 2872–2880.

Hawkins, M. E., and Balis, F. M. (2004). Use of pteridine nucleoside analogs as hybridization probes. *Nucleic Acids Res.* **32,** e62.

Heller, C. A., Henry, R. A., McLaughlin, B. A., and Bliss, D. E. (1974). Fluorescence spectra and quantum yields. Quinine, uranine, 9,10-diphenylanthracene, and 9,10-bis (phenylethynyl)anthracenes. *J. Chem. Eng. Data* **19,** 214–219.

Hess, S. T., and Webb, W. W. (2002). Focal volume optics and experimental artifacts in confocal fluorescence correlation spectroscopy. *Biophys. J.* **83,** 2300–2317.

Hill, J. J., and Royer, C. (1997). Fluorescence approaches to study of protein-nucleic acid complexation. *In* "Methods in Enzymology" (L. Brand, ed.). Vol. 278, pp. 410–411. Academic Press, New York.

Huang, S., Heikal, A. A., and Webb, W. W. (2002). Two-photon fluorescence spectroscopy and microscopy of NAD(P)H and flavoprotein. *Biophys. J.* **82,** 2811–2825.

Lakowicz, J. R. (1999). "Principles of Fluorescence Spectroscopy." Second edition ed. Plenum Publishers, New York.

Larsen, O. F. A., van Stokkum, I. H. M., Gobets, B., van Grondelle, R., and van Amerongen, H. (2001). Probing the structure and dynamics of a DNA hairpin by ultrafast quenching and fluorescence depolarization. *Biophys. J.* **81,** 1115–1126.

Lee, B. J., Barch, M., Castner, E. W., Jr., Volker, J., and Breslauer, K. J. (2007). Structure and dynamics in DNA looped domains: CAG triplet repeat sequence dynamics probed by 2-AMinopurine fluorescence. *Biochemistry* **46,** 10756–10766.

Liu, C., and Martin, C. T. (2001). Fluorescence characterization of the transcription bubble in elongation complexes of T7 RNA polymerase. *J. Mol. Biol.* **308,** 465–475.

Mukerji, I., and Williams, A. P. (2002). UV resonance raman and circular dichroism studies of a DNA duplex containing an A(3)T(3) tract: Evidence for a premelting transition and three-centered H-bonds. *Biochemistry* **41,** 69–77.

Myers, J. C., Moore, S. A., and Shamoo, Y. (2003). Structure-based incorporation of 6-methyl-8-(2-deoxy-beta-ribofuranosyl)isoxanthopteridine into the human telomeric repeat DNA as a probe for UP1 binding and destabilization of G-tetrad structures. *J. Biol. Chem.* **278,** 42300–42306.

Nordlund, T. M., Andersson, S., Nilsson, L., Rigler, R., Graslund, A., and McLaughlin, L. W. (1989). Structure and dynamics of a fluorescent DNA oligomer containing the EcoRI recognition sequence: Fluorescence, molecular dynamics, and NMR studies. *Biochemistry* **28,** 9095–9103.

Ou, C. Y., Kwok, S., Mitchell, S. W., Mack, D. H., Sninsky, J. J., Krebs, J. W., Feorino, P., Warfield, D., and Schochetman, G. (1988). DNA amplification for direct detection of HIV-1 in DNA of peripheral blood mononuclear cells. *Science* **239,** 295–297.

Park, Y.-W., and Breslauer, K. J. (1991). A spectroscopic and calorimetric study of the melting behaviors of a 'bent' and 'normal' DNA duplex: [d(GA4T4C)]2 versus [d(GT4A4C)]2. *Proc. Natl. Acad. Sci. USA* **88,** 1551–1555.

Rai, P., Cole, T. D., Thompson, E., Millar, D. P., and Linn, S. (2003). Steady-state and time-resolved fluorescence studies indicate an unusual conformation of 2-aminopurine within ATAT and TATA duplex DNA sequences. *Nucleic Acids Res.* **31,** 2323–2332.

Ramreddy, T., Rao, B. J., and Krishnamoorthy, G. (2007). Site-specific dynamics of strands in ss- and dsDNA as revealed by time-domain fluorescence of 2-aminopurine. *J. Phys. Chem. B* **111,** 5757–5766.

Rist, M. J., and Marino, J. P. (2002). Fluorescent nucleotide base analogs as probes of nucleic acid structure, dynamics and interactions. *Curr. Org. Chem.* **6,** 775–793.

Sanabia, J. E., Goldner, L. S., Lacaze, P.-A., and Hawkins, M. E. (2004). On the feasibility of single-molecule detection of the guanosine-analogue 3-MI. *J. Phys. Chem. B* **108,** 15293–15300.

Schurr, J. M., and Fujimoto, B. S. (1988). The amplitude of local angular motions of intercalated dyes and bases in DNA. *Biopolymers* **27,** 1543–1569.

Seibert, E., Chin, A. S., Pfleiderer, W., Hawkins, M. E., Laws, W. R., Osman, R., and Ross, J. B. A. (2003). pH-dependent spectroscopy and electronic structure of the guanine analogue 6,8-Dimethylisoxanthopterin. *J. Phys. Chem. A* **107,** 178–185.

Sischka, A., Toensing, K., Eckel, R., Wilking, S. D., Sewald, N., Ros, R., and Anselmetti, D. (2005). Molecular mechanisms and kinetics between DNA and DNA binding lignands. *Biophys. J.* **88,** 404–411.

Stanley, R. J., Hou, Z., Yang, A., and Hawkins, M. E. (2005). The two-photon excitation cross section of 6MAP, a fluorescent adenine analogue. *J. Phys. Chem. B* **109,** 3690–3695.

Tanaka, M., Srinivas, R. V., Ueno, T., Kavlick, M. F., Hui, F. K., Fridland, A., Driscoll, J. S., and Mitsuya, H. (1997). *In vitro* induction of human immunodeficiency virus type 1 variants resistant to 2′-beta-Fluoro-2′,3′-dideoxyadenosine. *Antimicrob. Agents Chemother.* **41,** 1313–1318.

Thompson, N. L. (1991). Fluorescence correlation spectroscopy. *In* "Topics in Fluorescence Spectroscopy" (J. R. Lakowicz, ed.), **1,** p. 337. Plenum Press, New York.

Unruh, J. R., Gokulrangan, G., Lushington, G. H., Johnson, C. K., and Wilson, G. S. (2005a). Orientational dynamics and dye-DNA interactions in a dye-labeled DNA aptamer. *Biophys. J.* **88,** 3455–3465.

Unruh, J. R., Gokulrangan, G., Wilson, G. S., and Johnson, C. K. (2005b). Fluorescence properties of fluorescein, tetramethylrhodamine and texas red linked to a DNA aptamer. *Photochem. Photobiol.* **81,** 682–690.

Vamosi, G., Gohlke, C., and Clegg, R. M. (1996). Fluorescence characteristics of 5-carboxytetramethylrhodamine linked covalently to the 5′ end of oligonucleotides: Multiple conformers of single-stranded and double-stranded dye-DNA complexes. *Biophys. J.* **71,** 972–994.

Wiegand, T. W., Williams, P., Dreskin, S. C., Jouvin, M.-H., Kinet, J.-P., and Tasset, D. (1996). High-affinity oligonucleotide ligands to human IgE inhibit binding to Fcε receptor I. *J. Immunol.* **157,** 221–230.

Wojtuszewski, K., Hawkins, M., Cole, J. L., and Mukerji, I. (2001). HU binding to DNA: Evidence for multiple complex formation and DNA bending. *Biochemistry* **40,** 2588–2598.

Xu, C., and Webb, W. W. (1996). Measurement of two-photon excitation cross sections of molecular fluorphores with data from 690 to 1050 nm. *J. Opt. Soci. Am. B: Opt. Phys.* **13,** 481–491.

Xu, C., Willliams, R. M., Zipfel, W., and Webb, W. W. (1996). Multiphoton excitation cross-sections of molecular fluorophores. *Bioimaging* **4,** 198–207.

Xu, Y., and Sugiyama, H. (2006). Formation of the G-quadruplex and i-motif structures in retinoblastoma susceptibility genes (Rb). *Nucleic Acids Res.* **34,** 949–954.

Zander, C., and Enderlein, J. (2002). *In* "Single Molecule Detection in Solution, Methods and Applications" (R. A. Keller, ed.). Wiley-VCH, Berlin.

CHAPTER ELEVEN

SINGLE-MOLECULE FLUORESCENCE METHODS FOR THE ANALYSIS OF RNA FOLDING AND RIBONUCLEOPROTEIN ASSEMBLY

Goran Pljevaljčić *and* David P. Millar

Contents

1. Introduction	233
2. Labeling Methods	236
3. Single-Molecule Fluorescence Detection Methods	238
4. Diffusion Single-Pair FRET for RNA Based Systems	238
5. Total Internal Reflection Fluorescence (TIRF) for RNA Based Systems	240
6. Application 1: Folding of the Hairpin Ribozyme	243
7. Application 2: Assembly of the Rev–RRE Complex	247
Acknowledgments	249
References	249

1. INTRODUCTION

In addition to its role as an information carrier in gene expression, RNA is now recognized to carry out a broad range of biological functions, and novel activities continue to emerge on a regular basis. RNA molecules can act as catalysts in reversible phosphodiester-cleavage reactions (Fedor, 2000; Hutchins *et al.*, 1986; Shih and Been, 2002), mediate splicing of premessenger RNA (Collins and Guthrie, 2000; Valadkhan, 2007), silence specific genes (Zamore and Haley, 2005), and act as molecular switches to sense metabolites and regulate the translation of genes into proteins (Barrick and Breaker, 2007; Winkler and Breaker, 2005), to name just a few

Department of Molecular Biology, The Scripps Research Institute, La Jolla, California

Methods in Enzymology, Volume 450
ISSN 0076-6879, DOI: 10.1016/S0076-6879(08)03411-3

© 2008 Elsevier Inc.
All rights reserved.

important examples. Perhaps the most impressive (and unexpected) function of RNA is the ability to catalyze peptide bond formation on the ribosome (Puglisi *et al.*, 2000). RNA molecules can achieve these biological activities independently, as in the case of small ribozymes, or by associating with one or more proteins to form ribonucleoproteins (RNPs), such as the spliceosome, ribosome, or signal recognition particle. In all cases, the RNA chain must fold into a defined tertiary structure that creates a catalytic center or a specific ligand recognition site. Consequently, it is important to understand the mechanisms governing the tertiary folding of RNA and the assembly of biologically active RNPs.

Similar to proteins, RNAs are synthesized as linear polymers that must fold into compact three-dimensional structures to attain their biological activity. However, there are fundamental differences between these two types of macromolecular folding processes. First, the chemical nature of the monomer units (nucleotides vs. amino acids) is distinctly different in each case. Second, RNA is a highly charged polyanion, compared to proteins that vary greatly in their net charge. Hence, cations of various types (especially divalent metal cations) play a critical role in RNA folding, acting to neutralize the negative charge of the RNA backbone and thereby allowing the chain to fold into a compact conformation that would otherwise be energetically unfavorable because of electrostatic repulsion.

In addition, RNA molecules fold much more slowly than do proteins, pointing to another key mechanistic difference (Sosnick and Pan, 2003; Treiber and Williamson, 1999, 2001). One reason for the slow folding is the propensity of RNA molecules to form nonnative secondary or tertiary structures that create kinetic traps. The escape from these kinetic traps is frequently rate-limiting (Treiber and Williamson, 1999). The preponderance of kinetic traps has obscured more fundamental aspects of the RNA folding process, such as conformational search and metal ion binding (Sosnick and Pan, 2003). These features of RNA folding have only begun to be explored recently.

While small RNAs, including many ribozymes, can fold autonomously, most large RNA molecules require one or more protein chaperones for efficient and correct folding. Proteins can prevent misfolding of large RNAs and/or accelerate the escape from kinetic traps. Additionally, proteins may guide proper folding by manipulating the structure of the RNA chain. For example, binding of one protein can induce an RNA conformational change that creates the binding site for a second protein or arranges the active site of a ribozyme. This coupling between protein binding and RNA folding events naturally leads to the orderly assembly of large RNPs. In addition, the biological function of many RNPs is linked to dynamic conformational changes that involve rearrangements in RNA tertiary structure. Hence, RNA folding and conformational dynamics are essential features of RNP assembly and function.

To gain an understanding of the general principles of RNA folding and RNP assembly, it is necessary to study the folding behavior of RNA, both in the absence and presence of protein chaperones. While a variety of methods have been used to study the folding of RNA molecules, few are capable of providing a complete picture of the pathways involved. Generally, a variety of methods have to be employed to generate a complete understanding of structure, function, and mechanism. Native gel electrophoresis (Pan et al., 1997) can detect changes in the overall conformation of RNA during folding, but the technique cannot provide specific structural details or kinetic information. X-ray crystallography can provide detailed structural information on large RNA molecules and complexes (Conn and Draper, 1998; Doudna and Cate, 1997; Ramakrishnan and Moore, 2001), and in favorable cases can provide structures of discrete intermediates during RNP assembly (Kuglstatter et al., 2002; Oubridge et al., 2002; Weichenrieder et al., 2000). However, crystallographic methods cannot readily capture dynamic conformational changes during folding or assembly. NMR spectroscopy can provide both structural and dynamic information on RNA molecules and RNA–protein complexes (Mollova and Pardi, 2000), although the size range is restricted. Hydroxyl radical footprinting can detect localized RNA conformational changes with high time resolution, but the technique is insensitive to alterations in the overall structure (Sclavi et al., 1998). Chemical cross-linking can capture RNA folding intermediates, but the technique provides only limited structural information (Narlikar and Herschlag, 1996).

Fluorescence spectroscopy can provide both structural and dynamic information and is emerging as a powerful tool for studies of RNA folding and RNP assembly processes. Fluorescence measurements can be performed in solution under physiologically relevant conditions, without restrictions arising from the size of the molecules under study. Moreover, the method provides dynamic information spanning a wide range of time scales, from picoseconds to minutes. Various types of fluorescence parameters can be recorded, including emission intensity, polarization, and lifetime, each of which reports different molecular properties. Förster resonance energy transfer (FRET) is another informative spectroscopic phenomenon. FRET refers to the long-range nonradiative coupling between two fluorophores with overlapping emission and absorption spectra. Since the efficiency of FRET is strongly dependent on the donor–acceptor distance in the range 20–80 Å (Stryer and Haugland, 1967), FRET measurements can provide information on the global structure of RNA. Moreover, the FRET phenomenon can be utilized in the design of assays that provide kinetic and thermodynamic information on RNA conformation and folding (Klostermeier and Millar, 2001a,b).

In the past decade, owing to technological advances in fluorescence detection methods, it has become possible to record fluorescence signals from individual molecules (Moerner and Orrit, 1999; Nie and Zare, 1997). Single-molecule measurements can directly reveal molecular subpopulations that are hidden in conventional (ensemble-averaged) measurements. Moreover, by observing the behavior of single molecules, it is possible to monitor kinetic processes that cannot be synchronized in a population of molecules. These capabilities are opening a new window into RNA folding and RNP assembly processes. Single-molecule fluorescence techniques have been applied to a variety of RNA systems in recent years and several exciting discoveries have been reported. For example, novel insights into RNA folding processes have emerged from studies of an RNA three-helix junction (Ha *et al.*, 1999; Kim *et al.*, 2002), the Tetrahymena group I intron (Lee *et al.*, 2007; Zhuang *et al.*, 2000), the hairpin ribozyme (Bokinsky *et al.*, 2003; Nahas *et al.*, 2004; Pljevaljčić *et al.*, 2004; Rueda *et al.*, 2004; Tan *et al.*, 2003; Zhuang *et al.*, 2002), RNase P RNA (Xie *et al.*, 2004), and the GAAA-tetraloop/receptor system (Fiore *et al.*, 2008; Hodak *et al.*, 2005). In other exciting applications, single-molecule fluorescence methods have been used to dissect the assembly pathway of the telomerase RNP (Stone *et al.*, 2007) and to reveal tRNA dynamics on the ribosome (Blanchard *et al.*, 2004a,b). These and other recent applications highlight the potential of single-molecule techniques to provide structural, dynamic, and mechanistic information across a range of biologically important RNA systems. The purpose of this review is to describe some of the key methodologies required and to illustrate two recent examples from our own laboratory. It is hoped that this brief review will encourage the use of single-molecule fluorescence techniques in an even broader range of RNA systems in the future.

2. Labeling Methods

For fluorescence-based studies of RNP assembly, it is useful to employ fluorophore-labeled proteins. Site-specific labeling of proteins is usually achieved by covalent modification of cysteine residues with maleimide derivatives of fluorescent dyes. Site-directed mutagenesis can be used to reduce the number of native cysteine residues in a protein to one, thereby ensuring site-specific labeling, or to introduce a cysteine residue into a desired location of an otherwise cysteine-less protein. It is important to confirm that the necessary mutations and dye attachments do not impair the biological function of the protein.

Site-specific incorporation of fluorescent labels into RNA is also essential for detailed studies of RNA folding or RNP assembly. Specific internal labeling of RNA oligonucleotides can be achieved by incorporation of amino- or thio-modified base analogs during solid phase chemical synthesis, after which the RNA is covalently labeled with dye NHS esters or dye maleimides, respectively. For FRET studies, both analogs can be incorporated into the same strand, allowing for orthogonal labeling with these dye derivatives. Alternatively, dye-labeled phosphoramidites can be directly introduced into an RNA oligonucleotide during solid phase synthesis. We have found that the latter method is more efficient, especially for oligonucleotides longer than ~30 nt. A limitation of solid phase RNA synthesis is the relatively short length of the synthetic oligonucleotides that can be produced. Generally, the RNA constructs do not exceed 50–60 bases in length.

Labeling of larger RNA molecules can be achieved by a combination of solid phase chemical synthesis, *in vitro* RNA transcription, and RNA ligation methodologies. Labeled RNA oligonucleotides generated as described earlier can be ligated to produce longer RNA strands containing donor and acceptor dyes for FRET studies. Alternatively, a synthetic RNA oligonucleotide (labeled) can be linked to a longer (unlabeled) RNA generated by *in vitro* transcription. There are three different enzymes available to ligate RNA, two RNA ligases and one DNA ligase. T4 RNA ligase joins two single stranded RNAs. The challenge is to align the ends of the two RNAs for ligation. Applying RNA ligase to nicks in hairpin loops solves the problem of bringing the single strands together and exposing the 3′-hydroxy end of the first RNA to the 5′-phosphate of the second RNA (England and Uhlenbeck, 1978). Recently, a DNA splint was used to form hairpin loop-like structures for RNA ligation with T4 RNA ligase (Stark *et al.*, 2006). T4 DNA ligase joins 3′ and 5′ ends in double stranded DNA. However, the RNA in a DNA–RNA hybrid can also be ligated. This splint ligation method (Fig. 11.1) was introduced by M. Moore and co-workers (Moore and Query, 2000). Recently, T4 RNA ligase 2 became commercially available, which can also be used for the splint ligation approach. Compared to DNA ligase, it links RNA strands ~100 times faster (Bullard and Bowater, 2006) and is our enzyme of choice.

Figure 11.1 Joining two RNA strands using the DNA splint ligation method. The two RNA strands are aligned by hybridization with a common DNA template. T4 DNA ligase or T4 RNA ligase 2 catalyzes the covalent joining of the 3′ hydroxyl group of one RNA strand and the 5′ phosphate group of the other strand. The DNA template is subsequently degraded with DNase I.

3. SINGLE-MOLECULE FLUORESCENCE DETECTION METHODS

Single-molecule measurements hold great promise for the study of complex RNA folding processes, because of the ability to resolve multiple species and parallel folding pathways under conditions of thermodynamic equilibrium. To record the fluorescence emitted by just a single molecule, it is essential to eliminate the background fluorescence arising from surrounding molecules and light scattering from solvent molecules. This requires that the detection volume is severely restricted, which can be achieved using either of two different optical configurations. In one approach, the detection volume is minimized by placing a confocal pinhole in the detection path or by using a femtosecond laser pulse to achieve two-photon excitation (confocal microscopy). Alternatively, molecules contained within a very shallow layer (∼100 nm) adjacent to an optical interface are illuminated using an evanescent field (total internal reflection microscopy). The two methods are complementary and both have been applied in studies of RNA folding and RNP assembly processes, as described later.

4. DIFFUSION SINGLE-PAIR FRET FOR RNA BASED SYSTEMS

Diffusion single-pair FRET is a single-molecule technique that combines the spectroscopic phenomenon of FRET with confocal fluorescence microscopy (Deniz et al., 2001). The method allows observation of single molecules freely diffusing in solution. For studies of RNA folding processes, the RNA of interest is labeled with donor and acceptor probes, as described earlier. The fluorescence of donor and acceptor are simultaneously recorded using separate avalanche photodiodes as individual RNA molecules traverse the excitation volume in a confocal microscope (Fig. 11.2). Very dilute solutions are used (typically 100–200 pM concentration) to minimize the probability that two RNA molecules occupy the detection volume simultaneously. For each single-molecule event, the FRET efficiency (E) is calculated using the equation $E = (1 + \gamma I_d/I_a)^{-1}$, where I_d and I_a are the donor and acceptor intensities, respectively, and γ is a correction factor that accounts for differences in quantum yield and detection efficiency of the donor and acceptor. The FRET efficiencies from many single-molecule observations (typically thousands of events are recorded during an acquisition time of a few minutes) are plotted in the form of a histogram. The resulting histogram can directly reveal conformational subpopulations that exist in equilibrium. These subpopulations appear as separate peaks in the

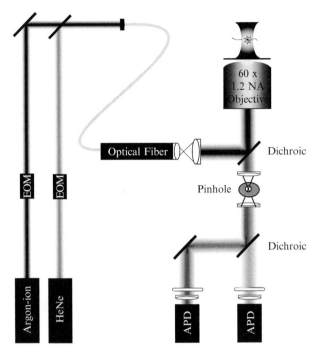

Figure 11.2 Experimental set up for diffusion spFRET measurements. Freely diffusing RNA molecules or RNA–protein complexes, labeled with donor and acceptor dyes, are excited with a green (argon-ion) or red (He–Ne) laser. The fluorescence of donor and acceptor are collected through the same objective used for laser illumination, transmitted through a small pinhole, spectrally separated with a dichroic mirror and detected on separate avalanche photodiodes (APDs). Electro-optical modulators (EOMs) are used to rapidly switch between the green and red laser sources.

FRET efficiency histogram, and are usually fitted with Gaussian functions to obtain the mean FRET efficiency, the width of the distribution, and the area under each peak. The mean FRET efficiencies are related to the average donor–acceptor distance (R) for each conformer, through the well known Förster equation, $E = (1 + (R/R_0)^6)^{-1}$, where R_0 is the Förster distance for the donor–acceptor pair. The peak areas reflect the fractional populations of each conformer. Hence, changes in the conformer populations can be followed in response to changes in solution conditions, such as metal ion concentration, ionic strength, temperature, or the presence of specific RNA-binding proteins (see Application 1). In contrast, all subpopulations are averaged in conventional ensemble FRET measurements. In addition, kinetic information on subpopulations interconverting on the microsecond to millisecond time scale can be obtained from detailed analysis of the histogram peak shapes.

There are several advantages of the spFRET technique for studies of RNA folding processes. First, the optical configuration is relatively simple. Second, there is no need to immobilize the RNA molecules at a surface, which can potentially perturb intrinsic molecular properties. Third, the use of fast avalanche photodiodes for fluorescence detection provides high time resolution (microseconds). However, there are a number of potential disadvantages that need to be considered as well. The throughput of spFRET measurements is relatively low because molecules are observed one at a time. Moreover, the observation period is limited by the diffusional transit time of the RNA through the confocal volume, typically 1–10 ms. Hence, the diffusion spFRET method is most appropriate for studies of fast folding processes. Another limitation is that subpopulations must have markedly different FRET efficiencies in order to be well resolved. Often this can be arranged through judicious placement of the donor and acceptor probes. Finally, a peak at zero FRET efficiency is usually observed in the FRET efficiency histograms. This peak arises from incomplete acceptor labeling or RNA molecules in which the acceptor probe has been destroyed by photobleaching. The zero peak can obscure a subpopulation with low FRET efficiency, such as an unfolded population of RNA molecules with a long donor–acceptor distance. While it is possible to reduce the rate of photobleaching through the addition of oxygen scavengers and triplet state quenchers such as propyl gallate, it is difficult to completely suppress the zero peak by these means. However, the zero peak can be eliminated through the use of alternating laser excitation (ALEX) (Kapanidis et al., 2005). In this method, the sample is alternately illuminated with separate lasers to excite the donor or acceptor. Electro-optical modulators (EOMs) are used to rapidly switch between the two laser sources (Fig. 11.2). Molecules that do not contain an acceptor are readily identified by the absence of any emission following direct acceptor excitation, allowing these molecules to be excluded from the analysis of the data obtained with donor excitation.

5. Total Internal Reflection Fluorescence (TIRF) for RNA Based Systems

TIRF microscopy (Axelrod et al., 1983) can be used in combination with different fluorescence observables to monitor both RNA folding and RNP assembly processes. In TIRF-based measurements, surface-immobilized RNA molecules or RNP complexes are excited by an evanescent field created by total internal reflection of the excitation laser at a quartz–water interface. Generally, the RNA molecule (labeled or unlabeled) is immobilized on the surface, as described later, and proteins (labeled or unlabeled) may be present in the solution phase. A prism in optical contact with the sample cover slip is typically used for TIR illumination (Fig. 11.3). Alternatively, TIR conditions can be achieved by off-axis illumination through an objective (Ha, 2001). In either case, fluorescence emitted by excited molecules is collected through an objective and recorded by an intensified CCD camera. This optical system provides a spatially resolved image of the surface in which individual immobilized molecules or complexes appear as discrete (diffraction-limited) fluorescent spots. Software is used to identify

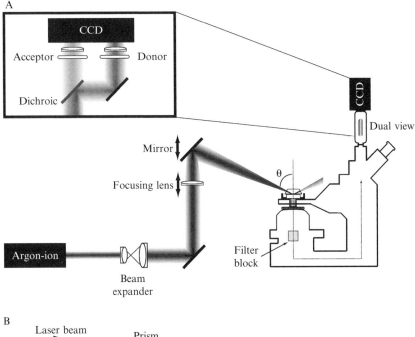

Figure 11.3 Experimental set up for TIRF measurements. (A) Overall view of the apparatus. An argon-ion laser beam enters a prism mounted on the stage of an inverted microscope and is totally internally reflected at a quartz–water interface. The evanescent field created by total internal reflection excites labeled molecules immobilized in the interfacial region. The resulting fluorescence is collected with an objective, separated into donor and acceptor components with a dual view system and detected on separate segments of an intensified CCD camera. (B) Close up view of the prism, sample cell and microscope objective.

single molecules in the images and to integrate the signal from all pixels corresponding to a given molecule.

The TIRF system can be used for either single- or dual-color (FRET) measurements. For single-color experiments, the fluorescence intensity or polarization of a probe attached to the RNA or protein is recorded over time. RNA folding transitions may cause changes in the local fluorophore environment, which are reported as jumps in the intensity or polarization of the probe. Alternatively, single-color TIRF imaging can be used to monitor the binding of labeled protein molecules to immobilized RNA during RNP assembly. In this case, intensity jumps are observed as proteins bind to or dissociate from the RNA (see Application 2). For two-color FRET measurements, doubly labeled RNA molecules are immobilized on the surface. The fluorescence of donor and acceptor is separated with a dichroic mirror and simultaneously recorded on separate segments of the CCD camera (Fig. 11.3). Software is used to match corresponding spots in the donor and acceptor channels. Once the donor and acceptor are identified, each is monitored over time and the FRET efficiency at each time point is calculated using the formula given in the preceding section. This procedure yields a FRET trajectory (time trace) for each donor–acceptor pair in the field of view. These trajectories can be analyzed using hidden Markov modeling to deduce the number of distinguishable FRET states and the transition rates connecting them (Mckinney et al., 2006). The transitions provide a direct readout of RNA structural changes at the single-molecule level during RNA folding or RNP assembly.

An advantage of this system is that hundreds of individual molecules can be monitored over time in parallel. Hence, TIRF measurements are more efficient than diffusion spFRET in terms of maximizing the data throughput. In addition, it is possible to discern transitions between conformers with similar FRET efficiencies as small but distinct steps in the FRET trajectories. In contrast, the spFRET method requires a much larger difference in FRET efficiency in order to resolve conformers. Another advantage of TIRF measurements is that the total observation period is only limited by the photobleaching time of the fluorophores, which can be extended to several minutes through the addition of oxygen scavengers and triplet state quenchers (Rasnik et al., 2006). However, the time resolution of these recordings is limited by the integration time of the CCD camera, typically on the order of 20–100 ms. Generally, the signal to noise ratio of the CCD camera dictates the shortest possible integration time. Hence, this system is most suitable for studying relatively slow RNA folding or RNP assembly processes.

RNA molecules can be readily immobilized on quartz cover slips by means of biotin–streptavidin interactions. A biotin moiety can be introduced at the $3'$ or $5'$ terminus of an RNA oligonucleotide during solid phase chemical synthesis. Alternatively, biotinylated RNA can be conveniently obtained by modifying the 3′-OH group with periodate, followed by treatment with biotin–hydrazide (Proudnikov and Mirzabekov, 1996). Two types of surface treatments can be used for immobilization of RNA. In one method, the surface is treated with a mixture of BSA and biotinylated-BSA that adsorb nonspecifically to the silicate surface. Subsequently the surface is exposed to streptavidin, which binds strongly to biotin via one of its four binding sites. The remaining binding sites are then available to capture biotinylated RNA, which is added last. For studies of RNA folding, the layer of adsorbed BSA molecules is usually sufficient to eliminate undesired surface-induced perturbations. However, for studies of RNP assembly, nonspecific adsorption of proteins to the surface can be a significant problem. In this case, it is more effective to passivate the surface by covalent attachment of polyethylene glycol (PEG) groups. As a first step, the hydroxyl groups of the quartz surface are amino functionalized using 3-aminopropyltriethoxysilane. In a second step, the free amino groups are reacted with an NHS ester of PEG. A small fraction of the PEG groups also contain a biotin moiety. The resulting surface-bound biotin groups are then exposed to streptavidin, as before, after which the biotinylated RNA is captured in a final step. This procedure is effective in suppressing nonspecific adsorption of proteins that are subsequently introduced into the sample chamber during studies of RNA-protein binding and RNP assembly. A more general discussion of surface treatment and immobilization procedures is presented elsewhere (Rasnik et al., 2005; Roy et al., 2008).

6. Application 1: Folding of the Hairpin Ribozyme

The hairpin ribozyme is an excellent system for detailed studies of autonomous RNA folding. The ribozyme has a relatively simple secondary structure, consisting of two internal loops carried within adjacent helical segments. In the natural form of the ribozyme, these helices constitute two arms of a four-way helical junction. To attain catalytic activity, the two loop regions must come into close proximity, allowing them to dock by forming a network of tertiary hydrogen bonding interactions (Fedor, 2000; Walter and Burke, 1998). The crystal structure of the docked ribozyme has been solved at high resolution, revealing the end point of the tertiary folding process (Rupert and Ferré-D'Amaré, 2001).

To monitor docking of the hairpin ribozyme by means of FRET, donor (Cy3) and acceptor (Cy5) dyes were attached to the end of the arms containing loops A and B, respectively (Fig. 11.4). The labeled ribozymes were examined using the diffusion spFRET method (Pljevaljčić et al., 2004). The FRET efficiency histograms revealed two well resolved peaks

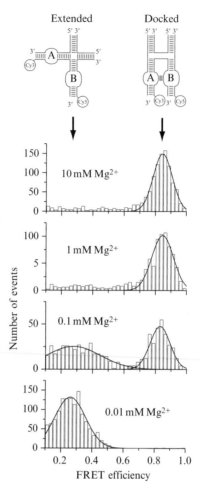

Figure 11.4 FRET efficiency histograms obtained for a hairpin ribozyme labeled with donor (Cy3) and acceptor (Cy5) dyes. spFRET data were recorded over a range of magnesium ion concentrations, as indicated. The peaks observed at $E \sim 0.25$ and $E \sim 0.85$ correspond to extended and docked ribozyme conformers, respectively, as depicted in schematic form at the top of the figure.

(in addition to the zero peak), one centered at $E \sim 0.25$ and another centered at $E \sim 0.85$, assigned to the ribozyme in extended (undocked) and compact (docked) conformations, respectively (Fig. 11.4). These assignments were confirmed using a ribozyme containing a G_{11} to inosine mutation, which was previously known to impair docking. The high FRET peak was markedly reduced in the mutant ribozyme and the low FRET peak was correspondingly enhanced (not shown). The observation of two well resolved peaks in the FRET histogram indicates that the extended and docked conformers must interconvert more slowly than the average photon burst duration for a single-molecule event (~300 µs). FRET histograms recorded over a range of magnesium ion concentrations showed that magnesium favored the docked conformation, as reflected in the respective peak areas of the extended and docked conformers (Fig. 11.4), consistent with the known role of magnesium cations in stabilizing tertiary structure formation in RNA. Notably, the ribozyme was in a fully docked conformation at a physiological concentration of magnesium (1 mM, Fig. 11.4). Similar spFRET measurements were carried out with ribozyme variants in which the central four-way junction was replaced with either a three- or two-way junction. These variants required much higher concentrations of magnesium in order to dock and never attained a completely docked state (Pljevaljčić et al., 2004). These results highlight the importance of the natural four-way helical junction in facilitating the folding of the ribozyme into the docked conformation required for catalytic activity.

The importance of the four-way junction in guiding proper ribozyme folding was manifested even more dramatically in a ribozyme variant in which loop A was replaced by fully complementary duplex RNA. This single-loop (SL) construct is incapable of forming tertiary docking interactions. Regardless, the SL construct was observed to fold to a compact conformation in the presence of magnesium ions. Moreover, the mean FRET efficiency was similar to the docked conformation of the natural ribozyme, indicating a similar arrangement of arms A and B. Hence, this species was termed a "quasi-docked" conformation. However, in contrast to the natural ribozyme, the FRET histogram of the SL construct revealed just a single, broad peak with a highly asymmetric shape (Fig. 11.5A). These features suggested that the SL ribozyme was rapidly exchanging between extended and compact conformations on the submillisecond timescale (Fig. 11.5B). The peak shape could be reproduced using a two-state kinetic model with particular values for the forward and reverse rate constants (Fig. 11.5A). These results reveal that the four-way helical junction folds in such a manner as to naturally position arms A and B in close proximity, facilitating formation of tertiary interactions between the two internal loops. Hence, the quasi-docked conformation is likely to be an intermediate species on the tertiary folding pathway of the natural ribozyme. Moreover, the results illustrate how kinetic information on fast RNA folding processes can be obtained from analysis of the spFRET histogram peak shapes. More sophisticated models used for simulation of the histogram peak shape have been developed by others (Gopich and Szabo, 2005).

The spFRET method has proven to be an excellent tool to obtain new information about the hairpin ribozyme folding pathway. However, the method is not limited to small ribozymes. The same approach could be applied to much larger RNA molecules labeled at specific sites with donor

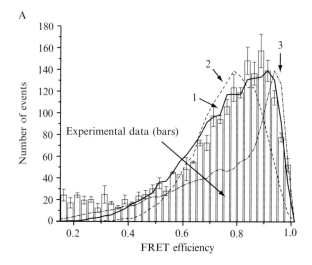

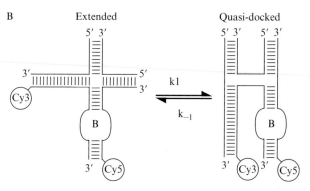

Figure 11.5 A single-loop variant of the hairpin ribozyme can fold to a compact conformation in the absence of tertiary docking interactions. (A) FRET efficiency histogram in the presence of 10 mM MgCl$_2$ (bars). The broad, asymmetric peak shape is consistent with rapid exchange between extended and compact (quasi-docked) conformers, illustrated schematically in part B. The lines in part A are two-state kinetic simulations with different assumed rate constants k_1 and k_{-1}. Line 1 ($k_1 = 2.0 \times 10^4$ s^{-1} and $k_{-1} = 6.8 \times 10^3$ s^{-1}), line 2 ($k_1 = 4.1 \times 10^4$ s^{-1} and $k_{-1} = 1.4 \times 10^4$ s^{-1}), line 3 ($k_1 = 1.0 \times 10^4$ s^{-1} and $k_{-1} = 3.4 \times 10^3$ s^{-1}). Line 1 is the best simulation of the observed histogram peak shape.

and acceptor dyes, using the ligation strategy described earlier. Folded and unfolded conformers could be resolved and information about the global arrangement of helices and other structural elements could be inferred from the corresponding donor–acceptor distances, especially when a variety of constructs with different labeling sites are employed. In addition, the conformer populations and exchange rates could be determined under various solution conditions, as described here for the hairpin ribozyme. Moreover, similar measurements could be performed in the presence of specific protein cofactors that bind to the RNA and alter its three-dimensional structure. These possibilities are illustrated by a recent spFRET study of the spliceosomal U4 snRNA in free and protein-bound forms (Woźniak et al., 2005). We anticipate that the spFRET method will provide a great deal of information about RNA folding and RNP assembly processes across a range of systems.

7. Application 2: Assembly of the Rev–RRE Complex

Rev is a key regulatory protein of HIV-1 that promotes the export of unspliced and partially spliced viral mRNAs from the nucleus to the cytoplasm of an HIV-1 infected cell (Pollard and Malim, 1998). The partially spliced mRNAs encode the structural proteins Gag, Pol, and Env, while the unspliced mRNA encodes the genomic RNA. Since the structural proteins and genomic RNA are needed to produce new virions, Rev-mediated RNA export is an essential step in the viral replication cycle and a potential therapeutic target for the treatment of HIV/AIDS.

Rev binds to a highly conserved region of the viral mRNA known as the Rev response element (RRE). The RRE contains a single high-affinity binding site for Rev, although the 1:1 complex is not active in mRNA export. Instead, multiple Rev molecules must assemble on a single RRE molecule in order to activate mRNA export (Malim and Cullen, 1991; Malim et al., 1989). The mechanism of oligomeric Rev–RRE complex assembly is not well understood. In principle, multiple Rev monomers could bind sequentially to the RRE. Alternatively, Rev might oligomerize in solution prior to binding the RRE. In fact, evidence has been presented for both assembly mechanisms in previous studies (Malim and Cullen, 1991; Malim et al., 1989; Olsen et al., 1990; Zapp et al., 1991).

To resolve this question, we devised a TIRF-based method to visualize Rev–RRE complex assembly at the single-molecule level in real-time (Pond et al.). A large fragment of the RRE capable of binding up to four Rev monomers was immobilized on a PEG-treated quartz surface by means of biotin–streptavidin interactions. The Rev protein was mutated so that only one of the native cysteine residues was available for covalent labeling with Alexa-Fluor 555. Binding of labeled Rev to individual RRE molecules immobilized on the surface was monitored over time by single-color TIRF microscopy.

A typical intensity trajectory reveals discrete and abrupt transitions between different intensity states, reflecting individual Rev binding and dissociation events (Fig. 11.6A). Notably, the jumps correspond to the

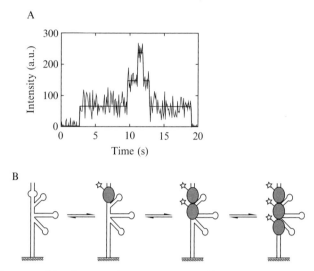

Figure 11.6 Assembly of Rev on the RRE monitored by TIRF microscopy. (A) Representative fluorescence intensity trajectory showing sequential binding of three Rev monomers to the immobilized RRE (upward jumps), followed by three spontaneous dissociation events (downward jumps) (Pond *et al.*). Each jump corresponds to binding or dissociation of a single Rev monomer labeled with Alexa-Fluor 555. (B) Assembly pathway of the Rev–RRE complex revealed by the single-molecule TIRF measurements.

emission intensity of a single labeled Rev molecule under the experimental conditions, indicating that Rev monomers bind to and dissociate from the RRE one at a time. The jump size was calibrated using a monomeric Rev mutant that is incapable of oligomerizing on the RRE. These results provide definitive evidence for the sequential monomer binding pathway of Rev–RRE complex assembly (Fig. 11.6B).

Kinetic information on each assembly and dissociation step was obtained from dwell time analysis of the fluorescence trajectories. An example of the analysis is shown in Fig. 11.7. In the example shown, the histogram of individual dwell times prior to binding of the first Rev monomer is fitted with an exponential function to extract the on-rate for the transition. This analysis can be performed for all of the transitions observed in the intensity trajectories, thereby providing the elementary rate constants for each step of complex assembly or dissociation. It is very difficult, if not impossible, to extract such detailed kinetic information from bulk measurements.

The TIRF imaging system could be extended to study even more complex RNP assembly processes. For example, multicolor imaging could be used to monitor the assembly of two or more proteins, each labeled with a different color dye, on the same RNA molecule. In addition, FRET measurements with doubly labeled RNA could report RNA

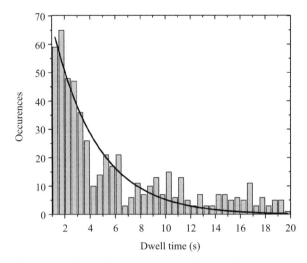

Figure 11.7 Kinetic analysis of a single step of Rev-RRE complex assembly. The histogram of dwell times of the 1:1 Rev:RRE complex prior to binding of the second Rev monomer is shown (Pond et al.). The histogram is fitted with an exponential function (solid line) to obtain the on-rate for binding. The first-order rate ($k_{on} = 0.28 \pm 0.02$ s^{-1}) corresponds to a bimolecular association rate constant of $2.8 \pm 0.2 \times 10^8$ M^{-1} s^{-1} at the Rev concentration used (1 nM).

conformational changes as the proteins bind. Recent studies of the assembly of the bI5 group I intron with its protein cofactor CBP2 (Bokinsky et al., 2006) and of the telomerase RNP (Stone et al., 2007) illustrate the prospects for using single-molecule TIRF methods to study RNP assembly processes. With these approaches, we envisage that novel discoveries will be made in understanding the assembly mechanisms of a variety of biologically important RNPs.

ACKNOWLEDGMENTS

We thank our colleagues Ashok Deniz, Stephanie Pond, William Ridgeway, Rae Robertson, Jun Wang, and Edwin Van der Schans for their contributions to the studies described here. We also apologize to those scientists whose work has not been cited because of space limitations. The research in our laboratory was supported by the National Institutes of Health (grants GM58873 and GM66669).

REFERENCES

Axelrod, D., Thompson, N. L., and Burghardt, T. P. (1983). Total internal reflection fluorescence microscopy. *J. Microsc.* **129**, 19.
Barrick, J. E., and Breaker, R. R. (2007). The distributions, mechanisms, and structures of metabolite-binding riboswitches. *Genome Biol.* **8**, R239.

Blanchard, S. C., Gonzalez, R. L., Kim, H., Chu, S., and Puglisi, J. (2004a). tRNA selection and kinetic proofreading in translation. *Nat. Struct. Mol. Biol.* **11,** 1008.

Blanchard, S. C., Kim, H., Gonzalez, R. L., Puglisi, J., and Chu, S. (2004b). tRNA dynamics on the ribosome during translation. *Proc. Natl. Acad. Sci. USA* **101,** 12893.

Bokinsky, G., Rueda, D., Misra, V. K., Rhodes, M. M., Gordus, A., Babcock, H. P., Walter, N., and Zhuang, X. (2003). Single-molecule transition-state analysis of RNA folding. *Proc. Natl. Acad. Sci. USA* **100,** 9302.

Bokinsky, G., Nivón, L. G., Liu, S., Chai, G., Hong, M., Weeks, K. M., and Zhuang, X. (2006). Two distinct binding modes of a protein cofactor with its target RNA. *J. Mol. Biol.* **361,** 771.

Bullard, D. R., and Bowater, R. P. (2006). Direct comparison of nick-joining activity of the nucleic acid ligases from bacteriophage T4. *Biochem. J.* **398,** 135.

Collins, C. A., and Guthrie, C. (2000). The question remains: Is the spliceosome a ribozyme? *Nat. Struct. Biol.* **7,** 850.

Conn, G. L., and Draper, D. E. (1998). RNA structure. *Curr. Opin. Struc. Biol.* **8,** 278.

Deniz, A. A., Laurence, T. A., Dahan, M., Chemla, D. S., Schultz, P. G., and Weiss, S. (2001). Ratiometric single-molecule studies of freely diffusing biomolecules. *Ann. Rev. Phys. Chem.* **52,** 233.

Doudna, J. A., and Cate, J. H. (1997). RNA structure: Crystal clear? *Curr. Opin. Struc. Biol.* **7,** 310.

England, T. E., and Uhlenbeck, O. C. (1978). Enzymatic oligoribonucleotide synthesis with T4 RNA ligase. *Biochemistry* **17,** 2069.

Fedor, M. J. (2000). Structure and function of the hairpin ribozyme. *J. Mol. Biol.* **297,** 269.

Fiore, J. L., Hodak, J. H., Piestert, O., Downey, C. D., and Nesbitt, D. J. (2008). Monovalent and Divalent Promoted GAAA-Tetraloop-Receptor Tertiary Interactions from Freely Diffusing Single-Molecule Studies. *Biophys. J.* **95,** 3892.

Gopich, I., and Szabo, A. (2005). Theory of photon statistics in single-molecule Förster resonance energy transfer. *J. Chem. Phys.* **122,** 14707.

Ha, T. (2001). Single-molecule fluorescence resonance energy transfer. *Methods* **25,** 78.

Ha, T., Zhuang, X., Kim, H. D., Orr, J. W., Williamson, J. R., and Chu, S. (1999). Ligand-induced conformational changes observed in single RNA molecules. *Proc. Natl. Acad. Sci. USA* **96,** 9077.

Hodak, J. H., Downey, C. D., Fiore, J. L., Pardi, A., and Nesbitt, D. J. (2005). Docking kinetics and equilibrium of a GAAA tetraloop-receptor motif probed by single-molecule FRET. *Proc. Natl. Acad. Sci. USA* **102,** 10505.

Hutchins, C. J., Rathjen, P. D., Forster, A. C., and Symons, R. H. (1986). Self-cleavage of plus and minus RNA transcripts of avocado sunblotch viroid. *Nucleic Acids Res.* **14,** 3627.

Kapanidis, A., Laurence, T. A., Lee, N. K., Margeat, E., Kong, X., and Weiss, S. (2005). Alternating-laser excitation of single molecules. *Acc. Chem. Res.* **38,** 523.

Kim, H., Nienhaus, G. U., Ha, T., Orr, J. W., Williamson, J. R., and Chu, S. (2002). Mg2+-dependent conformational change of RNA studied by fluorescence correlation and FRET on immobilized single molecules. *Proc. Natl. Acad. Sci. USA* **99,** 4284.

Klostermeier, D., and Millar, D. P. (2001a). Time-resolved fluorescence resonance energy transfer: A versatile tool for the analysis of nucleic acids. *Biopolymers* **61,** 159.

Klostermeier, D., and Millar, D. P. (2001b). RNA conformation and folding studied with fluorescence resonance energy transfer. *Methods* **23,** 240.

Kuglstatter, A., Oubridge, C., and Nagai, K. (2002). Induced structural changes of 7SL RNA during the assembly of human signal recognition particle. *Nat. Struct. Biol.* **9,** 740.

Lee, T., Lapidus, L. J., Zhao, W., Travers, K. J., Herschlag, D., and Chu, S. (2007). Measuring the folding transition time of single RNA molecules. *Biophys. J.* **92,** 3275.

Malim, M. H., Böhnlein, S., Hauber, J., and Cullen, B. R. (1989). Functional dissection of the HIV-1 Rev trans-activator–derivation of a trans-dominant repressor of Rev function. *Cell* **58,** 205.

Malim, M. H., and Cullen, B. R. (1991). HIV-1 structural gene expression requires the binding of multiple Rev monomers to the viral RRE: Implications for HIV-1 latency. *Cell* **65,** 241.

McKinney, S., Joo, C., and Ha, T. (2006). Analysis of single-molecule FRET trajectories using hidden Markov modeling. *Biophys. J.* **91,** 1941.

Moerner, W. E., and Orrit, M. (1999). Illuminating single molecules in condensed matter. *Science* **283,** 1670.

Mollova, E. T., and Pardi, A. (2000). NMR solution structure determination of RNAs. *Curr. Opin. Struc. Biol.* **10,** 298.

Moore, M. J., and Query, C. C. (2000). Joining of RNAs by splinted ligation. *Meth. Enzymol.* **317,** 109.

Nahas, M. K., Wilson, T. J., Hohng, S., Jarvie, K., Lilley, D. M., and Ha, T. (2004). Observation of internal cleavage and ligation reactions of a ribozyme. *Nat. Struct. Mol. Biol.* **11,** 1107.

Narlikar, G. J., and Herschlag, D. (1996). Isolation of a local tertiary folding transition in the context of a globally folded RNA. *Nat. Struct. Biol.* **3,** 701.

Nie, S., and Zare, R. N. (1997). Optical detection of single molecules. *Annu. Rev. Biophys. Biomol. Struct.* **26,** 567.

Olsen, H. S., Cochrane, A. W., Dillon, P. J., Nalin, C. M., and Rosen, C. A. (1990). Interaction of the human immunodeficiency virus type 1 Rev protein with a structured region in env mRNA is dependent on multimer formation mediated through a basic stretch of amino acids. *Genes Dev.* **4,** 1357.

Oubridge, C., Kuglstatter, A., Jovine, L., and Nagai, K. (2002). Crystal structure of SRP19 in complex with the S domain of SRP RNA and its implication for the assembly of the signal recognition particle. *Mol. Cell* **9,** 1251.

Pan, J., Thirumalai, D., and Woodson, S. A. (1997). Folding of RNA involves parallel pathways. *J. Mol. Biol.* **273,** 7.

Pljevaljčić, G., Millar, D., and Deniz, A. (2004). Freely diffusing single hairpin ribozymes provide insights into the role of secondary structure and partially folded states in RNA folding. *Biophys. J.* **87,** 457.

Pollard, V. W., and Malim, M. H. (1998). The HIV-1 Rev protein. *Ann. Rev. Microbiol.* **52,** 491.

Pond, S. J., Ridgeway, W., Robertson, R., Wang, J., and Millar, D. P., unpublished results.

Proudnikov, D., and Mirzabekov, A. (1996). Chemical methods of DNA and RNA fluorescent labeling. *Nucleic Acids Res.* **24,** 4535.

Puglisi, J. D., Blanchard, S. C., and Green, R. (2000). Approaching translation at atomic resolution. *Nat. Struct. Biol.* **7,** 855.

Ramakrishnan, V., and Moore, P. B. (2001). Atomic structures at last: The ribosome in 2000. *Curr. Opin. Struc. Biol.* **11,** 144.

Rasnik, I., McKinney, S., and Ha, T. (2005). Surfaces and orientations: Much to FRET about. *Acc. Chem. Res.* **38,** 542.

Rasnik, I., McKinney, S., and Ha, T. (2006). Nonblinking and long-lasting single-molecule fluorescence imaging. *Nat. Methods* **3,** 891.

Roy, R., Hohng, S., and Ha, T. (2008). A practical guide to single-molecule FRET. *Nat. Methods* **5,** 507.

Rueda, D., Bokinsky, G., Rhodes, M. M., Rust, M., Zhuang, X., and Walter, N. (2004). Single-molecule enzymology of RNA: essential functional groups impact catalysis from a distance. *Proc. Natl. Acad. Sci. USA* **101,** 10066.

Rupert, P. B., and Ferré-D'Amaré, A. R. (2001). Crystal structure of a hairpin ribozyme-inhibitor complex with implications for catalysis. *Nature* **410**, 780.

Sclavi, B., Sullivan, M., Chance, M. R., Brenowitz, M., and Woodson, S. A. (1998). RNA folding at millisecond intervals by synchrotron hydroxyl radical footprinting. *Science* **279**, 1940.

Shih, I. H., and Been, M. D. (2002). Catalytic strategies of the hepatitis delta virus ribozymes. *Annu. Rev. Biochem.* **71**, 887.

Sosnick, T. R., and Pan, T. (2003). RNA folding: Models and perspectives. *Curr. Opin. Struc. Biol.* **13**, 309.

Stark, M. R., Pleiss, J. A., Deras, M., Scaringe, S. A., and Rader, S. D. (2006). An RNA ligase-mediated method for the efficient creation of large, synthetic RNAs. *RNA* **12**, 2014.

Stone, M. D., Mihalusova, M., O'connor, C. M., Prathapam, R., Collins, K., and Zhuang, X. (2007). Stepwise protein-mediated RNA folding directs assembly of telomerase ribonucleoprotein. *Nature* **446**, 458.

Stryer, L., and Haugland, R. P. (1967). Energy transfer: A spectroscopic ruler. *Proc. Natl. Acad. Sci. USA* **58**, 719.

Tan, E., Wilson, T. J., Nahas, M. K., Clegg, R. M., Lilley, D. M., and Ha, T. (2003). A four-way junction accelerates hairpin ribozyme folding via a discrete intermediate. *Proc. Natl. Acad. Sci. USA* **100**, 9308.

Treiber, D. K., and Williamson, J. R. (1999). Exposing the kinetic traps in RNA folding. *Curr. Opin. Struc. Biol.* **9**, 339.

Treiber, D. K., and Williamson, J. R. (2001). Beyond kinetic traps in RNA folding. *Curr. Opin. Struc. Biol.* **11**, 309.

Valadkhan, S. (2007). The spliceosome: A ribozyme at heart? *Biol. Chem.* **388**, 693.

Walter, N. G., and Burke, J. M. (1998). The hairpin ribozyme: Structure, assembly and catalysis. *Curr. Opin. Chem. Biol.* **2**, 24.

Weichenrieder, O., Wild, K., Strub, K., and Cusack, S. (2000). Structure and assembly of the Alu domain of the mammalian signal recognition particle. *Nature* **408**, 167.

Winkler, W. C., and Breaker, R. R. (2005). Regulation of bacterial gene expression by riboswitches. *Annu. Rev. Microbiol.* **59**, 487.

Wooniak, A. K., Nottrott, S., Kühn-Hölsken, E., Schröder, G. F., Grubmüller, H., Lührmann, R., Seidel, C. A., and Oesterhelt, F. (2005). Detecting protein-induced folding of the U4 snRNA kink-turn by single-molecule multiparameter FRET measurements. *RNA* **11**, 1545.

Xie, Z., Srividya, N., Sosnick, T. R., Pan, T., and Scherer, N. F. (2004). Single-molecule studies highlight conformational heterogeneity in the early folding steps of a large ribozyme. *Proc. Natl. Acad. Sci. USA* **101**, 534.

Zamore, P. D., and Haley, B. (2005). Ribo-gnome: The big world of small RNAs. *Science* **309**, 1519.

Zapp, M. L., Hope, T. J., Parslow, T. G., and Green, M. R. (1991). Oligomerization and RNA binding domains of the type 1 human immunodeficiency virus Rev protein: A dual function for an arginine-rich binding motif. *Proc. Natl. Acad. Sci. USA* **88**, 7734.

Zhuang, X., Bartley, L. E., Babcock, H. P., Russell, R., Ha, T., Herschlag, D., and Chu, S. (2000). A single-molecule study of RNA catalysis and folding. *Science* **288**, 2048.

Zhuang, X., Kim, H., Pereira, M. J., Babcock, H. P., Walter, N., and Chu, S. (2002). Correlating structural dynamics and function in single ribozyme molecules. *Science* **296**, 1473.

CHAPTER TWELVE

Using Fluorophore-Labeled Oligonucleotides to Measure Affinities of Protein–DNA Interactions

Brian J. Anderson,[*,†] Chris Larkin,[*,‡] Kip Guja,[*,§] *and* Joel F. Schildbach[*]

Contents

1. Introduction	254
2. Definitions of Fluorescence Anisotropy and Intensity	254
3. Advantages of Fluorescence Measurements	256
4. Disadvantages of Fluorescence Measurements	257
5. Designing the Oligonucleotide	258
6. Designing the Experiment	266
7. Competition Assays for Determining Specificity	269
8. Conclusions	270
Acknowledgments	270
References	270

Abstract

Changes in fluorescence emission intensity and anisotropy can reflect changes in the environment and molecular motion of a fluorophore. Researchers can capitalize on these characteristics to assess the affinity and specificity of DNA-binding proteins using fluorophore-labeled oligonucleotides. While there are many advantages to measuring binding using fluorescent oligonucleotides, there are also some distinct disadvantages. Here we describe some of the relevant issues for the novice, illustrating key points using data collected with a variety of labeled oligonucleotides and the relaxase domain of F plasmid TraI. Topics include selection of a fluorophore, experimental design using a fluorometer equipped with an automatic titrating unit, and analysis of direct binding and competition assays.

---

[*] Department of Biology, Johns Hopkins University, Baltimore, Maryland
[†] George Washington University School of Medicine, Washington, District of Columbia
[‡] Food and Drug Administration, Rockville, Maryland
[§] School of Medicine at Stony Brook University Medical Center, Stony Brook, New York

1. Introduction

DNA-binding proteins perform or regulate myriad cellular activities. To fully appreciate the function of these proteins, we need to determine their affinities and specificities for DNA. Numerous methods have been developed and employed to characterize the interactions between proteins and DNA. For example, in a filter-binding assay, varying protein concentrations are combined with a radiolabeled oligonucleotide, and bound oligonucleotide is separated from free by filtration through nitrocellulose disks. In the conceptually similar electrophoretic mobility shift assay (EMSA), bound and free DNA are separated owing to their differing mobilities in a nondenaturing polyacrylamide gel. Using isothermal titration calorimetry (ITC), binding is observed by heat release or absorption occurring when protein and DNA are combined. While these methods are still widely used, researchers are increasingly following binding by observing changes in fluorescence emission intensity and anisotropy. The attraction is due in part to the diversity of available fluorophores, and the decreased cost of commercial syntheses.

Fluorescence as a technique to measure binding is powerful and useful, but also has limitations and some inherent complications. These potential problems may be obvious to the experienced fluorescence spectroscopist, but they could be troublesome for the unsuspecting novice. Here we hope to point out some of the strengths, weaknesses, and pitfalls of these methods for those starting out. Those desiring more detailed discussions of the topics in fluorescence introduced here are referred to *Principles of Fluorescence Spectroscopy* (Lakowicz, 2006).

2. Definitions of Fluorescence Anisotropy and Intensity

Frequently, the binding of protein to a fluorophore-labeled oligonucleotide is monitored by changes in fluorescence anisotropy. For anisotropy measurements, the solution is excited by a polarized beam of light at an appropriate wavelength. Use of a polarized excitation beam photoselects the subset of the fluorophores that is oriented properly in solution relative to the incoming light. If the excitation beam is vertically polarized, the vertically ($I_\parallel$) and horizontally ($I_\perp$) polarized emissions are measured and used to calculate the anisotropy, r, according to

$$r = \frac{I_\parallel - I_\perp}{I_\parallel + 2I_\perp} \qquad (12.1)$$

There is a directionality to the fluorescent emission as well as the excitation, and the directionality of the excitation and emission are related. Therefore, a fluorophore that rotates slowly in solution relative to its fluorescent lifetime (τ, the time required for the excited population to decay to $1/e$ of its initial value) will have a large $I_\parallel$ value and small $I_\perp$ value and a relatively large anisotropy. For fluorophores that tumble rapidly relative to their fluorescent lifetime, $I_\parallel$ and $I_\perp$ will be similar and the anisotropy will be near zero. Fluorescence anisotropy is a useful measure for binding because the rotation of biological macromolecules in solution occurs on the same nanosecond time scale as the fluorescent lifetime of many commonly used fluorophores. Frequently the binding of a protein to a fluorescently labeled oligonucleotide will increase the mass sufficiently to significantly decrease the rotational diffusion of the fluorophore. Thus, binding can be followed by the change in anisotropy. One disadvantage to using anisotropy to follow binding is that the use of polarimeters reduces the intensity of the excitation beam, and higher concentrations of the fluorophore-labeled oligonucleotide may be required to obtain sufficient signal.

The fluorescence emission intensity of a fluorophore-labeled oligonucleotide also can change upon binding of a protein. The fluorescent properties of dyes can vary considerably with changing environment, and a proximally bound protein can enhance or decrease fluorescence emission. The change in emission intensity as protein is titrated into the solution can be a convenient measure of protein binding, but it can also complicate anisotropy measurements. The conditions that cause these intensity changes frequently change the fluorescent lifetime of the fluorophore. Anisotropy and fluorescent lifetime are related, as is evident in the Perrin equation:

$$r = \frac{r_0}{1 + (\tau/\theta)} \qquad (12.2)$$

where r is the expected anisotropy, r_0 is the anisotropy in absence of rotational diffusion, τ is the fluorescent lifetime, and θ is the rotational correlation time. One assumption inherent in the analysis of binding curves is that the change in signal is proportional to the change in the distribution of molecules between the unbound and bound states. When binding is accompanied by changes in both fluorescence anisotropy and emission intensity, this proportionality starts to break down. Although this effect may be minor in most cases, it can be dealt with by fitting both intensity and anisotropy data with the program SPECTRABIND (Toptygin and Brand, 1995a,b). Use of this program has the added benefit that each calculated binding constant results from fitting two data sets, potentially improving the quality of the fits. Alternatively, if the change in fluorescence intensity upon binding is sufficient, intensity changes alone can be used to follow binding.

In summary, fluorescence anisotropy is expected to increase when a protein binds a fluorophore-labeled oligonucleotide. Fluorescence emission intensity may also change, and if used in the analysis, this additional data can be considered a bonus. Alternatively, the intensity data may be used alone. The effect of binding on both intensity and anisotropy should be assessed as part of the characterization of the system.

3. Advantages of Fluorescence Measurements

There are several advantages of fluorescent measurements over those using radioactivity. Unlike use of radioactivity, use of fluorophores does not require special licensing and incurs no costs for disposal of radioactive waste. The 2-week half-life of ^{32}P limits the useful lifetime of an end-labeled oligonucleotide, which may necessitate multiple costly and time-consuming labelings. In contrast, a commercially synthesized and purified fluorophore-labeled oligonucleotide can be aliquoted and stored indefinitely in a freezer (repeated freeze–thaw cycles, however, should be avoided). In addition, although these syntheses generate only nanomole quantities of the oligonucleotides, the sensitivity of fluorescence and fluorometers is sufficient to allow the yield to last for dozens of measurements.

Using fluorescence, affinities can be measured in solution without a step such as electrophoresis or filtration designed to separate free and bound oligonucleotide. Thus fluorescence measurements may yield true equilibrium binding constants. Furthermore, results from solution-based assays may be more accurate for measurements of lower affinity interactions, which are more likely to be affected by dissociation during the separation step than high-affinity interactions. Another advantage is that studies in which the influence of pH or salt concentrations on binding are examined to explore the intermolecular forces involved may be easier to perform because there is no worry that the conditions in the binding reaction and in the separation step differ meaningfully. As discussed later, however, these studies may be complicated by the behavior of some fluorophores as solution conditions are altered.

ITC, like fluorescence intensity and anisotropy, may be used to follow equilibrium binding in solution without the need for a separation step. ITC has the added benefit that it also provides information on the entropic and enthalpic contributions to binding (Oda and Nakamura, 2000). Calorimetric titrations, however, often require considerably more material and higher concentrations than comparable fluorescence experiments. These requirements can present particular problems if the protein under study is difficult to purify or has low solubility. In addition, the high concentrations used in ITC experiments can make it difficult to accurately measure binding when dissociation constants drop much below the micromolar range.

Fluorescence has considerable advantages when the kinetics of binding is being measured. Many fluorometers, even the most basic models, can collect intensity data in a kinetic mode, allowing continuous collection of data points. Scores or hundreds of time points can be collected over the course of a binding reaction, yielding well-defined binding curves. In contrast, other methods can be adapted to collect kinetic data, but are not nearly as amenable to collecting large data sets.

4. Disadvantages of Fluorescence Measurements

Fluorescence measurements have several disadvantages relative to EMSAs using ^{32}P-end-labeled oligonucleotides. One is the cost of equipment. One can obtain useful information from an EMSA with a nondenaturing gel, buffers, a power supply, a gel dryer, film, and developer. In contrast, fluorescence measurements require a comparatively expensive fluorometer. Another disadvantage is the lower sensitivity of fluorescence, relative to radioactivity. Picomolar concentrations of ^{32}P-labeled oligonucleotides provide ample signal for an EMSA, while nanomolar concentrations of a fluorophore-labeled oligonucleotide may be necessary for a suitable signal-to-noise ratio. This can present a problem for high-affinity ($K_D \sim$ subnanomolar) reactions because essentially all protein titrated into the reaction will bind the oligonucleotide, thus making it difficult to obtain a useful binding constant from the binding isotherm. If this poses a problem, however, binding can be measured under different conditions (e.g., higher ionic strength) that will reduce the affinity to a more accessible range.

Another problem with fluorescence measurements is that they may not report well on other than two-state reactions. Fluorescence measurements work well when observing a single transition between the unbound and bound states that is accompanied by a substantial change in anisotropy or intensity. Fluorescence also can work well for more complex systems, at least in some cases. If, for example, there are well-separated transitions between different bound states, the situation should be immediately apparent and an appropriate binding model developed. If multiple protein molecules bind with high cooperativity, this situation may be evident by poor fitting of a simple (stoichiometry 1:1) binding model to the data, with the slope of data points through the transition being greater than the slope of the fit through the region. If, however, there are multiple bound states and poorly separated transitions, it is likely that the binding model selected for analysis will be too simple. Unfortunately, it may still be possible to obtain reasonable quality fits to these data with a simple binding model, leaving the investigator with few indications that the studies are flawed. In contrast,

EMSA data can provide additional information in the form of shifted bands. The presence of multiple shifted bands indicates that different binding stoichiometries are present, and these are useful in determining an appropriate binding model. If an appropriate gel imaging system is available, such as a GE Healthcare Typhoon 9410, and an appropriate fluorophore is used, a researcher can use the same oligonucleotide for solution and EMSA studies, with the complementary approaches providing additional insight into the protein–DNA interaction.

Another disadvantage of using fluorophore-labeled oligonucleotides is that the fluorophores are large enough to potentially interfere with or otherwise influence binding (Wang et al., 1998). In contrast, ITC experiments can be performed with unmodified DNA oligonucleotides, and EMSA experiments can use oligonucleotides end-labeled with ^{32}P-phosphate, a comparatively minor modification. To check whether the fluorophore is affecting binding, routine characterization of an oligonucleotide should include comparing the measured K_D of binding with the measured K_I of an unlabeled version of the same oligonucleotide measured by competition experiments (Stern and Schildbach, 2001; Titolo et al., 2003b). (Competition assays are discussed later in this review.) Agreement between the K_D and K_I values indicates that the fluorophore has little effect on the binding.

In summary, while measuring binding using fluorophore-labeled oligonucleotides requires access to a fluorometer, the method can measure binding in solution using less material than ITC and without the radioactive waste generated by EMSA. EMSA, however, can more directly report on stoichiometry of binding through the presence of multiple bands that indicate different bound species. To make sure an appropriate binding model is selected, an investigator should pay careful attention to any systematic deviation of the fit from the data points, and if possible should use multiple experimental approaches to confirm any proposed model.

5. Designing the Oligonucleotide

In the last decade or so, preparing fluorophore-labeled oligonucleotides has progressed from a largely do-it-yourself endeavor involving fluorescein, rhodamine, or one of their derivatives, to one where an oligonucleotide labeled with one of dozens of different fluorophores can be commercially synthesized, purified, and delivered for a reasonable price. Available fluorophores have a variety of excitation and emission spectra with maxima above 500 nm, well above the absorbance maxima of soluble protein and DNA, reducing the possibility of inner filter effect (interference caused by light absorbance by the solutions). Many of these fluorophores are relatively insensitive to photobleaching, have large extinction coefficients (reflecting their ability to absorb photons) and have reasonable quantum yields (the fraction of photon emitted per photon absorbed), all of which are beneficial.

While several fluorophores that can be linked to oligonucleotides are available, their differing properties may make one optimal for one assay and a poor choice for another. To illustrate this point, we provide curves generated by the binding of TraI36 to single-stranded DNA (ssDNA) oligonucleotides 3′-end-labeled with different fluorophores. TraI36, a 36 kDa N-terminal fragment of the TraI protein of the conjugative bacterial plasmid F factor, binds with sequence specificity and subnanomolar K_D to a single-stranded oligonucleotide containing a plasmid sequence (Stern and Schildbach, 2001). All oligonucleotides were synthesized and HPLC-purified by Integrated DNA Technologies. While we believe that the curves presented are useful examples, the reader should keep in mind that the results are to some extent situational and cannot necessarily be extrapolated to another experimental system.

Figure 12.1 depicts the changes in fluorescence anisotropy and intensity that occur upon binding of the TraI36 protein to 17-base ssDNA oligonucleotides labeled with various fluorophores. All three fluorophores show significant increases in fluorescent anisotropy, with the increase of an indocarbocyanine-3 (Cy3)-labeled oligonucleotide being the greatest. All three also show changes in fluorescence emission intensity, but the intensity of the Cy3-labeled oligonucleotide falls, while the intensities of carboxytetramethylrhodamine (TAMRA) and 6-carboxyfluorescein (6-FAM)-labeled oligonucleotides rise.

The linker used to connect the fluorophore to the oligonucleotide can also affect results. Shown in Fig. 12.2 is a comparison of two 22-base oligonucleotides 3′-end-labeled with TAMRA, one synthesized using a TAMRA-modified controlled pore glass (CPG) column and the other linked through a succinimidyl (NHS) ester. The CPG-derived oligonucleotide, which has the fluorophore attached via a linker to a modified thymine, shows significantly greater increases in fluorescence emission intensity and anisotropy than the NHS oligonucleotide. In addition, the NHS-linked oligonucleotide, which has the fluorophore attached to the 3′-phosphate of the oligonucleotide via a 6-carbon linker, also yields a significantly sloped upper baseline for both intensity and anisotropy plots. The combination of small signal change and sloped upper baseline makes for data to which it is difficult to fit a model.

Sometimes the sequence of the oligonucleotide can affect the properties of the attached fluorophore. Time-resolved fluorescence studies of various tetramethylrhodamine (TMR)- or TAMRA-labeled oligonucleotides demonstrate that the fluorophore can have multiple fluorescent lifetimes, suggesting that the probe exists in multiple environments (Edman *et al.*, 1996; Eggeling *et al.*, 1998; Harley *et al.*, 2002; Vamosi *et al.*, 1996). One of these lifetimes may involve an interaction between fluorophore and DNA

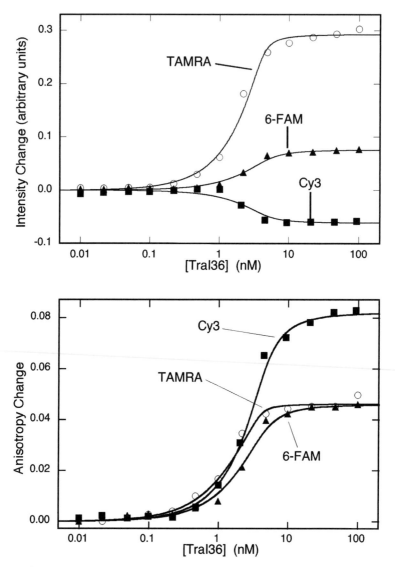

Figure 12.1 Binding of TraI36 to a fluorophore-labeled single-stranded oligonucleotide has different effects on different fluorophores. The changes in fluorescence intensity (top) and anisotropy (bottom) are shown for a 17-base oligonucleotide (5'-TTTGCGTGGGGTGTGTG-3') 3'-labeled with TAMRA (open circles), 6-FAM (filled triangles), or Cy3 (filled squares). The solid lines depict simultaneous fits to intensity and anisotropy data using SPECTRABIND (Toptygin and Brand, 1995a,b) as described (Stern and Schildbach, 2001).

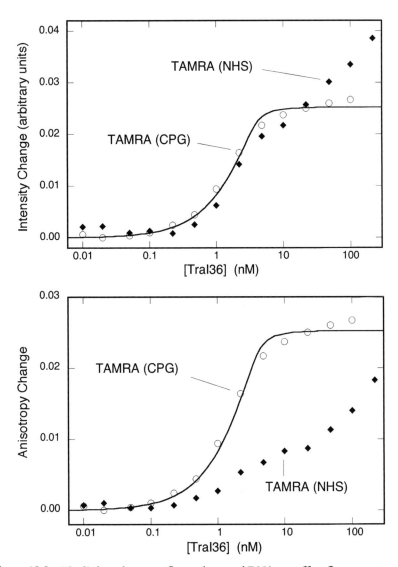

Figure 12.2 The linkage between fluorophore and DNA can affect fluorescent properties. TraI36 titrations into solutions of 22-base (5'-TTTGCGTGGGGTGTGT GCTTTT-3') single-stranded oligonucleotide 3'-labeled with TAMRA using either a CPG (open circles) or NHS (filled diamonds) linkage are shown. The solid lines depict simultaneous fits to intensity and anisotropy data using SPECTRABIND (Toptygin and Brand, 1995a,b) as described (Stern and Schildbach, 2001).

bases, which can potentially quench fluorescence emission or otherwise affect the measured values (Sauer *et al.*, 1995; Seidel *et al.*, 1996; Sevenich *et al.*, 1998). Fluorophore–nucleic acid interactions have also been proposed for Cy3 and fluorescein (Nazarenko *et al.*, 2002; Nelson *et al.*, 1993). Because these interactions can influence behavior of the fluorophore, two labeled oligonucleotides having the same recognition sequence placed within different contexts may have significantly different properties. Some fluorophores, notably TMR and TAMRA, which are more sensitive to their environment than other dyes, may show a greater context dependence than others.

The length of the oligonucleotide, and more specifically the distance between the fluorophore and the bound protein, can affect the results (Titolo et al., 2003a). We found that TraI36 binds 3′-TAMRA-labeled 17-base and 22-base ssDNA oligonucleotides (the latter contains five extra bases at the 3′ end) with similar affinities, but the shorter oligonucleotide exhibits greater increases in fluorescence anisotropy and intensity upon binding (Harley et al., 2002). This is a general phenomenon in our system, with 17-base oligonucleotides 3′-labeled with Cy3, 6-FAM, and NHS or CPG-linked TAMRA showing greater changes in intensity and anisotropy upon TraI36 binding than 22-base 3′-labeled oligonucleotides (Fig. 12.3 and data not shown). The differences in anisotropy can be explained by noting that anisotropy is a composite of the individual motions of the fluorophore and the global rotational diffusion of the macromolecule to which it is conjugated. A greater length of highly flexible ssDNA between dye and bound protein likely translates into a greater contribution of local motions to anisotropy in both bound and unbound states, and a smaller change in anisotropy upon binding. To what degree these observations with ssDNA can be generalized to double-stranded DNA (dsDNA) is not clear, especially given the more rigid nature of dsDNA relative to ssDNA.

As mentioned previously, one advantage to using fluorescence to follow formation of protein–DNA complexes is that it is possible to determine the effect of changing pH or salt conditions on binding without worrying whether the conditions for the binding reaction can be maintained while bound and free DNA are separated. Properties of fluorophores, however, are often sensitive to pH and salt. For example, the fluorescence intensity of a TAMRA-labeled oligonucleotide can decrease with increasing NaCl concentrations, while its fluorescence anisotropy increases (Vamosi et al., 1996). Both fluorescence intensity and anisotropy of a TAMRA-labeled oligonucleotide decrease with increasing temperature (Vamosi et al., 1996). The fluorescence emission intensity of fluorescein falls dramatically below pH 7. As shown in Fig. 12.4 and previously (Stern et al., 2004), while the fluorescence anisotropy of 22-base ssDNA oligonucleotides 3′-labeled with Cy3 or TAMRA generally increases upon binding by TraI36, at pH 10.5 and above the anisotropy decreases. Provided that binding causes a significant change in fluorescence anisotropy or emission intensity, the effect of pH, salt, or temperature on the behavior of the fluorophore probably will not affect data interpretation in most cases. It is possible, however, that a reduction in signal relative to noise could reduce the quality of data fits and, in the worst case, bury the transition in the noise.

Measuring Binding Using Fluorescent Oligos

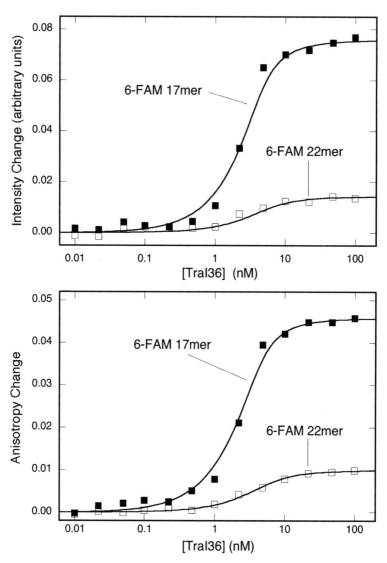

Figure 12.3 The proximity of fluorophore to bound protein can affect fluorescent properties. Single-stranded 22-base (5′-TTTGCGTGGGGTGTGTGCTTTT-3′; open squares) or 17-base (5′-TTTGCGTGGGGTGTGTG-3′; filled squares) oligonucleotides were synthesized with a 3′-6-FAM probe. Intensity (top) or anisotropy (bottom) changes occurring upon titration of TraI36 are shown. The solid lines depict simultaneous fits to intensity and anisotropy data using SPECTRABIND (Toptygin and Brand, 1995a,b) as described (Stern and Schildbach, 2001).

Fluorophores can potentially interact with the protein, altering their fluorescent properties. Often this is advantageous, causing a greater increase in anisotropy or a change in intensity than would have occurred in absence

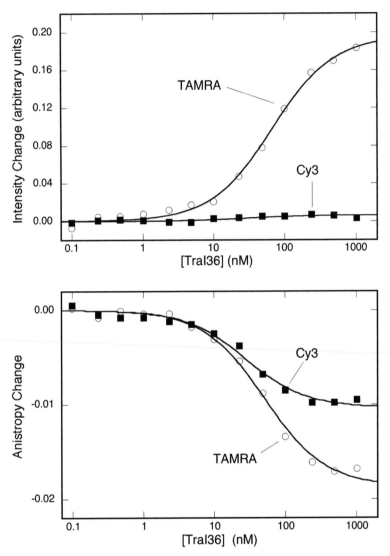

Figure 12.4 Changing solution conditions can alter fluorescent behavior of probes. TraI36 binding to 22-base oligonucleotides 3′-labeled with TAMRA (open circles) or Cy3 (filled squares) at pH 10.5 causes less significant changes to intensity (top) and decreases instead of increases in anisotropy (bottom) (compare to Fig. 12.1). The solid lines depict simultaneous fits to intensity and anisotropy data using SPECTRABIND (Toptygin and Brand, 1995a,b) as described (Stern and Schildbach, 2001).

of the fluorophore-protein interaction. The situation, however, can complicate analysis when examining the effects of amino acid substitutions. As part of a mutagenesis study designed to examine the contributions of various amino acids to TraI36 DNA recognition, we generated a series of protein variants with Ala substituting for contact residues (Harley and Schildbach, 2003; Larkin et al., 2005). Usually, the effects of variant proteins on anisotropy and intensity of a 3′-TAMRA-labeled 22-base oligonucleotide were similar to those caused by binding of the wild type protein. Two exceptions, R150A and E187A, affect fluorescence differently than the wild type protein and other variants. As shown in Fig. 12.5, R150A causes a smaller fluorescence anisotropy change when it binds, and E187A causes a larger change, than the change caused by binding of the wild type protein. In addition, both variant proteins show a steeply sloped upper baseline on the anisotropy plot, complicating analysis. Because R150 and E187 have opposite charges and Ala substitutions for them cause distinct effects on the fluorophore's behavior, one possibility is that altered electrostatic interactions between fluorophore and protein may explain the altered fluorescent characteristic. An examination of the TraI36 structure (Datta et al., 2003), however, shows that residues 150 and 187 are located nearly 30 Å apart, making a model involving direct charge–charge interactions seem less likely.

Given the differing properties of the fluorophores and the effects that altered conditions can have on them, how should an oligonucleotide be selected for binding studies? First, proceed with caution. Time and money

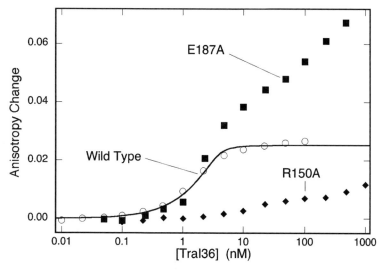

Figure 12.5 Amino acid substitutions in TraI36 can alter fluorophore behavior. Shown are the increases in fluorescent anisotropy of a 3′-TAMRA-labeled 22-base oligonucleotide upon binding of wild type TraI36 (open circles) or the TraI36 variant proteins E187A (filled squares), or R150A (filled diamonds). The solid line depicts simultaneous fits to intensity and anisotropy data using SPECTRABIND (Toptygin and Brand, 1995a,b) as described (Stern and Schildbach, 2001).

spent at the beginning in an effort to select and characterize a fluorophore is well spent. Second, the greater information available for certain dyes may give them an edge. For example, David Lilley and colleagues determined the NMR structure of a Cy3-labeled oligonucleotide (Norman *et al.*, 2000). They found that that 5′-labeled Cy3, attached by a 3-carbon-tether, stacked over the terminal C–G pair. This interaction may account for the stabilizing effect that Cy3 can have on an oligonucleotide duplex (Moreira *et al.*, 2005). This Cy3 characteristic may benefit anisotropy studies because if the local motions of the fluorophore are limited, such as by stacking over the terminal base pair of the oligonucleotide, then the anisotropy will more directly report on the binding by the protein and potentially will lead to greater changes in anisotropy. Another advantage is that if the fluorophore is stacked over the terminal base pair, it is less likely to be interacting directly with the protein and thus causing unexpected results. Because the behavior of Cy3 is better understood than many fluorophores, Cy3 is one of the more attractive choices of fluorophore. Finally, the equipment may in part dictate the fluorophore used. If one wishes to characterize the binding by both fluorometer and gel imaging system, the fluorophore has to be compatible with both systems. The allowed excitation wavelengths may be restricted by the light sources in, or the filters available for, the equipment to be used.

6. Designing the Experiment

Once the oligonucleotide is designed and the protein purified, a binding curve can be generated. Prior to collecting data, though, a number of parameters should be assessed. Perhaps the first item to check is the absorbance of the protein solution and binding buffer at the excitation and emission wavelengths. If the solution absorbs significantly (>0.1 optical density units at the excitation wavelength), inner filter effect can interfere with measurements. If binding is to be followed by observing a change in fluorescence emission, emission spectra of unbound and bound oligonucleotide should be collected to determine the optimal emission wavelength for the experiment. In addition, simple measurements of the association and dissociation kinetics should be made to determine reasonable equilibration periods that should elapse between addition of protein to the fluorometer cell and collection of data.

Regarding the data collection itself, an accurate fit requires well-defined upper and lower baselines, in addition to several data points through the transition. If performing a manual titration, such as those depicted in Figs. 12.1–12.5, a decent curve can be collected given a bit of planning, a rough estimate of affinity, and a pair of pipettors. If using a fluorometer outfitted with an automatic titrator, such as one in the Hamilton Microlab 500 series, the task is a little more complicated. While automatic titrators deliver the protein to the fluorometer cell with greater precision than is possible manually and save the user considerable time, the simple models are not particularly well designed to deliver protein over the three to four logs of concentration required to generate a good curve. Again, though, with some planning, generating a high-quality binding curve is possible. Shown in Fig. 12.6 is a binding curve of a TraI36 variant binding to an ssDNA sequence from the R100 plasmid (4 nM oligonucleotide, $K_D = 7$ nM). Using a titrator equipped with a 50 μl syringe, a protein solution of 250 nM, and a program using three concentration step sizes (0.08, 3, and 65 nM), we can cover the entire binding curve well. Titration schemes usually have to be optimized for each protein studied to make sure that the concentration jumps that occur when step sizes change do not occur at positions that would cause a less accurate fit, such as at the beginning of the transition.

Once the data are collected, a model may be fit to them to yield the binding constant. If fluorescence anisotropy was measured, the values may be used in a fit without further manipulation. If fluorescence emission intensity data were collected, the intensity values must be corrected for volume changes during the course of the experiment. As protein is titrated,

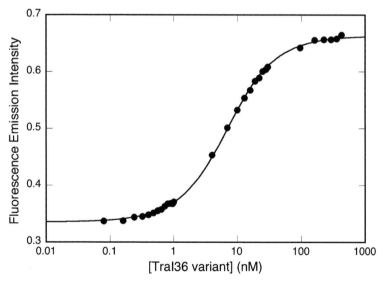

Figure 12.6 Titration scheme using an automatic titrator. Shown is a curve for the binding of an F TraI36 variant protein to an oligonucleotide containing a sequence from the R100 plasmid. For an interaction with $K_D = 7$ nM and fluorescent oligonucleotide concentration of 4 nM, three step sizes (0.08 nM step size to 0.96 nM, 3 nM step size to 30 nM, and 65 nM step size to 400 nM) were used to generate this binding curve. Note the breaks at the point at which the step size changes. The titration scheme has to be adjusted to prevent these from occurring at points in the curve where they could reduce the quality of the fit. The solid line depicts a fit to the data using KaleidaGraph as described (Larkin *et al.*, 2007).

the total volume increases and the fluorophore concentration, and thus emission intensity, decreases. To correct for the dilution, simply multiply the intensity value at protein concentration P by the ratio of the total volume in the fluorometer cell at P to the initial volume.

For the simple reaction having a 1:1 stoichiometry

$$P + D \leftrightarrow PD \quad (12.3)$$

where P is protein, D is DNA, and PD is the complex, the dissociation constant, K_D, is defined

$$K_D = \frac{[P_F][D_F]}{[PD]} \quad (12.4)$$

where $[P_F]$ and $[D_F]$ are the concentrations of free protein and free DNA respectively. Because the total DNA concentration ($[D_T]$) is the sum of $[D_F]$ and $[PD]$, and $[P_T]$ is the sum of $[P_F]$ and $[PD]$:

$$K_D = \frac{[P_F][D_F]}{[PD]} = \frac{([P_T] - [PD])([D_T] - [PD])}{[PD]} \quad (12.5)$$

Multiplying through and rearranging terms yields

$$[PD]^2 - ([PD] \times ([D_T] + [P_T] + K_D)) + ([D_T][P_T]) = 0 \quad (12.6)$$

Solving for $[PD]$ using the quadratic equation

$$[PD] = \frac{(D_T + P_T + K_D) - \sqrt{(D_T + P_T + K_D)^2 - (4[D_T][P_T])}}{2}$$

$$(12.7)$$

Using this definition of $[PD]$, we can fit data with the formula

$$\theta = \left((B_U - B_L) \times \frac{[PD]}{[D_T]} \right) + B_L \quad (12.8)$$

where θ is the fraction DNA bound, B_U is the upper baseline, and B_L is the lower baseline.

7. COMPETITION ASSAYS FOR DETERMINING SPECIFICITY

Frequently, determining the specificity of a protein–DNA interaction is as important as defining the affinity of the interaction. To measure specificity, one could, in principle, synthesize (or, more likely, pay to have synthesized) a series of variant oligonucleotides, all similarly fluorescently labeled, and measure the affinity of the protein for each. The expense of a large panel of labeled oligonucleotides, however, is prohibitive, limiting the sequence space that can be examined.

Alternatively, unlabeled oligonucleotides may be used in competition assays. One common method is to combine protein and labeled wild type oligonucleotide, then titrate unlabeled variant oligonucleotide to compete away binding of the labeled oligonucleotide. This approach has the advantage that less expensive unlabeled oligonucleotides, instead of fluorophore-labeled oligonucleotides, can be used to define binding specificity. We have had some problems with this approach, however. Even if we convert the IC_{50} value (concentration of unlabeled competitor oligonucleotide required to compete away 50% of the binding of labeled oligonucleotide) obtained in the competition experiment to a K_I value (the equivalent of a K_D for an inhibitor) using the method of Cheng and Prusoff (Cheng and Prusoff, 1973), the method underestimates our high-affinity interactions. We have also observed changes in fluorescence emission intensity of our labeled oligonucleotide in the presence of high concentrations (~ 1 μM) of an unlabeled competitor oligonucleotide, perhaps the result of interactions between the oligonucleotides. The resulting shift in the baseline makes fitting a model to the data challenging. While these apparent oligonucleotide interactions may be a problem specific to our system, or one that is more likely to plague ssDNA oligonucleotides than dsDNA oligonucleotides, they still are potential obstacles.

We have had greater success with a second competition method that assesses the binding of the protein to a labeled oligonucleotide in the presence of unlabeled competitor oligonucleotide. We then calculate the affinity of the protein for the competitor oligonucleotide based on the apparent affinity of the protein for the labeled oligonucleotide. The approach has several advantages. It requires less competitor oligonucleotide than a complete competition with the competitor, even if it does require more protein. If, as in our case, interactions between oligonucleotides at high concentration are a problem, these interactions may be minimized by carefully selecting the concentrations of oligonucleotides. Finally, because the assay does not require all binding of the labeled oligonucleotide to be competed away to obtain a complete curve and a good fit to the data, it may be more suitable for low-affinity interactions.

Wang (1995) published an expression that can be fit to data produced by this type of competition assay. (Because the derivation is detailed, we refer the reader to the paper rather than recapitulate it here.) We have found that using this method in test cases, the binding constant measured for an unlabeled competitor correlates well with the binding constant for the labeled version of that oligonucleotide. The method has limitations, however. Because the assay is indirect, it requires that the concentration of unlabeled competitor be accurately known. Well-defined baselines and numerous points through the transition are essential to obtaining a high-quality fit to the data. Finally, because two oligonucleotides with potentially different binding characteristics are used in the reaction, the equilibration times must be chosen carefully to ensure that true equilibrium measurements are made.

8. Conclusions

Using fluorophore-labeled oligonucleotides to measure the affinity and specificity of DNA recognition has many advantages. If a researcher has access to appropriate equipment, especially if that equipment is a fluorometer outfitted with an automatic titrator, a relatively modest investment in setup time can yield a wealth of binding data. As for any measurement, however, understanding the behavior of the system is crucial for subsequent successful measurements. In addition to carefully selecting the length and sequence of the DNA to be used in the assay and to determining the stoichiometry of the complex, a suitable fluorophore must be selected. As shown here, different fluorophores can demonstrate dramatically different behavior when attached to the same oligonucleotide, underscoring the importance of initial characterization to successful measurements.

ACKNOWLEDGMENTS

The authors thank Prof. Lenny Brand, and Drs. Dima Toptygin and Mike Rodgers for their generosity with their time, equipment, and expertise. This manuscript is based on work supported by National Institutes of Health grant GM61017 to J.F.S.

REFERENCES

Cheng, Y., and Prusoff, W. H. (1973). Relationship between the inhibition constant (KI) and the concentration of inhibitor which causes 50 per cent inhibition (I50) of an enzymatic reaction. *Biochem. Pharmacol.* **22**, 3099–3108.

Datta, S., *et al.* (2003). Structural insights into single-stranded DNA binding and cleavage by F factor TraI. *Structure (Camb.)* **11**, 1369–1379.

Edman, L., et al. (1996). Conformational transitions monitored for single molecules in solution. Proc. Natl Acad. Sci. USA **93**, 6710–6715.
Eggeling, C., et al. (1998). Monitoring conformational dynamics of a single molecule by selective fluorescence spectroscopy. Proc. Natl Acad. Sci. USA **95**, 1556–1561.
Harley, M. J., and Schildbach, J. F. (2003). Swapping single-stranded DNA sequence specificities of relaxases from conjugative plasmids F and R100. Proc. Natl Acad. Sci. USA **100**, 11243–11248.
Harley, M. J., et al. (2002). R150A mutant of F TraI relaxase domain: Reduced affinity and specificity for single-stranded DNA and altered fluorescence anisotropy of a bound labeled oligonucleotide. Biochemistry **41**, 6460–6468.
Lakowicz, J. R. (2006). "Principles of Fluorescence Spectroscopy." Springer Science + Business Media, LLC, New York.
Larkin, C., et al. (2005). Inter- and intramolecular determinants of the specificity of single-stranded DNA binding and cleavage by the F factor relaxase. Structure (Camb.) **13**, 1533–1544.
Larkin, C., et al. (2007). Roles of active site residues and the HUH motif of the F plasmid TraI relaxase. J. Biol. Chem. **282**, 33707–33713.
Moreira, B. G., et al. (2005). Effects of fluorescent dyes, quenchers, and dangling ends on DNA duplex stability. Biochem. Biophys. Res. Commun. **327**, 473–484.
Nazarenko, I., et al. (2002). Effect of primary and secondary structure of oligodeoxyribo-nucleotides on the fluorescent properties of conjugated dyes. Nucleic Acids Res. **30**, 2089–2195.
Nelson, W. C., et al. (1993). Characterization of the Escherichia coli F factor traY gene product and its binding sites. J. Bacteriol. **175**, 2221–2228.
Norman, D. G., et al. (2000). Location of cyanine-3 on double-stranded DNA: Importance for fluorescence resonance energy transfer studies. Biochemistry **39**, 6317–6324.
Oda, M., and Nakamura, H. (2000). Thermodynamic and kinetic analyses for understanding sequence-specific DNA recognition. Genes Cells **5**, 319–326.
Sauer, M., et al. (1995). New fluorescent dyes in the red region for biodiagnostics. J. Fluorescence **5**, 247–261.
Seidel, C. A. M., et al. (1996). Nucleobase-specific quenching of fluorescent dyes. 1. Nucleo-base one-electron redox potentials and their correlation with static and dynamic quenching efficiencies. J. Phys. Chem. **100**, 5541–5553.
Sevenich, F. W., et al. (1998). DNA binding and oligomerization of NtrC studied by fluorescence anisotropy and fluorescence correlation spectroscopy. Nucleic Acids Res. **26**, 1373–1381.
Stern, J. C., and Schildbach, J. F. (2001). DNA recognition by F Factor TraI36: Highly sequence-specific binding of single-stranded DNA. Biochemistry **40**, 11586–11595.
Stern, J. C., et al. (2004). Energetics of the sequence-specific binding of single-stranded DNA by the F factor relaxase domain. J. Biol. Chem. **279**, 29155–29159.
Titolo, S., et al. (2003a). Characterization of the minimal DNA binding domain of the human papillomavirus E1 helicase: Fluorescence anisotropy studies and characterization of a dimerization-defective mutant protein. J. Virol. **77**, 5178–5191.
Titolo, S., et al. (2003b). Characterization of the DNA-binding properties of the origin-binding domain of simian virus 40 large T antigen by fluorescence anisotropy. J. Virol. **77**, 5512–5518.
Toptygin, D., and Brand, L. (1995a). Analysis of equilibrium binding data obtained by linear-response spectroscopic techniques. Anal. Biochem. **224**, 330–338.
Toptygin, D., and Brand, L. (1995b). "Spectrabind User's Guide." The Johns Hopkins University, Baltimore, MD.

Vamosi, G., *et al.* (1996). Fluorescence characteristics of 5-carboxytetramethylrhodamine linked covalently to the 5′ end of oligonucleotides: Multiple conformers of single-stranded and double-stranded dye-DNA complexes. *Biophys. J.* **71,** 972–994.

Wang, Z. X. (1995). An exact mathematical expression for describing competitive binding of two different ligands to a protein molecule. *FEBS Lett.* **360,** 111–114.

Wang, K., *et al.* (1998). Fluorescence study of the multiple binding equilibria of the galactose repressor. *Biochemistry* **37,** 41–50.

CHAPTER THIRTEEN

IDENTIFYING SMALL PULSATILE SIGNALS WITHIN NOISY DATA: A FLUORESCENCE APPLICATION

Michael L. Johnson,[*,†] Leon S. Farhy,[†] Paula P. Veldhuis,[*,†] and Joseph R. Lakowicz[‡]

Contents

1. Introduction	274
2. Methods	274
2.1. The simulated data	275
3. Results	280
4. Discussion	284
Acknowledgments	286
References	286

Abstract

One of the most challenging scientific data analysis quandaries is the identification of small intermittent irregularly spaced pulsatile signals in the presence of large amounts of heteroscedastic experimental measurement uncertainties. We present an application of the use of *AutoDecon* to a typical fluorescence and/or spectroscopic data sampling paradigm, which is to detect a single fluorophore in the presence of high background emission. Our calculations demonstrate that single events can be reliably detected by *AutoDecon* with a signal-to-noise ratio of 3/20. *AutoDecon* was originally developed for the analysis of pulsatile hormone-concentration time-series data measured in human serum. However, *AutoDecon* has applications within many other scientific fields, such as fluorescence measurements where the goal is to count single analyte molecules in clinical samples.

[*] Department of Pharmacology, University of Virginia Health System, Charlottesville, Virginia
[†] Department of Medicine (Endocrine Division), University of Virginia Health System, Charlottesville, Virginia
[‡] Department of Biochemistry, University of Maryland at Baltimore, Baltimore, Maryland

1. INTRODUCTION

There are numerous cases in the scientific literature where the ability to identify small pulsatile signals of a magnitude comparable to the concomitant measurement errors is critical. One typical example of this would be detection of a low concentration of fluorophores in a flow cell or detection of a small number of fluorophores on a surface. The fluorescence might also originate from monitoring a single fluorescence molecule within the sample. The flow cell might be attached to the output of a chromatography column or any of a near infinity number of other types of experimental apparatus. Additionally, a wide variety of bioaffinity surfaces are used in genomic and proteomic analysis.

The basic data processing concepts presented here do not depend upon the specifics of the experiment other than the size and shape of the pulsatile signal or the properties of the concomitant measurement errors. Consequently, the current discussion will emphasize irregularly spaced approximately Gaussian-shaped pulsatile events of variable size and the typical Poisson distributed measurement uncertainties that originate from photo counting experiments. One example of this approach to signal analysis could be the counting of single analyte molecules in biological samples which may display high background emission. Single molecule counting (SMC) assays would represent the ultimate in high sensitivity detection. However, the basic mathematical approach presented here is not specific to this application nor to Gaussian events or Poisson noise. This work presents a fluorescence monitoring application of a data analysis algorithm that was originally developed for identifying pulsatile events in the hormone-concentration time-series that are observed in human serum (Johnson *et al.*, 2009).

2. METHODS

The general approach to be presented here is to generate a large number of simulated data sets with specific signal and measurement error characteristics. The advantage of simulated data is that the locations and sizes of the simulated pulsatile events within the data series are known *a priori*. The ability of the *AutoDecon* algorithm to identify small, irregularly timed events within noisy experimental data can then be evaluated by how well it can identify the events within these simulated data time-series. One example could be the photon counts from a single fluorophore flowing past a detector.

The specific event detection criteria utilized here are the: *True-Positives*, the fraction of identified events that correctly identified; *False-Positives*, the fraction of identified events that it incorrectly identified; *False-Negatives*, the fraction of simulated events that were not identified; and *Sensitivity*, the fraction of simulated events that were correctly identified. For an event to be correctly identified it must occur within a specific time window of a corresponding simulated event in the specific simulated data sets.

2.1. The simulated data

The data is simulated to mimic typical data that could be obtained from many different types of experiments. Specifically, each simulated data set consists of 1000 data points equally spaced in time units with three additive parts: (1) The background number of photons, typically 20 per unit time; (2) Several sparsely occurring Gaussian-shaped events of various heights and widths; and (3) Poisson distributed pseudo random measurement errors based upon the total number of counts within each unit of time. One thousand data sets were simulated for each example presented in this work. The results are independent of the specific units of time.

The number of Gaussian-shaped events within each data set was randomly selected between 1 and 5 with an even distribution. The temporal locations of these events were also selected based upon an even distribution with two caveats: no events were assigned within the first and last hundred data points, and the events could not occur within two standard deviations of the Gaussians. The first of these was intended to exclude the end effects of partial events at either end of the simulated data set. The second was to enforce the restriction that multiple events cannot be nearly simultaneous in time and thus indistinguishable.

The Poisson distributed measurement errors, that is, the experimental noise, were generated from the total number of background and event counts by the POIDEV procedure (Press *et al.*, 1986).

2.1.1. *AutoDecon* procedure

AutoDecon was originally developed for the analysis of the pulsatile hormone-concentration secretory events that are observed in serum. This deconvolution procedure functions by developing a mathematical model for the time course of the pulsatile event and then fitting this mathematical model to the experimentally observed time-series data with a weighted nonlinear least-squares algorithm. It implements a rigorous statistical test for the existence of pulsatile events. The algorithm automatically inserts presumed pulsatile events, then tests the significance of presumed events, and removes any events that are found to be nonsignificant. This automatic algorithm combines three modules: a parameter *fitting* module, an *insertion* module that automatically adds presumed events, and a *triage* module which automatically removes events which are deemed to be statistically nonsignificant. No user intervention is required subsequent to the initialization of the algorithm.

It is interesting to note that the mathematical form of hormone-concentration time-series data is the same mathematical form as expected in time-domain fluorescence lifetime measurements. Specifically, this mathematical model (Johnson and Veldhuis, 1995; Veldhuis *et al.*, 1987) is:

$$C(t) = \int_{-\infty}^{t} S(\tau) E(t-\tau) \, d\tau \qquad (13.1)$$

where $C(t)$ is the hormone-concentration as a function of time t, $S(\tau)$ is the secretion into the blood as a function of time and is typically modeled as the sum of Gaussian-shaped events, and $E(t-\tau)$ is the one- or two-exponential elimination of the hormone from the serum as a function of time. For the time-domain fluorescence lifetime experimental system, the $S(t)$ function corresponds to the lamp function (i.e., the instrument response function) and the $E(t-\tau)$ corresponds to the fluorescence decay function.

For the current study it was assumed that the elimination half-time was short compared with the width of the Gaussian-shaped events to be detected. In this limit, the shape of the events detected by *AutoDecon* will approach a Gaussian profile. This allowed the existing algorithm and software to detect the simulated Gaussian-shaped events without modification.

2.1.2. *AutoDecon* fitting module

The fitting module performs weighted nonlinear least-squares parameter estimations by the Nelder–Mead Simplex algorithm (Nelder and Mead, 1965; Straume *et al.*, 1991). In the present example, it fits Eq. (1) to the experimental data by adjusting the parameters of the Gaussian-shaped events function such that the parameters have the highest probability of being correct. The module is based upon the *Amoeba* routine (Press *et al.*, 1986) which was modified such that convergence is assumed when both the variance-of-fit and the individual parameter values do not change by more than 2×10^{-5} or when 15,000 iterations have occurred. This is essentially the original *Deconv* algorithm (Johnson and Veldhuis, 1995; Veldhuis *et al.*, 1987) with the exception that the Nelder–Mead Simplex parameter estimation algorithm (Nelder and Mead, 1965; Straume *et al.*, 1991) is used instead of the damped Gauss–Newton algorithm which was previously utilized as the Nelder–Mead algorithm simplifies the software since it does not require derivatives.

The *AutoDecon* fitting module constrains all of the events to have a positive amplitude by fitting to the logarithm of the amplitude instead of the amplitude. The fitting module also constrains all of the events to have the same standard deviation.

2.1.3. *AutoDecon* insertion module

The insertion module inserts a presumed event at the location of the maximum of the Probable Position Index (PPI).

$$\text{PPI}(t) = \begin{cases} -\dfrac{\partial[\text{Variance-of-fit}]}{\partial H_z} & \text{if } \dfrac{\partial[\text{Variance-of-fit}]}{\partial H_z} < 0 \\ 0 & \text{if } \dfrac{\partial[\text{Variance-of-fit}]}{\partial H_z} \geq 0 \end{cases} \quad (13.2)$$

The parameter H_z is the amplitude of a presumed Gaussian-shaped event at time z. The index function PPI(t) will have a maximum at the data point position in time where the insertion of an event will result in the largest negative derivative in the variance-of-fit versus event size. It is important to note that the partial derivatives of the variance-of-fit with respect to an event size can be evaluated without any additional weighted nonlinear least-squares parameter estimations or without even knowing the size of the presumed event, H_z. Using the definition of the variance-of-fit given in Eq. (3), the partial derivative with respect to the addition of an event at time z is shown in Eq. (4) where the summation is over all data points,

$$[\text{Variance-of-fit}] = \sum_i \left(\frac{Y_i - C(t_i)}{w_i}\right)^2 = \sum_i R_i^2 \quad (13.3)$$

$$\begin{aligned}\frac{\partial[\text{Variance-of-fit}]}{\partial H_z} &= \sum_i \left[\frac{2}{w_i^2}\left(Y_i - C(t)\right)\frac{\partial C(t)}{\partial H_z}\right] \frac{\partial C(t)}{\partial H_z} \\ &= \left[\exp\left(-\frac{1}{2}\left(\frac{t-z}{\text{Secretion SD}}\right)^2\right)\right] E(t)\end{aligned} \quad (13.4)$$

where w_i corresponds to the weighting factor for the ith data point, and R_i corresponds to the ith residual. The inclusion of these weighting factors is the statistically valid method to compensate for the heteroscedastic properties of the experimental data. For the present case it was assumed that these weighting factors were equal to the square-root of the number of photon counts at any specific time.

2.1.4. *AutoDecon* triage module

The triage module performs a statistical test to ascertain whether or not a presumed secretion event should be removed. This test requires two weighted nonlinear least-squares parameter estimations, one with the presumed event present and one with the presumed event removed. The ratio of the variance-of-fit resulting from these two parameter estimations is related to the probability that the presumed event does not exist, P, by an F-statistic, as in Eq. (5). For most of the present examples a probability level of 0.01 was used.

$$\frac{\text{Variance-of-fit}_{\text{removed}}}{\text{Variance-of-fit}_{\text{present}}} = 1 + \frac{2}{ndf} F\text{-statistic}(2, ndf, 1 - P) \quad (13.5)$$

This is the F-test for an additional term (Bevington, 1969) where the additional term is the presumed event. The 2's in Eq. (5) are included since each additional event increases the number of parameters being estimated by 2, specifically the location and the amplitude of the event. The number of degrees of freedom, ndf, is the number of data points minus the total number of parameters being estimated when the event is present. Each cycle of the triage module performs this statistical test for every event in an order determined by size, from smallest to largest. If an event is found to be not statistically significant it is removed and the triage module is restarted from the beginning (i.e., a new cycle starts). Thus, the triage module continues until all nonsignificant events have been removed. Each cycle of the triage module performs $m + 1$ weighted nonlinear least-squares parameter estimations where m is the current number of events for the current cycle: one where all of the events are present and one where each of the events has been removed and individually tested.

2.1.5. *AutoDecon* combined modules

The *AutoDecon* algorithm iteratively adds presumed events, tests the significance of all events, and removes nonsignificant events. The procedure is repeated until no additional events are added. The specific details of how this is accomplished with the insertion, fitting, and triage modules are outlined here.

For the present study, *AutoDecon* was initialized with the background set equal to zero, the elimination half-life set to a small value (specifically three time units), the standard deviation of the events set to the specific simulated value, and zero events. It is possible, but neither required nor performed in this study, that initial presumed event locations and sizes be included in the initialization. Initializing the program with event position and amplitude estimates might produce faster convergence and thus decrease the amount of computer time required.

The next step in the initialization of the *AutoDecon* algorithm is for the fitting module to estimate only the background. The fitting module then estimates all of the model parameters except for the elimination half-life and the standard deviation of the secretion events. If any secretion events have been included in the initialization, the second fit will also refine the locations and sizes of these secretion events. Next, the triage module is utilized to remove any nonsignificant events. At this point the parameter estimations that are performed within the triage module will estimate all of the current model parameters except for the elimination half-life and the standard deviation of the secretion events.

The *AutoDecon* algorithm next proceeds with Phase 1 by using the insertion module to add a presumed event. This is followed by the triage module to remove any nonsignificant events. Again, the parameter estimations performed within the triage module during Phase 1 will estimate all of the current model parameters with the exception of the elimination half-life and the standard deviation of the secretion events. If during this phase the triage module does not remove any secretion events, Phase 1 is repeated to add an additional presumed event. Phase 1 is repeated until no additional events are added in the insertion followed by triage cycle.

For the current example, Phase 2 repeats the triage module with the fitting module estimating all of the current model parameters but this time including the standard deviation of the secretion events and not the elimination half-life.

Phase 3 will repeat Phase 1 (i.e., insertion and triage) utilizing the parameter estimations that are performed by the fitting module within the triage module estimating all of the current model parameters again including the standard deviation of the secretion events but not the elimination half-life. For the present example, the elimination half-life is never estimated. Phase 3 is repeated until no additional events have been added in the insertion followed by triage cycle.

2.1.6. Concordant secretion events

Determining the operating characteristics of the algorithms requires a comparison of the apparent event positions from the *AutoDecon* analysis of a simulated time-series with the actual known event positions upon which the simulations were based. This process must consider whether the concordance of the peak positions is statistically significant or whether it is a consequence of a simply random position of the apparent and simulated events. Specifically, the question could be posed: given two time-series with n and m distinct events, what is the probability that j coincidences (i.e., concordances) will occur based upon a random positioning of the distinct events within each of the time-series? The resulting probabilities are dependent upon the size of the specific time window employed for the definition of coincidence. This question can easily be resolved utilizing a Monte-Carlo approach. One hundred thousand pairs of time-series are generated with the n and m distinct randomly timed events, respectively. The distribution of the expected number of concordances can then be evaluated by scanning these pairs of random event sequences for coincident peaks where coincidence is defined by any desired time interval.

Obviously, as the coincidence interval increases so will the expected number of coincident events. The expected number of coincident events will also increase with the numbers of distinct events, n and m. Thus, the coincidence interval should be kept small.

3. RESULTS

The procedure to demonstrate the functionality of *AutoDecon* for the detection of small events within noisy time-series involves simulating and then analyzing one thousand time-series which have between one and five events temporally assigned at random between data points 100 and 900. These are analyzed via *AutoDecon* and the results compared with the known answers upon which the simulations were based. If an observed event is within a concordance window of 20 data points from a simulated event then it is considered a true-positive, if an observed event is not within 20 data points of an actual simulated event it is a false-positive, etc. However, an initial question is how many true-positive events are expected based simply upon the random locations of the events. For a specific data set containing one simulated event there is a 0.05 probability that a coincidence will be randomly coincident to within ±20 data points. For each specific data set with five simulated events randomly located within 800 data points there is a 0.71 probability that one or more coincident events will be randomly coincident to within ±20 data points, a 0.26 probability of two or more coincident events, a 0.04 probability of three or more coincident events, a 0.0024 probability of four or more coincident events, and less than a 0.0001 probability that five or more will be randomly located. However, when combining 1000 independent data sets these probabilities are raised to the 1000th power. Thus, the probability of observing five or more randomly coincident events in 1000 data sets is less that 10^{-4000}.

Figure 13.1 presents the analysis of the first of a 1000 data sets that were simulated with between 1 and 5 noncoincident events between data points 100 and 900 of a total of 1000 data points in the time sets. The simulated number of photons at each data point was a Poisson distributed random deviate with a mean based upon Gaussian-shaped randomly occurring noncoincident events with a standard deviation (i.e., half-width/2.354) of 20 data points and a height of 10 photons and a background, that is, dark current, of 20 photons. The probability of randomly observing these four events correctly to within ±20 data points is less than 0.0001.

Table 13.1 presents the operating characteristics of the *AutoDecon* algorithm when applied to 1000 time-series analogous to Fig. 13.1. The mean,

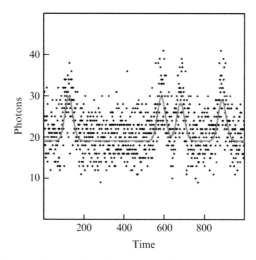

Figure 13.1 A simulated case where four simulated events were correctly identified. Background is 20 photons, simulated Gaussian event standard deviations are 20 time units and the event heights are 10 photons. The points are the simulated noisy photon counts and the curve represents the best estimated values.

Table 13.1 Operating characteristics of *AutoDecon* for 1000 data sets simulated as in Fig. 13.1

	Median	Mean ± SEM	Interquartile range
±20 Data point concordance			
True-positive%	100.0	98.6 ± 0.2	0.0
False-positive%	0.0	1.4 ± 0.2	0.0
False-negative%	0.0	0.1 ± 0.1	0.0
Sensitivity%	100.0	99.9 ± 0.1	0.0
±15 Data point concordance			
True-positive%	100.0	98.3 ± 0.2	0.0
False-positive%	0.0	1.7 ± 0.2	0.0
False-negative%	0.0	0.4 ± 0.1	0.0
Sensitivity%	100.0	99.6 ± 0.1	0.0
±10 Data point concordance			
True-positive%	100.0	96.8 ± 0.4	0.0
False-positive%	0.0	3.2 ± 0.4	0.0
False-negative%	0.0	2.0 ± 0.3	0.0
Sensitivity%	100.0	98.0 ± 0.4	0.0

median, and interquartile distances are given because the operating characteristic distributions do not always follow Gaussian distributions. For this simulation the sensitivity is ∼98% for finding the event locations to within ±10 time units when the half-width of the simulated events is ∼50 time units (~2.354 SD). Obviously this simulation, where the events heights are 50% of the dark current and event standard deviation is equal to the dark current, represents a comparatively easy case. For an analogous simulation (not shown) where the standard deviation of the event is decreased to half of the dark current sensitivity increases to 99.1 ± 0.2% and the false-positive % increases slightly to 4.0 ± 0.04%.

Figure 13.2 and Table 13.2 present a somewhat more difficult case. This simulation is identical to the previous simulation except that the simulated event height is only 5 photons. Here the events are half the size of the previous simulations. For this simulation the *AutoDecon* algorithm was still able to accurately find the event locations to within ±15 time units.

It is instructive to examine the median values found in Tables 13.1 and 13.2. In every case the medians are either 100% or 0%. This indicates that the *AutoDecon* algorithm correctly located all of the simulated events to within ±10 time units for more than half of the simulated data sets. Furthermore, for the instances where the interquartile distance is also 0.0 the *AutoDecon* algorithm correctly located all of the simulated events in more than 75% of the simulated data sets.

Figure 13.3 and Table 13.3 takes the simulations presented in Figs. 13.1 and 13.2 and Tables 13.1 and 13.2 to the next logical step where the standard deviation is 20 data points and a height of 3 photons and a background of 20 photons. In Fig. 13.3, five events were correctly identified within ±20 data points. When all 1000 simulated time-series are examined the false-positive% for event identification to within ±20 data points is 6.8 ± 0.6%

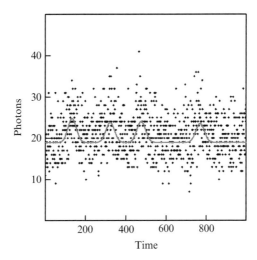

Figure 13.2 A second simulated case where four simulated events were correctly identified. Background is 20 photons, simulated Gaussian event standard deviations are 20 time units and the event heights are 5 photons. The points are the simulated noisy photon counts and the curve depicts the best estimated values.

Table 13.2 Operating characteristics of *AutoDecon* for 1000 data sets simulated as in Fig. 13.2

	Median	Mean ± SEM	Interquartile range
±20 Data point concordance			
True-positive%	100.0	97.1 ± 0.3	0.0
False-positive%	0.0	2.9 ± 0.3	0.0
False-negative%	0.0	2.6 ± 0.3	0.0
Sensitivity%	100.0	97.4 ± 0.3	0.0
±15 Data point concordance			
True-positive%	100.0	94.7 ± 0.5	0.0
False-positive%	0.0	5.3 ± 0.5	0.0
False-negative%	0.0	4.8 ± 0.5	0.0
Sensitivity%	100.0	95.2 ± 0.5	0.0
±10 Data point concordance			
True-positive%	100.0	84.7 ± 0.8	25.0
False-positive%	0.0	15.3 ± 0.8	25.0
False-negative%	0.0	14.8 ± 0.8	25.0
Sensitivity%	100.0	85.2 ± 0.8	25.0

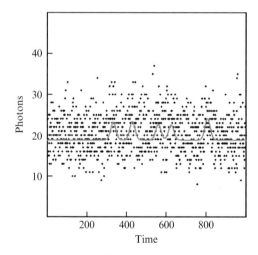

Figure 13.3 A third simulated case where five simulated events were correctly identified. Background is 20 photons, simulated Gaussian event standard deviations are 20 time units and the event heights are 3 photons. The points are the simulated noisy photon counts and the curve depicts the best estimated values.

and the corresponding false-negative% is 28.5 ± 1.0%. This is clearly a very difficult case where the *Autodecon* algorithm performs surprisingly well.

In Table 13.1, the false-positive percentages are substantially higher than the corresponding false-negative percentages. They are of a comparable size

Table 13.3 Operating characteristics of *AutoDecon* for 1000 data sets simulated as in Fig. 13.3

	Median	Mean ± SEM	Interquartile range
±20 Data point concordance			
True-positive%	100.0	89.0 ± 0.8	0.0
False-positive%	0.0	6.8 ± 0.6	0.0
False-negative%	20.0	28.5 ± 1.0	50.0
Sensitivity%	80.0	71.5 ± 1.0	50.0
±15 Data point concordance			
True-positive%	100.0	81.1 ± 1.0	33.3
False-positive%	0.0	14.7 ± 0.9	25.0
False-negative%	33.3	34.2 ± 1.0	60.0
Sensitivity%	66.7	65.8 ± 1.0	60.0
±10 Data point concordance			
True-positive%	66.7	65.5 ± 1.2	50.0
False-positive%	20.0	30.3 ± 1.1	50.0
False-negative%	50.0	46.1 ± 1.1	75.0
Sensitivity%	50.0	53.9 ± 1.1	75.0

in Table 13.2, while in Table 13.3 the false-positive percentages are substantially lower than the corresponding false-negative percentages. Clearly, these percentages are both expected and observed to be a function of the relative height of the events. But the false-negative percentages are more sensitive to the relative event heights than are the false-positive percentages. These percentages are also a function of the probability level within the triage module, see Eq. (5). Table 13.4 presents the results from the analysis of the same simulated data sets that were used for Table 13.3 where the only difference is that a probability level of 0.05 was used in Eq. (5). With the probability level of 0.05 the false-positive and false-negative percentages are approximately equal for simulated event sizes of three photons.

4. Discussion

The *Autodecon* algorithm functions by: (1) creating a mathematical model for the shape of an event, (2) using the derivative of the variance-of-fit with respect to the possible existence of events, predicts where the next presumptive event should be added, (3) performing a least-squares parameter estimation to determine the exact event locations and sizes, and (4) using a rigorous statistical test to determine if the events are actually statistically significant. Steps 2–4 are repeated until no additional events are added.

Table 13.4 Operating characteristics of *AutoDecon* for 1000 data sets simulated as in Fig. 13.3

	Median	Mean ± SEM	Interquartile range
±20 Data point concordance			
True-positive%	100.0	83.6 ± 0.8	33.3
False-positive%	0.0	15.6 ± 0.7	33.3
False-negative%	0.0	13.5 ± 0.7	25.0
Sensitivity%	100.0	86.5 ± 0.7	25.0
±15 Data point concordance			
True-positive%	83.3	76.8 ± 0.9	40.0
False-positive%	16.7	22.4 ± 0.8	40.0
False-negative%	0.0	20.1 ± 0.9	33.3
Sensitivity%	100.0	79.9 ± 0.9	33.3
±10 Data point concordance			
True-positive%	66.7	61.6 ± 1.0	56.3
False-positive%	33.3	37.6 ± 1.0	50.0
False-negative%	33.3	35.3 ± 1.1	60.0
Sensitivity%	66.7	64.7 ± 1.1	56.3

The probability level 0.05 was used in Eq. (5) for this table.

AutoDecon is a totally automated algorithm. For the current examples the algorithm was initialized with a probability level of 0.01 and an approximate standard deviation for the Gaussian-shaped events of 20 time units. The algorithm then automatically determined the number of events, their locations, and sizes.

The algorithm does not assume that the events occur at regular intervals. However, if they did occur at regular intervals the algorithm could be modified to include this assumption and would then perform even better.

Similarly the algorithm does not assume that the secretion events have the same height. If the events had a constant height this could also be included in a modified algorithm to obtain even better performance.

The *Autodecon* algorithm is also somewhat independent of the form of the noise distribution. In the present example, the simulated data contained Poisson distributed simulated measurement uncertainties while the current *Autodecon* software assumes that the measurement error distribution follows Gaussian distribution.

The *Autodecon* algorithm functions by performing a series of weighted nonlinear regressions. The weighting factors employed here are proportional to the square-root of the observed number of photons at each simulated data point. This is consistent with the procedure commonly employed for experimental photon counting protocols. A consequence of this is that the statistical weight assigned to a data point with n too many photons is less than the statistical weight of a data point with n too few photons. This introduces an asymmetrical bias into the weighting of the data points. The major consequence of this is that the baseline levels in Figs. 13.1–13.3 are ~19 photons instead of the simulated value of 20 photons.

The relative values of the false-positive and false-negative percentages are adjustable within the *Autodecon* algorithm. Tables 13.3 and 13.4 present an example of how the probability level within the algorithm's triage module can be manipulated so that the false-positive and false-negative percentages are approximately equal. The optimal value for the probability level is a complex function of the relative sizes of the events and the baseline values as well as many other variables. Thus, the optimal probability level cannot be predicted *a priori* but it can be determined by computer simulations which include the specific experimental details.

The *Concordance* and *Autodecon* algorithms are part of our hormone pulsatility analysis suite. They can be downloaded from www.mljohnson.pharm.virginia.edu/pulse_xp/.

ACKNOWLEDGMENTS

This work was supported in part by NIH grants RR-00847, HD28934, R01 RR019991, R25 DK064122, R21 DK072095, P30 DK063609, and R01 DK51562 to the University of Virginia. This work was also supported by grants at the University of Maryland, HG-002655, EB000682, and EB006521.

REFERENCES

Bevington, P. R. (1969). "Data Reduction and Error Analysis for the Physical Sciences." McGraw Hill, New York.

Johnson, M. L., and Veldhuis, J. D. (1995). Evolution of deconvolution analysis as a hormone pulse detection algorithm. *Methods Neurosci.* **28**, 1–24.

Johnson, M. L., Pipes, L., Veldhuis, P. P., Farhy, L. S., Nass, R., Thorner, M. O., and Evans, W. S. (2009). *AutoDecon*: A robust numerical method for the quantification of pulsatile events. *Methods Enzymology.* **453**, (in press).

Nelder, J. A., and Mead, R. (1965). A simplex method for function minimization. *Comput. J.* **7**, 308–313.

Press, W. H., Flannery, B. P., Teukolsky, S. A., and Vettering, W. T. (1986). "Numerical Recipies: The Art of Scientific Computing." Cambridge University Press, New York.

Straume, M., Frasier-Cadoret, S. G., and Johnson, M. L. (1991). Least-squares analysis of fluorescence data. *Topics Fluoresc. Spectrosc.* **2**, 171–240.

Veldhuis, J. D., Carlson, M. L., and Johnson, M. L. (1987). The pituitary gland secretes in bursts: Appraising the nature of glandular secretory impulses by simultaneous multiple-parameter deconvolution of plasma hormone concentrations. *Proc. Natl. Acad. Sci. USA* **84**(21), 7686–7690.

CHAPTER FOURTEEN

DETERMINATION OF ZINC USING CARBONIC ANHYDRASE-BASED FLUORESCENCE BIOSENSORS

Rebecca Bozym,[*,†] Tamiika K. Hurst,[‡] Nissa Westerberg,[*,§] Andrea Stoddard,[‡] Carol A. Fierke,[‡] Christopher J. Frederickson,[¶] and Richard B. Thompson[*]

Contents

1. Introduction	288
2. Principles of CA-Based Zinc Sensing	288
3. Transducing Zinc Binding as a Fluorescence Change	290
4. "Free" Versus Bound Zinc Ion: Speciation	295
5. Metal Ion Buffers	297
6. Kinetics	301
7. Applications: Ratiometric Determination of Free Zinc in Solution	302
8. Preparation of Apocarbonic Anhydrase	303
9. Intracellular Sensing with TAT Tag	304
10. Intracellular Sensing with an Expressible CA Sensor	305
References	307

Abstract

This chapter summarizes the use of carbonic anhydrase (CA)-based fluorescent indicators to determine free zinc in solution, in cells, and in subcellular organelles. Expression (both *in situ* and *in vitro*) and preparation of CA-based indicators are described, together with techniques of their use, and procedures to minimize contamination. Recipes for zinc buffers are supplied.

---

[*] Department of Biochemistry and Molecular Biology, University of Maryland School of Medicine, Baltimore, MD 21201
[†] Cellumen, Inc., 3180 William Pitt Way, Pittsburgh, Pennsylvania
[‡] Department of Chemistry, University of Michigan, Ann Arbor, Michigan
[§] U.S. Patent and Trademark Office, Alexandria, Virginia
[¶] NeuroBioTex, Inc., Galveston, Texas

1. Introduction

Over the past few years, fluorescent indicator systems based on apo–carbonic anhydrase (apo-CA) have demonstrated outstanding sensitivity, selectivity, and flexibility in determining free zinc and other metal ions in a variety of complex aqueous matrices (Bozym *et al.*, 2006; Fierke and Thompson, 2001; Zeng *et al.*, 2003, 2005). The heart of the technique lies in transducing the high affinity and selectivity for binding of Zn(II) (and a handful of other ions) as a change in fluorescence, which can be simply related to the fractional occupancy of the binding site and thus the concentration of the zinc in a "free" state. By "free" zinc we mean zinc ion in aqueous solution bound to labile, rapidly exchangeable ligands such as water or chloride (see Section 4). The most important advantage (and unique feature) of the CA-based zinc sensors is that the affinity, selectivity, kinetics, and signal change have all been improved for particular applications by relatively modest mutagenesis of the CA molecule. By comparison, it is often infeasible to improve the properties of organic synthetic fluorescent indicators without dramatic change in the molecules. The fact that these are fluorescence sensors confers the well-known advantages of fluorescence techniques in chemical analysis: high sensitivity, reduced susceptibility to interference, reduced or no need for separation (e.g., a homogenous assay), widely available instrumentation, speed, real-time readout, adaptability to microscopy of specimens, adaptability to remote sensing in some cases, and often ease of use. In this chapter the principles of CA-based zinc sensors will be described, together with their applications for particular determinations and details of ancillary procedures for their successful use.

2. Principles of CA-Based Zinc Sensing

While the catalytic role of zinc in carbonic anhydrases is well established (Lindskog *et al.*, 1971), from our standpoint the protein primarily represents a ligand that can bind the metal ion, albeit one with unique properties. Wild-type CA II from humans binds zinc ion with picomolar affinity, whereas other metal ions abundant in biological systems (Na, K, Mg, Ca) are not observed to bind to the active site ligands at all. This may be attributed to the positioning of the three histidine imidazole moieties which bind the metal in a tetrahedral geometry; the fourth ligand is water (Christianson and Fierke, 1996; Eriksson and Jones, 1988). The histidine imidazoles are held in place in the active site by a network of hydrogen bonds (Huang et al., 1996; Kiefer et al., 1995) which accounts in large measure for the selectivity. As O'Halloran has pointed out, the protein's ability to maintain the imidazoles in a fixed position contrasts with most small molecule chelators such as EDTA, whose metal-binding ligands enjoy substantial flexibility and are able to accommodate metal ions of widely differing size and charge. As a result, such small molecule chelators (and the indicators based on them) typically lack selectivity, even though they may bind metal ions tightly.

By comparison with small molecule indicators, the CA-based indicators offer high affinity and selectivity for zinc. While high affinity can often be achieved in small molecule indicators, selectivity is more important in sensors for use in biological systems and natural waters because these media are so complex. For instance, blood serum and sea water have concentrations of total calcium and magnesium which are in the millimolar range, and therefore fluorescent indicators intended to be used to determine free zinc at nanomolar levels in such media must have billion-fold selectivity against these potential interferents. In passing, we note that some fluorescent indicators bind nontarget metal ions without exhibiting a fluorescence change: they interfere with determination of the zinc or other target by occupying the binding site and effectively reducing the concentration of sites. The common practice of assessing selectivity by looking for fluorescence changes by adding a single concentration of the potential interferent to a solution of the indicator fails to detect this problem. A better method is to prepare metal ion buffers with a range of concentrations of the potential interferent and the target ion near its K_D (Thompson et al., 2002b). If the interfering ion binds, it will shift the apparent K_D of the indicator even if the interferent is spectroscopically silent. CA-based indicators do not exhibit competition for zinc in the binding site by calcium or magnesium at levels up to 10 and 50 mM, respectively. Wild-type human CA II binds Cu(II) more tightly than Zn(II) (K_D = 0.1 vs 4 pM) with Cd(II) (2.9 nM), Ni(II) (16 nM), and Co(II) (160 nM) all exhibiting substantially lower affinities (McCall, 2000). We note that in biological systems and natural waters free Co(II), Ni(II), and Cd(II) are generally expected to be present only at much lower levels.

The unique and powerful advantage of CA-based indicators is that their affinity and selectivity may be tuned over a broad range by subtle modification of the protein molecule. For instance, we have constructed variants with affinity for zinc varying over thirteen orders of magnitude (Ippolito et al., 1995), and with relative affinity for copper with respect to zinc varying from 10^7-fold tighter for copper, to essentially equal affinities for copper and zinc (Hunt et al., 1999; McCall and Fierke, 2004), which contravenes the well-known Irving–Williams series of metal ion affinities. These properties are evidently difficult to engineer into small, organic fluorescent indicators.

3. Transducing Zinc Binding as a Fluorescence Change

We have developed a number of means whereby the binding of zinc ion to CA is transduced as a change in fluorescence which can be measured and related to the free zinc concentration. While these changes typically include simple intensity changes, for a number of well-established reasons simple intensity changes are undesirable for quantitative determinations, and we eschew them. Rather, we prefer to transduce the zinc binding as a change in fluorescence spectrum (usually measured as a ratio of intensities at different wavelengths) (Thompson and Jones, 1993; Thompson et al., 2000a,b), fluorescence lifetime (Thompson and Patchan, 1995a,b; Thompson et al., 1998a), or fluorescence anisotropy (polarization) (Elbaum et al., 1996; Thompson et al., 1998a, 2000b). While most fluorometers and fluorescence microscopes can also perform wavelength-ratiometric determinations using filters, fewer laboratories possess instruments or microscopes capable of fluorescence anisotropy or lifetime measurements. Consequently, this chapter focuses mainly on the wavelength-ratiometric methods as being one of the broadest interest.

Most of the CA-based zinc sensors rely on the zinc-dependent binding of a second molecule to the protein, which results in a fluorescence change. These molecules are often fluorescent or colored themselves. They all (until now) are of a well-known class of inhibitors of CA described as aryl sulfonamides; these inhibitors have been used therapeutically to treat glaucoma and altitude sickness, so hundreds are known (Maren, 1977; Supuran and Scozzafava, 2000). The weakly acidic aryl sulfonamide proton ($pK_a \sim 10$) enables the zinc in the CA active site to stably bind the sulfonamide as the anion in place of the water normally found in the active site. If fluorescent, binding of the sulfonamide (with ionization) in the more rigid, somewhat hydrophobic-binding site on the protein often results in enhanced lifetime and quantum yield, as well as a blue shift in emission. The first example of such an aryl sulfonamide was dansylamide, whose fluorescence changes upon binding to CA were described by Chen and Kernohan (1967). We showed that the binding essentially required zinc be in the active site, and that the fluorescence changes could be related to zinc concentration (Thompson and Jones, 1993). Dansylamide's weak absorbance and requirement for UV excitation were undesirable in many applications, and so we developed other fluorescent aryl sulfonamides which exhibited shifts in their emission upon binding to zinc in the active site: they include ABDN, ABDM, and Dapoxyl sulfonamide (Thompson et al., 1998b, 2000a,b). The approach is shown schematically for ABDN in Fig. 14.1.

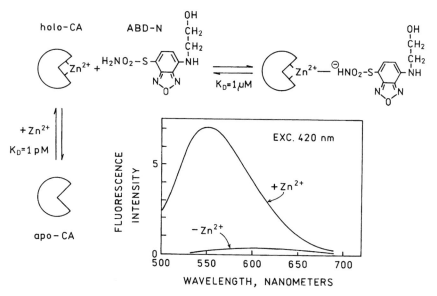

Figure 14.1 Principle of carbonic anhydrase-based zinc determination with ABDN; *inset* shows fluorescence emission spectra in the presence and absence of zinc.

For all these sulfonamides we described wavelength-ratiometric zinc determinations; an example data set is shown for the visibly excited ABDN Fig. 14.2. We have since made a polymeric form of ABDN which may be entrapped in a gel together with apo-CA for continuous monitoring, such as with a fiber optic (Zeng et al., 2005).

These fluorescent aryl sulfonamides also typically display changes in their fluorescence anisotropy (polarization) when they bind to holo-CA because the binding is usually accompanied by a substantial reduction in their rotational rate. Thus zinc concentration can be transduced as a change in fluorescence polarization (Elbaum et al., 1996). In some circumstances these determinations can exhibit a substantially enhanced dynamic range (Thompson et al., 1998b). An example is shown in Fig. 14.3. Less useful for this purpose are those which have low quantum yields in the free form like Dapoxyl sulfonamide, or substantial increases in lifetime upon binding, like dansylamide. Fluorescence anisotropy measurements can also be ratiometric (Weber, 1956) and thus very accurate and precise; however, relatively few fluorescence polarization microscopes have been constructed (Axelrod, 1989; Dix and Verkman, 1990) and we do not believe any are offered commercially.

We also have described different approaches to CA-based zinc measurement which rely on changes in fluorescence lifetime. While more complex from an instrumental or imaging standpoint, lifetime measurements are also largely independent of fluorescence intensity (and the things which affect it)

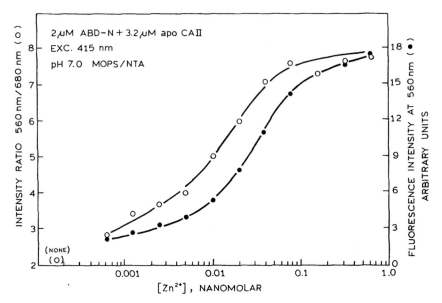

Figure 14.2 Fluorescence intensity at 560 nm (●) and ratio of emission intensity 560/680 nm (○) as a function of free zinc concentration. Reproduced from Thompson et al. (2000a,b) with permission.

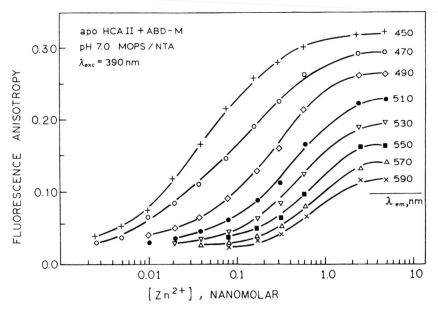

Figure 14.3 Fluorescence anisotropy is depicted as a function of free zinc for ABDM and apo-CA at emission wavelengths ranging from 450 to 590 nm. Reproduced from Thompson et al. (1998a,b) with permission.

(Szmacinski and Lakowicz, 1994). Another generic advantage of lifetime-based sensing is that with measurement of the complete decay one recovers the proportions of free and bound directly, even if it is otherwise infeasible to measure the fluorescent indicator in the fully bound form. These approaches involve (mostly) fluorescent aryl sulfonamides whose lifetimes change dramatically upon binding to *holo*-CA. Thus dansylamide exhibits an increase of lifetime from about 3 to 22 ns on binding (Chen and Kernohan, 1967), and we can directly determine the fraction of protein with bound dansylamide (and therefore bound zinc) by measuring the proportion of dansylamide (or ABDN) exhibiting the bound lifetime (Thompson and Patchan, 1995a; Thompson *et al.*, 2000a). Like other fluorescent indicators which exhibit lifetime changes as well as wavelength shifts (Szmacinski and Lakowicz, 1993), these indicators can exhibit dynamic ranges up to five orders of magnitude These measurements are not altogether independent of the sulfonamide amounts, but are reasonably robust if the sulfonamide concentrations are kept above the sulfonamide K_D. We also have developed a lifetime-based approach that relies on FRET (Forster resonance energy transfer; Forster, 1948) between a label attached covalently to the CA and a colored aryl sulfonamide (azosulfamide) whose binding is zinc-dependent (Thompson and Patchan, 1995b). The approach is depicted schematically in Fig. 14.4. The principle is based on the

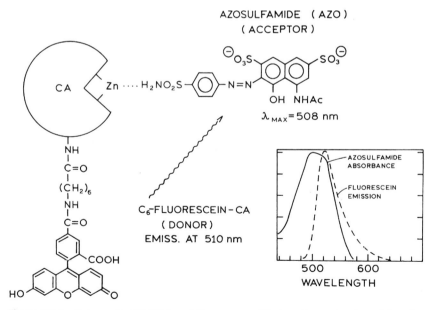

Figure 14.4 Scheme for FRET-based fluorescence lifetime determination of free zinc using fluorescent-labeled apo-CA and azosulfamide.

azosulfamide acting to partially quench (and thereby reduce the lifetime of) a fluorescent label on the CA upon the former's binding to *holo*-CA, which brings the azosulfamide into close proximity to the labeled CA. The proportion of CA with the reduced lifetime corresponds to the free zinc concentration, and is determined directly from the fluorescence decay. The azosulfamide is present at some low concentration (*ca.* 1 μM) so that on the average it is too distant from the protein to efficiently accept energy from the label on the protein. Finally, we have developed one example of a covalently fluorescent-labeled CA which exhibits lifetime and anisotropy changes upon binding zinc (Thompson *et al.*, 1999) but the label is only available on a custom basis.

Finally, we have also described an *excitation ratiometric* method for zinc determination (Thompson *et al.*, 2002a). One issue with the emission ratiometric methods described earlier is that they are dependent on the level of aryl sulfonamide present to accurately report the fractional saturation (and thus the free zinc level). We devised a FRET-based approach that utilizes the aryl sulfonamide as an energy transfer donor. The approach is diagrammed in Fig. 14.5. Briefly, the apo-CA has a covalent fluorescent label such as Alexa Fluor 594 chosen to serve as an acceptor for the sulfonamide. In the absence of zinc, the Dapoxyl sulfonamide does not bind to CA, and its UV-excited emission in the red at 600 nm is very weak, as is the UV-excited emission in the red from the label on the CA. If zinc ion is present and binds to the labeled apo-CA, the Dapoxyl sulfonamide will bind to it; however, when excited in the UV the sulfonamide does not

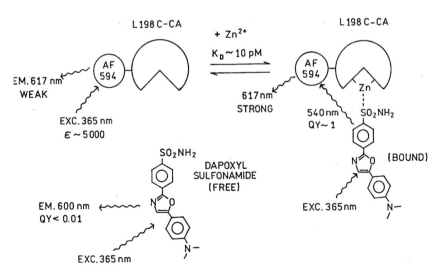

Figure 14.5 Principle of excitation-ratiometric determination of zinc using Dapoxyl sulfonamide and Alexa Fluor 594-labeled apo-CA.

emit but transfers energy almost quantitatively to the acceptor on the CA, which then strongly emits in the reddish orange spectral region. The proportion of CA with zinc bound is thus a simple function of the ratio of UV-excited emission to the emission of the label directly excited in the green.

Among the important advantages of this approach is its insensitivity to the amount of sulfonamide and (in equilibrium binding mode) fluorescent-labeled CA present. Moreover, although the Dapoxyl sulfonamide is known to bind to membranes (and may bind proteins as well) with much enhanced quantum yield, its emission in the blue-green is not observed and thus does not interfere. Similarly, because one is only observing emission from labeled CA, emission from Dapoxyl sulfonamide bound to CA present in the cell or medium does not interfere. The fluorescent label can be covalently attached at a cysteine residue inserted in the CA gene at a position selected to maximize energy transfer, or the CA can be labeled *in vivo* by constructing a fusion protein together with an appropriate green fluorescent protein (GFP) variant; we have used both approaches. The latter is of course an expressible indicator that may be targeted to particular cells or (as we have shown) organelles based on transfection mode and targeting sequences (see below).

4. "Free" Versus Bound Zinc Ion: Speciation

A central concept in zinc biochemistry (and metallobiochemistry generally) is the distinction between "free" and "bound" metal ions. In fact, there is probably no such thing as "free" zinc ion in physiologic solution: that is, zinc ion with no water molecules or anions such as chloride bound to it, even transiently. Rather, one can think of zinc ions bound with weak, rapidly exchangeable ligands such as these as being "free" in that they are free to bind to other ligands which may be present. Most cells have *total* zinc concentrations of 100 μM or greater, but the vast majority of these zinc ions are tightly bound to ligands such as proteins from which they dissociate rarely, if at all. The proportions of total zinc (or any metal ion) with various ligands bound is referred to as the *speciation* by marine chemists. Indeed, the question of binding to ligands in solution (or a cell) is not only a matter of affinity, but kinetics as well.

A corollary to the presence of ligands in physiological media such as cytoplasm, sera, cerebrospinal fluid, and growth media is that zinc added to media such as these is not all free; rather, the vast majority becomes bound to ligands present in the media. This is illustrated in Fig. 14.6, which depicts measured and calculated free zinc ion levels in growth media as zinc is added to it; the growth media are Dulbecco's modified Eagle's medium with

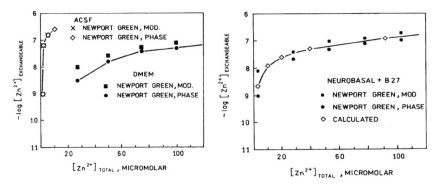

Figure 14.6 Measured (by phase and modulation fluorometry) and calculated (using MINEQL) free zinc are depicted as a function of total added zinc for (left panel) artificial cerebrospinal fluid ($\Diamond$, x) and Dulbecco's modified Eagle's medium ($\blacksquare$, $\bullet$) and (right panel) Neurobasal+B-27 ($\blacksquare$, $\bullet$, $\Diamond$).

serum (DMEM), an artificial cerebrospinal fluid (ACSF), and Neurobasal, whose composition is also known (Price and Brewer, 2001). For growth media which contain sera the composition is not so well known because of the variability of zinc levels in sera: normal human zinc levels in serum range from 14 to 24 μM (Berkow, 1992). For the ACSF the free Zn (expressed as pZn, the negative logarithm of the free zinc concentration) tracks very closely the added zinc because there are no tight-binding ligands in the solution (phosphate has been omitted). By comparison, DMEM starts out with a micromolar concentration of total zinc and an abundant, fairly strong ligand in the form of bovine serum albumin, providing less than 10 nM free zinc, and addition of zinc to a total of 100 μM only results in the free zinc rising to 50 nM. Because the composition of Neurobasal is known, we can predict the proportion of free zinc for a given level of total zinc, evidently with good accuracy by comparison with measurements using fluorescent indicators. The vital message is that free zinc is less than added zinc, usually by orders of magnitude, and varies dramatically depending on choice of the medium. For media with known composition this can be predicted to a degree, but measurement is preferable. The inverse experiment, identifying the concentrations and affinities of ligand(s) by titrating with zinc and measuring free zinc, can be done, but only a simplified approximation may be obtained (Moffett et al., 1997; van den Berg, 1984).

For our purposes we take the following labile complexes of zinc ion to be "free" or rapidly exchangeable: $ZnHCO_3^+$, $ZnSO_4$, $ZnCl^+$, $ZnOH^+$, $ZnOHCl$, $ZnCl_2$, and $Zn(SO_4)^{2-}$. $ZnCO_3$ seems to be on the borderline in terms of stability (log stability constant $\sim$ 5; Zirino and Yamamoto, 1972): the CA-based indicators do not respond to it, but other indicators might. In certain media other complexes may be present and also exchange rapidly with apo-CA even though they are otherwise stable, in the manner of dipicolinate.

5. Metal Ion Buffers

The presence of contaminating ligands together with zinc contamination from many sources generally makes it infeasible to reproducibly prepare solutions with low concentrations of free zinc by simple addition of small amounts of zinc to ordinary pH buffers. Indeed, the propensity of zinc to hydrolyse and form precipitates near neutrality makes it generally infeasible to prepare stocks of zinc salts near neutrality (Cotton and Wilkinson, 1988). For these reasons we prepare zinc ion "buffers" to maintain fixed low zinc ion concentrations in a solution. These buffers contain moderate concentrations of zinc (μM) together with a fairly tight-binding ligand such as nitrilotriacetic acid (NTA), citrate, or Bicine; the ligand binds the vast majority of the zinc, and this composition resists elevation of the free zinc concentration if contaminating zinc is present, or its decline if some process sequesters the zinc. Several authors have described procedures and computer programs for formulating such buffers (Aslamkhan *et al.*, 2002; Baker, 1988; McCall, 2000; Nuccitelli, 1994); we find the program MINEQL+ (Environmental Research Software, Hallowell, Maine) well suited for this purpose.

Tables 14.1–14.4 contain recipes for preparing zinc buffers to produce a range of free zinc concentrations in various media, including calibration buffers, ACSF, and Neurobasal+B-27 supplement (Price and Brewer, 2001). Note that phosphate is omitted from the ACSF due to the insolubility of its zinc salt ($K_{SP} = 9 \times 10^{-33}$); adding phosphate effectively precipitates free zinc out of solution. This is calculated in Fig. 14.7 for a medium under conditions under which cells are cultured (free exchange with 5% CO_2). At zinc concentrations near 1 μM total, the majority is present as labile complexes or the somewhat more stable carbonate complex, but as zinc is added it begins to precipitate as the phosphate, and by 10 μM nearly half the zinc has precipitated. The pH buffer used in the ACSF is MOPS or another Good's buffer chosen with low affinity for zinc. Neurobasal+B-27 supplement is a serum-free medium used to culture neurons and the rat pheochromocytoma line PC-12, which can differentiate to form processes in a neuron-like manner; for this reason PC-12s are used as a model for neurons. For studying the effects of varying zinc levels on cells this defined medium has the important advantage that it has known amounts of total zinc and zinc buffers; media incorporating sera perforce have unknown amounts of zinc and zinc buffers present, confounding interpretation unless free zinc is measured therein.

Table 14.1 2× 2 mM NTA 10 mM MOPS buffers pH 7.0, 25 °C for pZn 14.7–7.3

	[Zn], molar total	[Zn], free	2× [Zn], total	$ZnSO_4$ to add	NTA/MOPS to add (ml)	MOPS to add (ml)
0	Control	0	0	0	10	40
1	1.00×10^{-9}	2.13×10^{-15}	2.00×10^{-9}	$20.61\mu l^a$	10	39.97
2	5.00×10^{-9}	1.06×10^{-14}	1.00×10^{-8}	$98.61\mu l^a$	10	39.90
3	1.00×10^{-7}	2.13×10^{-13}	2.00×10^{-7}	1.97ml^a	10	38.03
4	2.50×10^{-7}	5.23×10^{-12}	5.00×10^{-7}	$49.31\mu l^b$	10	39.95
5	5.00×10^{-7}	1.06×10^{-12}	1.00×10^{-6}	$98.61\mu l^b$	10	39.90
6	2.00×10^{-6}	4.26×10^{-12}	4.00×10^{-6}	$394.47\mu l^b$	10	39.60
7	5.00×10^{-6}	1.07×10^{-11}	1.00×10^{-5}	$986.19\mu l^b$	10	39.01
8	2.50×10^{-5}	5.46×10^{-11}	5.00×10^{-5}	$493.09\mu l^c$	10	39.50
9	5.00×10^{-5}	1.12×10^{-10}	1.00×10^{-4}	$986.19\mu l^c$	10	39.01
10	2.00×10^{-4}	5.23×10^{-10}	4.00×10^{-4}	$394.47\mu l^d$	10	39.60
11	5.00×10^{-4}	2.13×10^{-9}	1.00×10^{-3}	$986.19\mu l^d$	10	39.01
12	7.50×10^{-4}	6.39×10^{-9}	1.50×10^{-3}	1.47ml^d	10	38.53
13	9.00×10^{-4}	1.92×10^{-8}	1.80×10^{-3}	1.77ml^d	10	38.23
14	9.60×10^{-4}	5.10×10^{-8}	1.92×10^{-3}	1.89ml^d	10	38.11

This is a 2× buffer: for example, dilute with equal volume ultrapure water to make final buffer. To make: prepare 10 mM MOPS alone (Sigma M5162; 2.09gl^{-1}), and 10 mM NTA (Sigma 398144; 1.91gl^{-1} final volume) in 10 mM MOPS. Add MOPS and NTA/MOPS per table to 50ml standup plastic centrifuge tube (Corning 430897). Make stock 10, 100, 1000, and 10,000-fold dilutions of the atomic absorption Zn standard (Aldrich cat. No. 207667) in ultrapure water (use immediately) or 1wt% HCl (dilute EM Omnipure high purity HCl (HX0607-1) 1ml into 30ml ultrapure water) (lasts indefinitely). Add listed volumes of stocks (note icons) to the NTA/MOPS mixture to give desired free zinc concentrations. Note that the listed volumes of stock dilutions (and MOPS volumes) are for a zinc AA standard of 0.0507 M concentration to give the desired total (and thus free) zinc concentrations. Other bottles may be slightly different requiring adjustment in the resulting free zinc concentrations (or recalculation of the volumes to use to get the same free zinc concentrations).

[a] Zn stock 5: 5.07×10^{-6} M.
[b] Zn stock 3: 5.07×10^{-4} M.
[c] Zn stock 2: 5.07×10^{-3} M.
[d] Zn standard: 5.07×10^{-2} M.

Table 14.1 is a set of free zinc calibration buffers made up with just MOPS as the pH buffer and NTA as the metal ion buffer. Table 14.2 is a set of buffers covering a smaller range of free zinc concentration in the micromolar range. Table 14.3 is a set of zinc buffers in ACSF, such as might be used to bathe a brain slice preparation. Table 14.4 is a set of Neurobasal+ B-27 buffers where free zinc is elevated by adding zinc sulfate, and reduced by the addition of NTA. The buffers are specified for a particular pH, temperature, and ionic strength. The ACSF is also specified for a particular

Table 14.2 $2\times$ NTA/MOPS buffers pH 7.0, 25 °C for pZn 5–7

	[Zn], molar total	[Zn], free	$.2\times$ [Zn], total	$ZnSO_4$ to add	NTA/MOPS to add (ml)	MOPS to add (ml)
0	Control	0	0	0	2.5	47.50
1	2.00×10^{-7}	1.01×10^{-7}	4.00×10^{-7}	$39.4\mu l^a$	2.5	47.46
2	3.00×10^{-7}	1.99×10^{-7}	6.00×10^{-7}	$59.17\mu l^a$	2.5	47.44
3	5.00×10^{-7}	3.96×10^{-7}	1.00×10^{-6}	$98.6\mu l^a$	2.5	47.40
4	7.00×10^{-7}	5.94×10^{-7}	1.40×10^{-6}	$13.8\mu l^b$	2.5	47.48
5	9.00×10^{-7}	7.92×10^{-7}	1.80×10^{-6}	$17.75\mu l^b$	2.5	47.48
6	1.50×10^{-6}	1.39×10^{-6}	3.00×10^{-6}	$29.5\mu l^b$	2.5	47.47
7	2.20×10^{-6}	2.08×10^{-6}	4.40×10^{-6}	$43.39\mu l^b$	2.5	47.45
8	4.20×10^{-6}	4.05×10^{-6}	8.40×10^{-6}	$82.84\mu l^b$	2.5	47.41
9	6.20×10^{-6}	6.03×10^{-6}	1.24×10^{-5}	$12.2\mu l$	2.5	47.48
10	8.20×10^{-6}	8.00×10^{-6}	1.64×10^{-5}	$16.17\mu l$	2.5	47.48
11	1.02×10^{-5}	9.98×10^{-6}	2.04×10^{-5}	$20.11\mu l$	2.5	47.47

^{a} Zn stock 3: 5.07×10^{-4} M.
^{b} Zn stock 2: 5.07×10^{-3} M.
This is a $2\times$ set of buffers for the micromolar range of free zinc concentration. The buffers are prepared as described in the footnote to Table 14.1, again assuming a concentration of 0.0507 M total zinc in the AA standard.

carbonate/bicarbonate status: whether there is significant added carbonate/bicarbonate, or the buffer is intended for use with tissue culture conditions (e.g., under a 95% O_2, 5% CO_2 atmosphere), or open to the atmosphere (0.3% CO_2) makes a difference in the formulation.

In general these buffers are made up as somewhat concentrated stock solutions; we find it convenient to make up 50 ml volumes as these can be stored in Corning 430897 50-ml self-standing centrifuge tubes, which in our experience are reasonably free of contamination. Glassware should be avoided as it will leach out metal ions even when acid cleaned. In general we use "ultrapure" water, for example, distilled water that has been further purified by passage through a system containing beds of activated carbon and ion exchange resin to further reduce contaminant levels. Such systems are offered by Bransted, Millipore, and other manufacturers. Buffers lacking added divalent cations can be freed of such contaminants by passage through beds of chelating resin containing immobilized iminodiacetate residues (e.g., Chelex-100, Bio-Rad). We also use uncolored (e.g., "natural") disposable pipette tips (e.g., Bio-Rad BR-41 or equivalent). We find it easiest to use a volumetric standard for atomic absorption spectroscopy (e.g., Aldrich cat. No. 207667) to accurately add zinc to buffers. Recipes given below are for a particular solution; since the product assay is given on each

Table 14.3 Artificial cerebrospinal fluid (ACSF) pH 7.4, 37 °C, with pZn from 13 to 6 (142 mM NaCl, 5 mM KCl, 10 mM glucose, 1 mM MgCl$_2$, 2 mM CaCl$_2$, 10 mM HEPES, 50 μM Bicine; open to the atmosphere)

Free [Zn], molar	Total [Zn], molar
6.94×10^{-14}	1.00×10^{-12}
2.08×10^{-13}	3.00×10^{-12}
4.86×10^{-13}	7.00×10^{-12}
6.94×10^{-13}	1.00×10^{-11}
2.08×10^{-12}	3.00×10^{-11}
4.86×10^{-12}	7.00×10^{-11}
6.94×10^{-12}	1.00×10^{-10}
2.08×10^{-11}	3.00×10^{-10}
4.86×10^{-11}	7.00×10^{-10}
6.94×10^{-11}	1.00×10^{-9}
2.08×10^{-10}	3.00×10^{-9}
4.86×10^{-10}	7.00×10^{-9}
6.94×10^{-10}	1.00×10^{-8}
2.08×10^{-9}	3.00×10^{-8}
4.87×10^{-9}	7.00×10^{-8}
6.96×10^{-9}	1.00×10^{-7}
2.09×10^{-8}	3.00×10^{-7}
4.92×10^{-8}	7.00×10^{-7}
6.35×10^{-8}	9.00×10^{-7}
2.20×10^{-7}	3.00×10^{-6}
7.38×10^{-7}	9.00×10^{-6}
2.03×10^{-6}	2.00×10^{-5}

This is a 1× ACSF; note there is no added phosphate or carbonate/bicarbonate and it is pH buffered by HEPES. To make combine NaCl, KCl, glucose, HEPES, and Bicine and remove divalent contaminants by passage over Chelex. Add Ca and Mg as the highest purity salts available: previously we had used "electronic purity" grades of MgCl$_2$ and CaCl$_2$, but since these are no longer available the best grades we are aware of are the 99.999% grades from Sigma-Aldrich or GFS Chemicals.

Table 14.4 Neurobasal A+B-27 supplement with varying pZn

Free [Zn], molar	Total [Zn], molar	[NTA] added, molar
2.0×10^{-11}	1.86×10^{-6}	7.5×10^{-4}
3.3×10^{-10}	1.86×10^{-6}	4.2×10^{-6}
1.2×10^{-9}	1.86×10^{-6}	
5.1×10^{-8}	3.00×10^{-5}	
6.57×10^{-7}	3.00×10^{-4}	
4.49×10^{-6}	5.00×10^{-3}	

This buffer series is made by adding ZnSO$_4$ or NTA to Neurobasal A+B-27 supplement (Price and Brewer, 2001), which are commercially available. As formulated, the Neurobasal already contains 1.86×10^{-6} molar total zinc. The free zinc concentrations given here do not include the carbonate/bicarbonate zinc complexes.

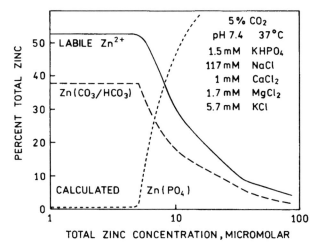

Figure 14.7 Effect of phosphate on zinc in solution. Labile zinc (—), zinc carbonate/bicarbonate (- - - -), and precipitated zinc phosphate (...) are depicted as a function of total added zinc for an artificial cerebrospinal fluid formulation.

bottle, it may be necessary to adjust the recipe slightly. Note that these standards are supplied in dilute acid to maintain solubility, and that it is occasionally necessary to adjust the pH of a buffer after it has been completed.

6. Kinetics

It should be noted that the kinetics of metal ion-binding reactions to proteins (such as CA) as well as indicator molecules and other solutes can play an important and sometimes dominant role in these reactions. For instance, let us consider the response time of a well-known fluorescent zinc indicator, ZinPyr-1, to a free zinc concentration near its K_D. The association rate constant k_{on} for ZinPyr-1 is $3.3 \pm 0.4 \times 10^6$ $M^{-1}s^{-1}$ at 25 °C, pH 7.0 (Nolan et al., 2005), and the dissociation rate constant k_{off} is $2.3 \pm 0.4 \times 10^{-3} s^{-1}$. We can calculate the time it will take for the indicator to come to equilibrium with a free zinc concentration of 1 nM (and presuming we have contrived that the ZinPyr-1 concentration is 0.1 μM, well in excess of the K_D of 0.7 nM) (Stumm and Morgan, 1996; Thompson et al., 2006). The rate at which water is exchanged from the zinc–water labile complex is fast ($\sim 10^8 s^{-1}$) and thus not rate limiting. The observed rate k_{obs} can be approximated by

$$k_{obs} = k_{on}[\text{ZinPyr} - 1] + k_{off}. \qquad (14.1)$$

Substituting, we find $k_{obs} \sim 0.33\ s^{-1}$, and since the half time of the reaction is $\ln 2/k_{obs}$, the half time will be approximately 1.83 s. The point is that this indicator (and ones like it that also use the di(2-picolyl)amine zinc-binding moiety) can respond fast enough for static zinc observations, but is unlikely to follow rapid (ms) changes like those associated with nerve transmission with adequate fidelity. Similarly, even with high-affinity indicators having near diffusion-controlled association rate constants for zinc binding ($\sim 10^8\ M^{-1}s^{-1}$ at room temperature in water), picomolar concentrations of free zinc will take hours to equilibrate unless catalysts are present (Bozym et al., 2006).

7. Applications: Ratiometric Determination of Free Zinc in Solution

While we have described several different means of free zinc determination in solution above using CA, it is necessary first to decide whether the determination is to be done in the "stoichiometric" or "equilibrium binding" regimes (Eftink, 1997; Weber, 1992). The former is well suited for determinations when the expected free zinc concentration is above the K_D of the CA variant, but somewhat lower than a convenient concentration of apo-CA and the fluorescent sulfonamide (in this example, ABDN). For instance, suppose the free zinc concentration is around 100 nM, and the wild-type apo-CA and ABDN concentrations can be made 1 μM by additions to the sample. Since the binding constant for zinc is 4 pM, the zinc ion will be bound stoichiometrically (hence the name) and the fractional occupancy of the protein with zinc (and thus ABDN) will be approximately 10%. Unlike the calibration curve for ABDN depicted earlier, the fractional occupancy is linear with zinc concentration until saturation of the protein. The ratio is measured conveniently in the fluorometer by measuring the intensities at 560 and 600 nm, with excitation at 430 nm. In this case the accuracy of the measurement is constrained by the accuracy with which the protein and ABDN concentrations are known. ABDN is synthesized from ABD-F (7-fluorobenz-2-oxa-1,3-diazole-4-sulfonamide; Molecular Probes F6053) and ethanolamine as previously described (Thompson et al., 2000a). A calibration curve is nearly linear (as opposed to the sigmoidal curve in Fig. 14.2) and the dynamic range is limited by how precisely low fractional saturations can be measured. We have found the stoichiometric assay to be very useful when measuring zinc release in the brain using a dialysis probe: the apo-CA is pumped through the dialysis tube and captures nearly all of the nanomolar concentrations of zinc that are released (Frederickson et al., 2006).

For concentrations of free zinc closer to the protein's Zn(II) K_D, the stoichiometric method ultimately becomes infeasible when the protein is dispersed uniformly throughout the solution because the protein concentration is so low that it is hard to discern the signal, and either the sulfonamide is also present at such low concentrations that it does not bind fully and underreports the level of zinc, or if at or above its K_D the signal from the bound form is only a small fraction of the total. Under these conditions, we switch to an "equilibrium binding" mode where the fractional saturation of the CA Zn-binding site is simply controlled by the law of mass action and the free zinc concentration. The problem arises again that if the protein and sulfonamide are at a convenient concentration dispersed throughout a finite sample: the fractional occupancy with zinc will be low because the protein and sulfonamide are in much greater excess. Expedients for avoiding this problem are to use a tiny amount of protein and sulfonamide localized at the foci of the excitation and emission beams (a fiber optic is a convenient way to achieve this—see accompanying chapter), or work in a naturally buffered system like serum, where other ligands release zinc in response to the uptake by the apo-CA of free zinc.

8. Preparation of Apocarbonic Anhydrase

We routinely use carbonic anhydrase isozyme II from humans and its variants to prepare our sensors; the gene is easily expressed in the BL 21 (DE3) strain of *E. coli* and purified as previously described (DiTusa *et al.*, 2001; Kiefer and Fierke, 1994; Lesburg *et al.*, 1997; McCall and Fierke, 2004). However, bovine CA II also is satisfactory for some applications and cheaper than the human variant; some preparations contain substantial amounts of the CA I isozyme, which differs slightly but functions similarly. Removal of the native zinc ion is performed by either of the two methods based on the use of 2,6-dipicolinic acid (DPA) described (Hunt *et al.*, 1977). The first method is simpler, slower, and requires less labor: the protein in some suitable buffer (e.g., 50 mM HEPES pH 7.5, 100 mM Na$_2$SO$_4$) is dialyzed (Spectrapor 3 dialysis tubing, Spectrum Medical Industries) versus two changes of 10 mM Tris pH 7.0, 50 mM DPA (Sigma P63808) in a plastic vessel; the DPA can then be removed by dialysis against a suitable buffer (treated with Chelex-100 resin, Bio-Rad) or gel filtration in a spin column (see below). For any dialysis of CA that has been "apoized" we add a few milligrams of Chelex resin to the dialysis buffer to bind contaminating metal ions that may emerge posttreatment with Chelex. The second method (developed in the Fierke lab) is faster and requires more labor. The protein solution at a concentration in the micromolar range (ε_{280} = 48,900 M^{-1} cm^{-1}) in some buffer (about 600 μl) is added to a Centriprep

10 concentrator tube (Amicon, Beverly, MA, 01915; (800) 343–0696). Fill to the mark with buffer (10 mM Tris pH 7.0, 50 mM DPA) and centrifuge at 3000 rpm for 25 min. Discard filtrate. Add more buffer and repeat centrifugation as earlier. Retentate from second centrifugation is put into a new Centriprep tube, which is filled to the mark with Chelexed 50 mM HEPES pH 7.3, 150 mM Na$_2$SO$_4$. Centrifuge at 3000 rpm for 25 min. The retentate can also be dialyzed or passed through a PD-10 gel filtration column against this buffer to remove DPA. Some fluorescent conjugates (e.g., L198C) are difficult to apoize, and it may be necessary to apoize the unconjugated protein, replace with Co(II), then conjugate the cobalt-bound protein and remove the cobalt last since cobalt binds much less tightly. Verification of apoization can be done by adding a stoichiometric amount of dansylamide (5-dimethylaminonaphthalene-1-sulfonamide; Fluka 39225), exciting at 330 nm and observing the emission at 450 and 560 nm; appearance of a shoulder at 450 nm indicates the presence of the holoprotein. The apoprotein is best used promptly and/or refrigerated (not frozen), but unfrozen fluorescent-labeled apoproteins have performed satisfactorily years after zinc removal.

Wild-type CA and many variants are such avid binders of zinc when in the apo-form that special precautions are necessary to prevent their contamination once the zinc is removed (Fierke and Thompson, 2001; Thompson et al., 2006). These precautions are akin to the "trace metal clean procedures" developed by marine geochemists. One begins by recognizing that zinc (and other metal ions which will also bind tightly to CA and small molecule fluorescent indicators, such as Cu(II)) is ubiquitous, being found in airborne dust, many plastic mold release agents, many paints, skin, and paper products such as Kimwipes. We wear gloves, use metal free disposable pipette tips (e.g., Bio-Rad BR-41), avoid using glassware or rinse it in acid if we must, and perform some manipulations inside laminar flow hoods to avoid contamination. Ordinary buffer salts (e.g., A.C.S. reagent grade) have tens of parts per million total contaminants, so millimolar buffer concentrations will have tens of nanomolar Zn, Cu, etc., above that present in even highly purified water. The ligands present in metal ion buffers limit the effect of contamination by most metal ions. Further details are given elsewhere (Thompson et al., 2006).

9. Intracellular Sensing with TAT Tag

Free zinc inside cells can be quantitatively imaged using fluorescent-labeled TAT-apo-CA and Dapoxyl sulfonamide in the excitation ratiometric approach described earlier (Thompson et al., 2002a). Ordinarily CA (like most proteins) is not taken up intact by most cells, so cannot be used alone for intracellular measurements. Fortunately, attachment of the small (12-residue) TAT peptide to the CA (which may also be a CA–GFP homolog fusion protein) enables CA to penetrate some cell membranes readily (Schwarze et al., 1999). The fluorescent-labeled apo-CA is added to the medium at low micromolar concentrations and is taken up within a few minutes. Dapoxyl sulfonamide is synthesized from Dapoxyl sulfonyl chloride (Invitrogen D10160) and ammonium hydroxide as previously described (Thompson et al., 2000b) then dissolved in DMSO (1 μM) and added to the cells; it enters within a minute. Images of the cells are taken in the fluorescence microscope with excitation at 365 nm (Omega D360–40×filter) and 540 nm (Omega D540–25×) with emission monitored through Chroma 570 LP dichroic and HQ 620–60 m barrier filters. Ratiometric images are constructed with IPLab and compared with calibration images of a 1536-well plate taken under identical conditions but with lower magnification. The image taken with excitation 365 is divided by the image taken with 540 nm excitation. A segment layer was then applied to the interior of the wells which allowed us to extract numerical data from each well (the mean pixel value, as well as the minimum and maximum pixel value with standard deviation for each well). The mean value for each well was then plotted against the free zinc concentration to determine the apparent K_D on the microscope. With PC-12 and CHO cells at least, the UV- and green-excitable red fluorescence background is low enough to be negligible. For cells (or phenomena therein) requiring reduced sensitivity to higher zinc concentrations that would readily saturate the wild-type apo-CA, variants with lower affinity have been substituted.

10. Intracellular Sensing with an Expressible CA Sensor

Conjugation of CA variants (having a cysteine residue substituted for another at a chosen point, such as H36C) with thiol-reactive fluorescent labels does not always proceed with high yields, and the reagents are costly. Replacing the fluorescent label Alexa Fluor 594 in Fig. 14.5 with a variant of the GFP exhibiting similar spectral properties by expressing it as a fusion protein with CA offers several advantages. First, expression and purification is simplified, in that one is producing the fluorescent-labeled protein with the same effort as the unlabeled variant, without the additional labor, expense, and inefficiency of purifying the unlabeled variant and labeling it. Indeed, the brightly colored fusion protein is simpler to quantify and follow through the purification process. Second, the fusion protein can in principle be expressed in any cell (or all of them) in a given organism, making *in vivo* sensing possible. Third, in principle different colors might be expressed in different cells or compartments for multisite experiments. Finally, the intracellularly expressed fusion protein may be "targeted" to particular organelles by fusing small peptides which direct the expression product. We note that the spectral overlap between the chosen acceptors and Dapoxyl sulfonamide, together with Dapoxyl sulfonamide's high quantum yield when bound and the relatively small size of the GFP variants and CA generally means that energy transfer between the Dapoxyl sulfonamide and the GFP variant acceptor is nearly quantitative.

For a variety of reasons it was of interest to measure resting free zinc levels in the mitochondrion, particularly under conditions of oxidative stress (Bozym *et al.*, submitted for publication). The existing mitochondrial zinc indicator (RhodZin-3) only offers affinity (K_D = 65 nM) to free zinc levels we would ordinarily consider to be toxic, and is not ratiometric. Thus we constructed and expressed a DsRed2–CA fusion protein and expressed it with a mitochondrial targeting sequence from a cytochrome oxidase (Complex IV) subunit (Rizzuto *et al.*, 1989, 1995). This assured the fusion protein would only be localized in the mitochondrion and enabled the free zinc level therein to be measured. We note that the Dapoxyl sulfonamide penetrates cells and mitochondria readily.

The expression vectors for CA-fluorescent protein fusion proteins for use in PC-12 cells were constructed and transfected using standard techniques. The red fluorescent protein used was a commercially available optimized form of the *Discosoma* protein DsRed (DsRed2, Clontech) which folds more quickly than the wild-type as well. Briefly, human CA II was PCR amplified using the two separate sense primers:

5′-CCGCGCGCCAAAGATCCATTCGTTGATGGCCCATCACTG GGGGTACGGC-3′ or 5′-ATCCGCTAGCATGGCCCATCACTG GGGGTACGGCAAACAC-3′, containing a *Bss*HII or *Nhe*I site, respectively, and the antisense primer.

5′-CATCGGATCCCCACCTTTGAAGGAAGCTTTG-3′. Each fragment was cloned into the pDsRed2-Mito vector (Clontech). The pDsRed2-Mito/CA vector contains the CA in frame with an upstream mitochondrial signal sequence (from subunit VIII of human cytochrome *c* oxidase and a downstream DsRed2 tag). This was achieved by cloning the CA into the *Bss*HII and blunted *Bam*HI sites of the pDsRed2-Mito vector. The inserts and the expression vector were sequenced to ensure everything was in frame and no sense mutations had occurred. The vector was transfected into the PC-12 cells using Lipofectamine and the manufacturer-supplied protocol. Imaging and calibration were performed as described for the TAT-tagged protein.

REFERENCES

Aslamkhan, A. G., Aslamkhan, A., and Ahearn, G. A. (2002). Preparation of metal ion buffers for biological experimentation: A methods approach with emphasis on iron and zinc. *J. Exp. Zool.* **292,** 507–522.

Axelrod, D. (1989). Fluorescence polarization microscopy. *In* "Methods in Cell Biology: Fluorescence Microscopy of Living Cells in Culture. Part B. Quantitative Fluorescence Microscopy—Imaging and Spectroscopy," (D. L. Taylor and Y.-L. Wang, eds.), Vol. 30, pp. 333–352. Academic Press, New York.

Baker, J. O. (1988). Metal-buffered systems. *Methods Enzymol.* **158,** 33–55.

Berkow, R. (1992). *In* "Merck Manual of Diagnosis and Therapy," p. 2844. Merck Research Laboratories, Rahway, NJ.

Bozym, R. A., Thompson, R. B., Stoddard, A. K., and Fierke, C. A. (2006). Measuring picomolar intracellular exchangeable zinc in PC-12 cells using a ratiometric fluorescence biosensor. *ACS Chem. Biol.* **1,** 103–111.

Chen, R. F., and Kernohan, J. (1967). Combination of bovine carbonic anhydrase with a fluorescent sulfonamide. *J. Biol. Chem.* **242,** 5813–5823.

Christianson, D. W., and Fierke, C. A. (1996). Carbonic anhydrase—Evolution of the zinc binding site by nature and by design. *Acc. Chem. Res.* **29,** 331–339.

Cotton, F. A., and Wilkinson, G. (1988). "Advanced Inorganic Chemistry," Wiley-Interscience, New York.

DiTusa, C. A., McCall, K. A., Christensen, T., Mahapatro, M., Fierke, C. A., and Toone, E. J. (2001). Thermodynamics of metal ion binding. II. Metal ion binding by carbonic anhydrase variants. *Biochemistry* **40,** 5345–5351.

Dix, J. A., and Verkman, A. S. (1990). Mapping of fluorescence anisotropy in living cells by ratio imaging: Application to cytoplasmic viscosity. *Biophys. J.* **57,** 231–240.

Eftink, M. R. (1997). Fluorescence methods for studying equilibrium macromolecule—Ligand interactions. *In* "Fluorescence Spectroscopy," (L. Brand and M. L. Johnson, eds.), Vol. 278, pp. 221–257. Academic Press, New York.

Elbaum, D., Nair, S. K., Patchan, M. W., Thompson, R. B., and Christianson, D. W. (1996). Structure-based design of a sulfonamide probe for fluorescence anisotropy detection of zinc with a carbonic anhydrase-based biosensor. *J. Am. Chem. Soc.* **118,** 8381–8387.

Eriksson, A. E., and Jones, T. A. (1988). Refined structure of human carbonic anhydrase II at 2.0A resolution. *Proteins* **4,** 274–282.

Fierke, C. A., and Thompson, R. B. (2001). Fluorescence-based biosensing of zinc using carbonic anhydrase. *BioMetals* **14,** 205–222.

Forster, T. (1948). Intermolecular energy migration and fluorescence (Ger.). *Ann. Phys.* **2,** 55–75.

Frederickson, C. J., Giblin, L. J., Krezel, A., McAdoo, D. J., Muelle, R. N., Zeng, Y., Balaji, R. V., Masalha, R., Thompson, R. B., Fierke, C. A., Sarvey, J. M., Valdenebro, M. d., Prough, D. S., and Zornow, M. H. (2006). Concentrations of extracellular free zinc (pZn)e in the central nervous system during simple anesthetization, ischemia, and reperfusion. *Exp. Neurol.* **198,** 285–293.

Huang, C.-c., Lesburg, C. A., Kiefer, L. L., Fierke, C. A., and Christianson, D. W. (1996). Reversal of the hydrogen bond to zinc ligand histidine-119 dramatically diminishes catalysis and enhances metal equilibration kinetics in carbonic anhydrase II. *Biochemistry* **35,** 3439–3446.

Hunt, J. B., Rhee, M. J., and Storm, C. B. (1977). A rapid and convenient preparation of apocarbonic anhydrase. *Anal. Biochem.* **79,** 614–617.

Hunt, J. A., Ahmed, M., and Fierke, C. A. (1999). Metal binding specificity in carbonic anhydrase is influenced by conserved hydrophobic amino acids. *Biochemistry* **38,** 9054–9060.

Ippolito, J. A., Baird, T. T., McGee, S. A., Christianson, D. W., and Fierke, C. A. (1995). Structure-assisted redesign of a protein–zinc binding site with femtomolar affinity. *Proc. Natl Acad. Sci. USA* **92,** 5017–5021.

Kiefer, L. L., and Fierke, C. A. (1994). Functional characterization of human carbonic anhydrase II variants with altered zinc binding sites. *Biochemistry* **33,** 15233–15240.

Kiefer, L. L., Paterno, S. A., and Fierke, C. A. (1995). Hydrogen bond network in the metal binding site of carbonic anhydrase enhances zinc affinity and catalytic efficiency. *J. Am. Chem. Soc.* **117,** 6831–6837.

Lesburg, C. A., Huang, C.-c., Christianson, D. W., and Fierke, C. A. (1997). Histidine to carboxamide ligand substitutions in the zinc binding site of carbonic anhydrase II alter metal coordination geometry but retain catalytic activity. *Biochemistry* **36,** 15780–15791.

Lindskog, S., Henderson, L. E., Kannan, K. K., Liljas, A., Nyman, P. O., and Strandberg, B. (1971). Carbonic anhydrase. *In* "The Enzymes," (P. D. Boyer, ed.), Vol. 5, pp. 587–665. Academic Press, New York.

Maren, T. H. (1977). Use of inhibitors in physiological studies of carbonic anhydrase. *Am. J. Physiol.* **232,** F291–F297.

McCall, K. A. (2000). *In* "Department of Biochemistry," p. 190. Duke University, Durham, NC.

McCall, K. A., and Fierke, C. A. (2004). Probing determinants of the metal ion selectivity in carbonic anhydrase using mutagenesis. *Biochemistry* **43,** 3979–3986.

Moffett, J. W., Brand, L. E., Croot, P. L., and Barbeau, K. A. (1997). Cu speciation and cyanobacterial distribution in harbors subject to anthropogenic Cu inputs. *Limnol. Oceanogr.* **42,** 789–799.

Nolan, E. M., Jaworski, J., Okamoto, K.-I., Hayashi, Y., Sheng, M., and Lippard, S. J. (2005). Qz1 and Qz2: Rapid, reversible quinoline-derivatized fluoresceins for sensing biological Zn(II). *J. Am. Chem. Soc.* **127,** 16812–16823.

Nuccitelli, R. (1994). A practical guide to the study of calcium in living cells. *In* "Methods in Cell Biology," (L. Wilson and P. Matsudaira, eds.), Vol. 40, p. 364. Academic Press, New York.

Price, P. J., and Brewer, G. J. (2001). Serum-free media for neural cell cultures. *In* "Protocols for Neural Cell Culture," (S. Federoff and A. Richardson, eds.), pp. 255–264. Humana Press, Totowa, NJ.

Rizzuto, R., Nakase, H., Darras, B., Francke, U., Fabrizi, G. M., Mengel, T., Walsh, F., Kadenbach, B., DiMauro, S., and Schon, E. A. (1989). A gene specifying subunit VIII of human cytochrome *c* oxidase is localized to chromosome 11 and is expressed in both muscle and non-muscle tissues. *J. Biol. Chem.* **264,** 10595–10600.

Rizzuto, R., Brini, M., Pizzo, P., Murgia, M., and Pozzan, T. (1995). Chimeric green fluorescent protein as a tool for visualizing subcellular organelles in living cells. *Curr. Biol.* **5,** 635–642.

Schwarze, S. R., Ho, A., Vocero-Akbani, A., and Dowdy, S. F. (1999). *In vivo* protein transduction: Delivery of a biologically active protein into the mouse. *Science* **285,** 1569–1572.

Stumm, W., and Morgan, J. J. (1996). "Aquatic Chemistry: Chemical Equilibria and Rates in Natural Waters," Wiley-Interscience, New York.

Supuran, C. T., and Scozzafava, A. (2000). Carbonic anhydrase inhibitors and their therapeutic potential. *Exp. Opin. Ther. Patents* **10,** 575–600.

Szmacinski, H., and Lakowicz, J. R. (1993). Optical measurements of pH using fluorescence lifetimes and phase-modulation fluorometry. *Anal. Chem.* **65,** 1668–1674.

Szmacinski, H., and Lakowicz, J. R. (1994). Lifetime-based sensing. *In* "Topics in Fluorescence Spectroscopy, Vol. 4: Probe Design and Chemical Sensing," (J. R. Lakowicz, ed.), pp. 295–334. Plenum, New York.

Thompson, R. B., and Jones, E. R. (1993). Enzyme-based fiber optic zinc biosensor. *Anal. Chem.* **65,** 730–734.

Thompson, R. B., and Patchan, M. W. (1995a). Fluorescence lifetime-based biosensing of zinc: Origin of the broad dynamic range. *J. Fluorescence* **5,** 123–130.

Thompson, R. B., and Patchan, M. W. (1995b). Lifetime-based fluorescence energy transfer biosensing of zinc. *Anal. Biochem.* **227,** 123–128.

Thompson, R. B., Maliwal, B. P., Feliccia, V. L., Fierke, C. A., and McCall, K. (1998a). Determination of picomolar concentrations of metal ions using fluorescence anisotropy: Biosensing with a "reagentless" enzyme transducer. *Anal. Chem.* **70,** 4717–4723.

Thompson, R. B., Maliwal, B. P., and Fierke, C. A. (1998b). Expanded dynamic range of free zinc ion determination by fluorescence anisotropy. *Anal. Chem.* **70,** 1749–1754.

Thompson, R. B., Maliwal, B. P., and Fierke, C. A. (1999). Fluorescence-based sensing of transition metal ions by a carbonic anhydrase transducer with a tethered fluorophore. *In* "SPIE Conference on Advances in Fluorescence Sensing Technology IV," (J. R. Lakowicz, S. A. Soper, and R. B. Thompson, eds.) Vol. 3602, pp. 85–92. SPIE, Bellingham, WA.

Thompson, R. B., Whetsell, W. O., Jr., Maliwal, B. P., Fierke, C. A., and Frederickson, C. J. (2000a). Fluorescence microscopy of stimulated Zn(II) release from organotypic cultures of mammalian hippocampus using a carbonic anhydrase-based biosensor system. *J. Neurosci. Methods* **96,** 35–45.

Thompson, R. B., Maliwal, B. P., and Zeng, H. H. (2000b). Zinc biosensing with multiphoton excitation using carbonic anhydrase and improved fluorophores. *J. Biomed. Opt.* **5,** 17–22.

Thompson, R. B., Cramer, M. L., Bozym, R., and Fierke, C. A. (2002a). Excitation ratiometric fluorescent biosensor for zinc ion at picomolar levels. *J. Biomed. Opt.* **7,** 555–560.

Thompson, R. B., Peterson, D., Mahoney, W., Cramer, M., Maliwal, B. P., Suh, S. W., and Frederickson, C. J. (2002b). Fluorescent zinc indicators for neurobiology. *J. Neurosci. Methods* **118,** 63–75.

Thompson, R. B., Frederickson, C. J., Fierke, C. A., Westerberg, N. M., Bozym, R. A., Cramer, M. L., and Hershfinkel, M. (2006). Practical aspects of fluorescence analysis of free zinc ion in biological systems: pZn for the biologist. *In* "Fluorescence Sensors and Biosensors," (R. B. Thompson, ed.), pp. 351–376. CRC Press, Boca Raton, FL.

van den Berg, C. (1984). Determining the copper complexing capacity and conditional stability constants of complexes of copper (II) with natural organic ligands in seawater by cathodic stripping voltammetry of copper–catechol complex ions. *Mar. Chem.* **15,** 1–18.

Weber, G. (1956). Photoelectric method for the measurement of the polarization of the fluorescence of solutions. *J. Opt. Soc. Am.* **46,** 962–970.

Weber, G. (1992). "Protein Interactions," Chapman & Hall, New York.

Zeng, H. H., Thompson, R. B., Maliwal, B. P., Fones, G. R., Moffett, J. W., and Fierke, C. A. (2003). Real-time determination of picomolar free Cu(II) in seawater using a fluorescence-based fiber optic biosensor. *Anal. Chem.* **75,** 6807–6812.

Zeng, H.-H., Bozym, R. A., Rosenthal, R. E., Fiskum, G., Cotto-Cumba, C., Westerberg, N., Fierke, C. A., Stoddard, A., Cramer, M. L., Frederickson, C. J., and Thompson, R. B. (2005). *In situ* measurement of free zinc in an ischemia model and cell culture using a ratiometric fluorescence-based biosensor. *In* "SPIE Conference on Advanced Biomedical and Clinical Diagnostic Systems III," (T. Vo-Dinh, W. S. Grundfest, D. A. Benaron, and G. E. Cohn, eds.), Vol. 5692, pp. 51–59. SPIE, San Jose, CA.

Zirino, A. and Yamamoto, S. (1972). A pH-dependent model for the chemical speciation of copper, zinc, cadmium, and lead in seawater. *Limnol. Oceanogr.* **17,** 661-671.

CHAPTER FIFTEEN

Instrumentation for Fluorescence-Based Fiber Optic Biosensors

Richard B. Thompson,* Hui-Hui Zeng,* Daniel Ohnemus,[†] Bryan McCranor,* Michele Cramer,* and James Moffett[‡]

Contents

1. Introduction and Rationale for Fluorescence-Based Fiber Optic Sensors — 312
2. Basic Principles of Fiber Optics for Fluorescence Sensors — 313
3. Optics and Mechanics of Fluorescence-Based Fiber Optic Sensors — 316
    3.1. Alignment of sensors — 316
4. Mounting and Alignment of the Instrument — 319
5. Standards for Ratiometric and Lifetime-Based Fiber Optic Fluorescence Sensors — 323
6. Construction of Fiber Optic Probes — 326
7. Zn^{2+} Probe — 329
8. Cu^{2+} Probe — 330
9. Use of Fiber Optic Probes for Discrete Samples — 332
10. Operating Issues: Noise, Background, Thermal Drift, and Mode Hopping — 333
11. Field and Shipboard Use — 334
Acknowledgments — 335
References — 336

Abstract

This chapter summarizes the construction principles, operation, and calibration of (single-fiber) fluorescence-based fiber optic sensors. These sensors transduce recognition of a chemical analyte by a transducer such as a protein molecule as a change in fluorescence wavelength or lifetime that can be measured remotely through a length of fiber optic. Examples are given of

* Department of Biochemistry and Molecular Biology, University of Maryland School of Medicine, Baltimore, MD 21201
[†] Woods Hole Oceanographic Institution, Woods Hole, Massachusetts
[‡] Department of Biological Sciences, University of Southern California, Los Angeles, California

determination of metal ions in aqueous solution by fluorescence ratio and lifetime. Included are descriptions of instruments, alignment procedures, identification of noise sources, use of calibration standards, factors in the use of long fibers for sensing, issues in field and shipboard operation, and probe preparation.

1. Introduction and Rationale for Fluorescence-Based Fiber Optic Sensors

Over the last 30 years, there has been substantial development of fiber optic sensors for many applications (Grattan and Zhang, 1994; Thompson, 1991, 2006; Wolfbeis, 1991). A significant proportion of these are sensors for the detection and determination of chemical analytes (as distinct from influence sensors for temperature, sound, magnetic fields, etc.); of these, a large fraction detects the analyte by a change in fluorescence. Some of the fluorescence-based chemical sensors employ the surface specificity of the "evanescent wave" propagating through waveguides including optical fibers to transduce binding events (such as antibody–antigen recognition) as changes in fluorescence (Golden *et al.*, 1992; Harrick and Loeb, 1973; Herron *et al.*, 1998; Hirschfeld, 1984; Kronick and Little, 1975). However, most fluorescence-based fiber optic chemical sensors simply employ the fiber optic as a light pipe, and it is on these we will focus. Some of the fluorescence-based fiber optic sensors discussed herein are *biosensors*, which we define as sensors that employ a biological (or biomimetic) molecule or larger assembly to recognize the analyte, and which transduce the presence or level of the analyte in this case as a change in fluorescence we can measure. In our jargon, a "sensor" comprises the entire instrument, fiber optic, transducer head, and provision for data collection and analysis: in Malmstadt's concept of data domains, the sensor ultimately communicates the analyte concentration as an output on a computer screen that humans can interpret. In our view a "sensor" can monitor the analyte concentration at least quasicontinuously over some period of time *in situ*, as distinct from an "assay" which makes a discrete determination on a sample which must be collected.

The advantages of fiber optic sensors for some applications were apparent to the first inventors, and still obtain today. Perhaps most important, they permit the analysis to be performed *in situ*: essentially, the analysis is brought to the sample, instead of vice versa. This is of particular interest if the sample is remote, is in a hazardous (toxic, flammable, high voltage, etc.) environment, is inside the body (e.g., the brain or digestive tract; Peterson *et al.*, 1980) or is otherwise difficult to sample, such as deep in the ocean. Fiber optic sensors are also attractive when an analyte must be determined simultaneously throughout an array of sites distributed over some space (Thompson et al., 2002). This is particularly true when the alternative is repeatedly sampling numerous sites, which may be prohibitively costly or infeasible altogether. Unlike electrically-based sensors including ion-sensitive electrodes, fiber optics are safe in inflammable or explosive media, are safe to use in humans, and are not interfered with by environmental electrical noise (e.g., near generators or radar transmitters) or caused by movement through the Earth's magnetic field because they are dielectric waveguides. Fiber sensors can be quite small: tens of micrometers in diameter, or much smaller in near field conditions. Finally, because the amount of transducer material in the sensor is typically very small, they are convenient to measure trace analytes in relatively small volumes (see below).

2. BASIC PRINCIPLES OF FIBER OPTICS FOR FLUORESCENCE SENSORS

Optical fibers are a subset of optical waveguides wherein light is confined while propagating by differences in refractive index. It will be necessary to consider a few of their optical properties which are most germane for sensing; comprehensive treatments may be found elsewhere (Marcuse, 1991; Snyder and Love, 1983) Classically, one can consider a fiber optic as a solid cylinder of transparent material called the *core*, surrounded by a layer of transparent material of slightly lower refractive index called the *cladding*. If light is launched into the core roughly parallel to the axis, it will tend to stay confined within the core because according to Snell's law it will be totally internally reflected at the core: cladding interface because of the differing refractive indices (Fig. 15.1):

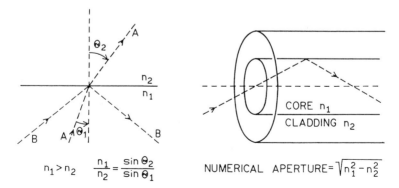

Figure 15.1 Principle of total internal reflection (Snell's law) (left panel) and light confinement within optical fiber (right panel).

$$(n_1^2 - n_2^2)^{1/2} = \text{NA} = \sin\theta, \tag{15.1}$$

where n_1 is the refractive index of the core, n_2 is the refractive index of the cladding, NA is the *numerical aperture* of the fiber, and θ is the critical angle. Several things are readily apparent: if light is launched into the core at an angle greater than θ, the light will not be reflected at the interface and will be lost into the cladding. Similarly, if the fiber optic is bent sharply enough the angle steepens and loss from the core will occur. If light in the core exits the end of the fiber it will not remain collimated, but will spread out to a degree controlled by the refractive index ratio and termed the *numerical aperture* of the fiber. The attenuation or *loss* of the fiber optic describes essentially how transparent it is, typically by measuring the reduction in light intensity of some source in passing through some length of fiber. The loss is expressed logarithmically in decibels per kilometer (dB/km), where 10 dB of loss corresponds exactly to an optical density (OD) of 1.0. Thus system loss is conveniently calculated by summing the losses of the components (see below). Loss is strongly wavelength dependent, mainly because Rayleigh scattering is a principal loss mechanism and scattering increases as the fourth power of the light frequency. Therefore working at longer wavelengths greatly reduces loss: telecommunications are sent through fiber optics at a wavelength of 1550 nm in the near infrared, such that the fiber attenuation is *circa* 1 dB/km.

This simplified theory breaks down if we consider fibers (or other waveguide geometries) with small cores a few micrometers in diameter whose refractive index is only slightly greater (~ 0.01) than that of the cladding. In this regime the fiber will only propagate light having particular distributions of electromagnetic energy in space and time; these distributions are called *modes* and are precisely analogous to the modes which a microwave waveguide or laser cavity will allow to propagate. A typical fiber with a core 100 μm in diameter and a core refractive index 0.05 greater than the cladding might support dozens of modes; such fibers are termed *multimode*. By comparison, fibers with very small cores and refractive index differences will allow only a single mode to propagate. These "single-mode fibers" are important in telecommunications (and time-resolved fluorescence-based sensors, see below) because different modes propagate through fibers at different velocities. If we simultaneously launch picosecond pulses of light into two modes of a multimode fiber, we will get a broadened pulse (or two pulses) out the other end. The single-mode fiber will only support the one mode, and the pulse emerges unbroadened; the multimode fiber suffers from modal dispersion. Single-mode fibers are harder to work with, however.

If we wish to measure fluorescence through optical fiber(s), additional considerations come into play. The first is whether to use separate fibers for emission and excitation, or a single fiber. Separate fibers have the virtue of simplicity: one launches the excitation through one fiber at the fluorescent transducer material, and collects the emission in a separate fiber to send it to the detector. Many excellent sensors have been made this way, particularly those developed by Barnard and Walt (1991). Additional advantages of this approach are that any fluorescence excited in the excitation fiber is not "seen" by the detector, it is easy to use with extended excitation sources such as lamps, and it lends itself to imaging application with fused fiber imaging bundles. However, they have drawbacks for many of the applications which interest us. Fiber bundles are bulky, costly, and lossy for distances longer than a few meters. The multimode fibers in these bundles have very large modal dispersion, an issue for time-resolved sensing. A bigger issue for us has been the issue of registration between the distribution of exciting light emitted by the excitation fiber in the transducer medium, and the distribution of light collected by the emission fiber (Fig. 15.2). Evidently fluorescence can only be observed where the two overlap, this is illustrated for two fibers each with NA = 0.30, typical for multimode fibers (Fig. 15.2). For a point close to the fiber end faces ("A" in the figure) a reasonable proportion of its emission (uniformly distributed throughout 2π-steradians) is directed such that it will be launched into and thus collected by the emission fiber core. By comparison, the solid angle into which the emission from the point "B" (farther from the end face) must be launched to be captured is obviously very small, and much less light will be

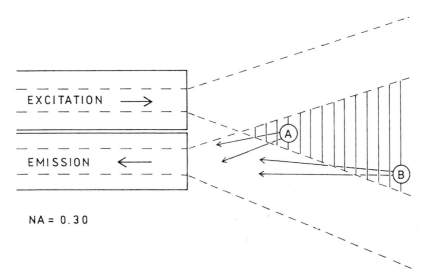

Figure 15.2 Principle of fluorescence measurement through a bifurcated fiber. Fluorescence is excited in the medium to the right from the upper fiber and collected through the lower fiber only from the crosshatched portion of the medium.

collected from a fluorophore at point "B" than point "A." Clearly the relative sizes of core and cladding and the numerical apertures are large factors in the efficiency of these devices, but the fundamental drawback still applies. This has been considered in detail by Papaioannou *et al.* (2003) and others.

For our purposes, we prefer a single fiber to conduct excitation to the sample and emission from it back to the detector. There is no difference between the volume excited and that from which the emission is collected; typically fluorescence cannot be collected from deeper than a couple of millimeters in the transducer (and most from within a few hundred micrometers of the end face), making the use of small transducers sensible. For distances longer than a few meters single fibers (even cabled) are cheaper, lighter, more flexible, and much easier to handle than bifurcated fibers or fiber bundles. The mechanical components such as connectors used for single fibers are manufactured with much greater precision, which is of value (see below); some devices can only be used with discrete fibers. At wavelengths in the orange-red portion of the spectrum and longer, the background photoluminescence of telecommunications fibers is quite low. While separation of emission from scattered excitation and other background (the *sine qua non* of sensitivity in fluorescence measurement) is more complex, it usually can be dealt with by approaches well known to the art (see below).

3. Optics and Mechanics of Fluorescence-Based Fiber Optic Sensors

3.1. Alignment of sensors

Figure 15.3 depicts a typical optical setup for a fluorescence-based fiber optic sensor. Essentially, the optics are designed to measure fluorescence of whatever is at the distal end of the optical fiber, basically in a front-face geometry. Fundamentally it is similar to an epifluorescence microscope and similar optics have been used by ourselves and other developers of fiber optic fluorometers for many years (Hirschfeld, 1984; Lippitsch *et al.*, 1988; Thompson *et al.*, 1990): excitation light (usually from a laser) is launched along the *x*-axis in the illustration through a dichroic mirror and an objective lens, which focuses the light onto the fiber core at its proximal end; the light propagates through the fiber and excites fluorescence in the transducer material at the distal end. The fluorescence is collected by the fiber and passed back to the proximal end, where it is collected by the objective and recollimated, reflects off the dichroic mirror, and is detected by the detector. This differs somewhat from the usual practice in fluorescence microscopes, where the excitation is reflected and the emission passes through the mirror. Previously, we had used a perforated off-axis parabolic

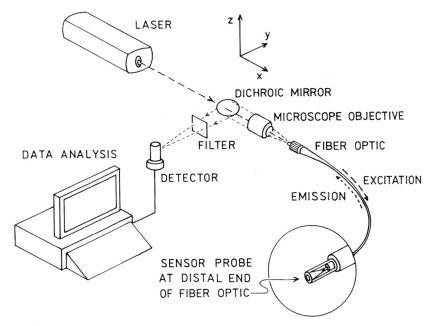

Figure 15.3 Schematic of single-fiber fluorescence-based fiber optic sensor.

mirror in place of the dichroic, which had the virtue of being achromatic, but it was difficult to use with any light source but a laser having a narrow, well-collimated beam (Thompson et al., 1990).

For convenience, the components are usually mounted on a small plate or breadboard in place of the cuvette holder in one of our ISS fluorometers (Chronos or K2). These fluorometers have a sizable sample chamber (16 × 16 × 20 cm) in which it is relatively convenient to mount the components; we note that other fluorometers have similarly sized sample chambers and should be usable as well. We found it most convenient to mount the dichroic, objective, and fiber holder (bulkhead connector) on their own xyz-translation mounts bolted to a 150 × 210 mm breadboard which is screwed into the chamber in place of the cuvette holder or turret; the breadboard is additionally drilled and tapped for 0.25–20 screws on 0.5-in. centers (Fig. 15.4). We have found that higher precision, high stiffness mounts are the easiest to use and provide the most reproducible results; other mounts will work, but at some cost in aggravation. The microscope objective holder (Newport 561-OBJ) and fiber holder (Newport 561-FCH) are mounted on stainless steel Ultralign stages (561-XYZ-LH and 561-XYZ-RH); the left- and right-hand models are selected to put the micrometers on the same side. The dichroic is less position sensitive (see below) and is mounted on a New Focus 9920 45° holder held by a Newport

Figure 15.4 Components for fiber optic sensor mounted on small plate for insertion into the instrument: left to right, dichroic mirror, microscope objective, proximal end of fiber optic pigtail.

P-100 optic holder mounted on a Newport MT-XYZ miniature translation stage; this in turn is mounted on a Newport EB-1 riser mount. The dichroic is essentially mounted where the cuvette would ordinarily be held in the fluorometer. Other configurations are doubtless possible, and may be desirable.

Of course, the excitation source to be used depends upon the fluorophore to be excited, but for this purpose lasers have many advantages. In addition to their intensity and highly collimated beam for ease of alignment, lasers are also quite monochromatic and the best sources for lifetime measurements (see below). For the most part we use diode lasers because they are small, light, efficient, less expensive, and simpler to modulate than other laser sources or lamps. However, mode hops can add noise to high frequency measurements, diode lasers are generally more poorly collimated than gas (e.g., Ar ion or HeCd) lasers, and thermal control is required for best results (see below). Gas lasers have highly collimated beams of small diameter, which can make them harder to launch into small fiber cores unless the beam is expanded. While the laser beam itself is highly monochromatic, it often is necessary to filter out other plasma lines emitted by the laser tube collinearly with the laser beam. These plasma lines are weak and uncollimated compared to the laser beam, but may mimic the fluorescence. For instance, the presence of a 442 nm plasma line in the emission of a HeCd laser operating at 326 nm might easily interfere with studies of a fluorophore in the same blue spectral regime. While modulated light-emitting diodes have been used for phase fluorometry (Lippitsch *et al.*, 1988; Sipior *et al.*, 1996), their lack of monochromaticity and uncollimated output makes them less desirable for fiber optic fluorometry. Finally, gas lasers exhibit noisier output unless they are stabilized, and must be externally modulated for phase fluorometry; such modulators are lossy.

4. MOUNTING AND ALIGNMENT OF THE INSTRUMENT

Good alignment and stability are crucial to successful measurements. If at all possible, the fluorometer, laser source, and other optical components should be mounted on a vibration-isolated optical table or breadboard. All pneumatically isolated tables we have used have been satisfactory, as well as breadboards isolated by air bladders or plastic foam from the benchtop. If using the ISS instruments their small perforated mounting plates are quite satisfactory for reproducibly and stably mounting the instruments on breadboards; if the plates must be removed for shipping, it is a good idea to mark their position on the tabletop unambiguously. It is essential that the instrument be parallel with the tabletop for ease in alignment; this is most easily accomplished if the tabletop and instrument are each leveled with a 1 m carpenter's spirit level. At this time it is convenient to measure the height above the tabletop of the instrument's optical axis. All fluorometers have an excitation focusing lens that must be removed. We have five lasers that are all used with the same ISS K2 in place of the laser in the illustration, and it is tedious to have to remount and realign a laser each time it must be changed. To avoid this, the gas lasers and Ti:sapphire are aimed at beam steerers (periscopes: e.g., paired, rod-mounted New Focus 9807/9920s) that are kinematically mounted (e.g., Newport BK-3). The kinematic mounts permit the periscope (one dedicated to each laser) to be very reproducibly (to within microns and milliarcseconds) removed and replaced when that laser is to be used. All lasers should also be stably mounted so that their beams are parallel to the optical table. The reason for this is that it is much easier to translate a beam or change its plane in increments of 90° using mirrors in 45° mounts: a beam hitting a periscope perpendicular to the periscope's vertical axis will emerge in a different plane parallel to the first. Moving the periscope to launch the beam into the instrument is much easier than moving the laser itself unless it is a diode laser. While ISS offers modulatable diode lasers that are easily attached to the instrument itself, we eschew directly attaching the lasers to the ISS chassis for two reasons. First, the diode lasers must be thermostatted using a Peltier junction or thermal jacket for stability (see below), and it is difficult to mount the jacket when the laser is attached to the ISS chassis. Second, the ISS focusing tool cannot be used with the laser mounted in this fashion. Thus the laser in its jacket is mounted on a xyz-translation stage (e.g., Newport 461-XYZ) in an L-bracket using rubber bands; the cylindrical jacket cannot fit too snugly because it changes dimensions with temperature and thus is ultimately unsatisfactory. The new CVI Melles Griot 56 RCS modulatable laser diodes are thermoelectrically cooled and thus avoid the need for a jacket; it is to be hoped they are worthy successors to the 56 DOL models now discontinued. The xyz-translation stage is itself conveniently mounted on a magnetic kinematic base (e.g., Newport 110). The laser can be reliably determined to be level by aiming the beam to pass through a ruler held normal to the optical table (clear plastic minimizes the reflection), and assuring that the beam height stays constant at the value measured for the instrument optical axis height at various distances along the tabletop. Once level and at the right height, the diode laser can usually be roughly boresighted into alignment with the ISS by aiming the beam through the iris in the excitation optical path to hit the "+" marked on the inside front of the ISS sample chamber.

Aligning the components mounted in the ISS sample chamber is relatively straightforward. With the laser beam entering the sample chamber along the x-axis (see Fig. 15.3) the dichroic holder and fiber connector holder should be centered on the beam. For safety's sake the beam power should be reduced to just few milliwatts if possible; note that for alignment one generally cannot wear protective laser goggles since one then cannot see the laser beam, and caution must be commensurately greater. The holes into which the mounts are bolted should be chosen such that the micrometers or screws on the mounts are near the centers of their ranges of travel for convenience. The objective holder should be mounted such that the beam is well centered in the objective; if the beam is well off-axis it will be necessary to shift the fiber connector holder a relatively large distance. For wavelengths in the visible to near infrared we have used "long working distance" objectives from Nikon and Olympus (e.g., Olympus LUCPlan FL N 20×/0.45 NA) these have been particularly convenient for seeing that the excitation beam is well centered in the objective, and being able to see where on the proximal face of the fiber connector the beam is focused. For this purpose a quality 7× or 10× magnifier (e.g., Edmund Scientific Hastings triplet cat. No. A38-452) is essential. For wavelengths in the UV and blue we have had good success with a Newport U13× UV objective. The laser beam (*in the absence of the objective*) should hit the proximal end of the fiber squarely if the beam and fiber holder are well aligned. When the objective is installed, the focused beam is likely to be slightly off-axis, and the objective should be translated along the y- and z-axes to bring the focal spot into position on the proximal face of the fiber using the 7× magnifier. Light output from the distal end of the fiber pigtail indicates approximate alignment, and maximizing power by use of the power meter (a bulkhead connector for the pigtail distal end designed to attach to the meter's sensor head is very convenient) will usually bring the alignment very close to optimal. A card is helpful in judging the focal point of the objective along the optical axis; usually it is easiest to translate the (connectorized) proximal end of the fiber along the optical axis to maximize the launched power; power and alignment are much less sensitive to position along the optical axis than the other two axes.

Criteria for good alignment include low loss (or attenuation) in launched power and good mode structure of the beam exiting a ~2 m fiber pigtail.

Low loss is measured at the entrance to the objective and the distal end of the pigtail. We have used a Newport Model 835 power meter with 818-UV sensor with satisfaction; it helps to be able to read the meter in the dark and that the meter probe be small and mountable on the optical table, since often one needs to measure the power continuously while adjusting the position of a component. Power loss is best expressed (and may be specified for components) in terms of decibels (abbreviated dB):

$$\text{Loss in dB} = -10(-\log(I_{in}/I_{out})), \quad (15.2)$$

where I_{in} is the power entering the component and I_{out} is the power leaving the component. Decibels obviously represent a logarithmic scale, like OD: thus -20 dB $=$ an OD of 2.0. For a given wavelength the loss in a component (objective, fiber optic, filter, etc.) is constant, and measurement of a greater loss is evidence of misalignment. A virtue of the decibel system is that cumulative optical losses are multiplicative, and more easily expressed as a sum of logarithms. Thus if we have one optical component with a loss of 90% (e.g., 10% transmission, or an OD of 1.0, or a loss of -10 dB) and a second component with 50% loss (50% transmission, OD $= 0.3$, or -3 dB), light passing through both would experience 95% loss (transmission is 5%, or 50% of 10%, or 1.3 OD, or -13 dB).

Good mode structure in the beam leaving the fiber pigtail is possible to measure but is more commonly judged by eye: the spot the beam makes should be circular, symmetrical, and uniformly increasing in intensity from the periphery to the center. A very reliable criterion for good alignment occurs when (as is often the case) the spot size of the focused laser beam on the proximal end of the fiber is significantly smaller than the core size. Spot radius r is primarily determined by the incoming (collimated) laser beam radius d, the wavelength λ, and the focal length f of the objective:

$$r \sim 4\lambda f/\pi d. \quad (15.3)$$

If the spot is well centered on the core (of the short pigtail), most of the power will be present in the axial, low-order modes of the fiber (an underfilled launch condition). When this occurs the output spot from the pigtail is not only well formed but slightly smaller than if the modes in the fiber are uniformly excited. Thus when translating the fiber connector holder slightly along the Y- or Z-axes one observes the output spot size to increase significantly, and optimum alignment is easily found by minimizing the spot size.

Currently for optical fiber pigtails we use gradient index multimode fiber of either 100/140 μm core/cladding diameter (for ratiometric sensors) or 50/125 fiber for lifetime-based sensors. We had used plastic-clad silica fibers (especially for ease in accessing the core by stripping the cladding), but these do not seem to mount well in connectors. Also, such large core, high numerical aperture fibers have very low bandwidth for lifetime-based sensing. Finally, the plastic cladding is often quite fluorescent and contributes to background. Telecommunication fibers have the advantages that they are all glass (the higher refractive index of the core is usually achieved by doping with elements such as germanium), they are relatively inexpensive, and they are available in cabled form, essential for field use and desirable in the laboratory. Finally, it is straightforward to cleanly cleave the all-glass fiber using cleavers designed for the purpose (e.g., Fujikura CT-04B) whereas the simple hand cleavers usable with plastic-clad silica fiber require the cladding be stripped first and require skill to avoid hackle on cleaved faces.

We find it essential to be able to repeatably mount a fiber sensor probe to the system with high precision, particularly if it is at the end of a fiber pigtail. The best way is to use fibers terminated with connectors. We had used SMA connectors from various manufacturers with both plastic clad and all-glass fibers, but these are manufactured to tolerances that are unacceptably loose with our system. If using SMA connectors, it is imperative to insert them into the receptacle or union with the fiber oriented around the optical axis the same way each time (mark the ferrule with a marker). Most recently, we have used FC connectors with ceramic ferrules, which are very reproducible. They offer another advantage in alignment: if the beam is not launching into the fiber well the translucent ferrule lights up due to scattered excitation, and the focused spot is easy to see on the proximal face of the ferrule. When the excitation is well launched into the proximal end of the fiber, the glow of the ferrule is minimized. While it is certainly feasible to cleave and connectorize fiber pigtails oneself, we have found it cost-effective to buy premade pigtails, especially for sensors (see below).

While alignment of the system for launching light into the fiber is important, it is equally vital to collect the returning fluorescence and direct it into the detector, and alignment of the apparatus for this purpose must be checked. We find it convenient to use a pigtail connectorized on both ends and launch light from a red or green HeNe laser into the *distal* end of the pigtail; since the intensity required is modest, perfect alignment of the launching light is unnecessary and the narrow beam of the HeNe lasers may make a focusing lens to launch usable light in the distal end unnecessary. The light emerging from the *proximal* end of the fiber is collected by the objective and recollimated into a spot a few millimeters across, and reflected toward the detector (Fig. 15.3). In general there is a collection lens in the fluorometer which will focus this light on the detector. We check that the light coming out of the back of the objective is reasonably well centered on the dichroic and the collection lens. We use a dentist's mirror or a suitably cut business card to observe the beam in the optical path to the detector. In the ISS instruments there is a filter holder halfway between the detector and sample chamber which can be removed for this purpose. In general the laser light is too intense for the detector without substantial attenuation with neutral density filters. Since the image of the returning beam is similar in extent to the fluorescence emitted from a cuvette mounted in the sample chamber, the optical train leading to the detector is close to optimum for its collection and measurement. If a laser is unavailable, a suitable fluorescent card may be substituted if the excitation wavelength is short enough. In this case to get maximum light back the distal end of the fiber pigtail is placed on the card and usually a bright spot appears on the dichroic.

5. STANDARDS FOR RATIOMETRIC AND LIFETIME-BASED FIBER OPTIC FLUORESCENCE SENSORS

We have developed standards for ratiometric and lifetime-based fluorescence-based fiber optic sensors. These standards are distinct from the calibration standards used for the sensors themselves to relate the measured ratios or lifetime data to the analyte concentrations. Rather, they are designed to test instrument function and alignment with samples that are bright and readily reproducible. For instance, the ratiometric standards should give intensities at the two emission wavelengths that are comparable to previous results obtained using the same excitation power, optical filtration, and detector gain setting. Similarly, the lifetime standards give a reproducible range of lifetimes (or, in the frequency domain, frequency-dependent phases and modulations) with readily available, relatively inexpensive specimens. In both cases, the standards can be used simply by dipping the (cleaved) distal end of a fiber pigtail in the solution, rather than having to use an intact fiber optic sensor probe (see below), and enables one to quickly identify whether performance issues arise in the instrument or the probe, and provide guidance as to their resolution. Also in both cases, the standards are chosen to be intense, to minimize the influence of relatively weak noise sources (shot noise, dark current, $1/f$ noise) that become important when signal levels are low.

The ratiometric standards simulate the response of poly-ABDN used with our ratiometric Zn sensors (Thompson *et al.*, 2000) (see below) with a set of mixtures of two fluorescent dyes in different proportions: Sulforhodamine 101 (Molecular Probes/Invitrogen catalog No. S-359) and Lucifer Yellow CH (L-682). Emission spectra of the mixtures in MOPS buffer are shown in Fig. 15.5, and their proportions in Table 15.1. Of course, other concentrations can be chosen to give different ratio values, or for different optical conditions. We find that the mixtures can be made up in some

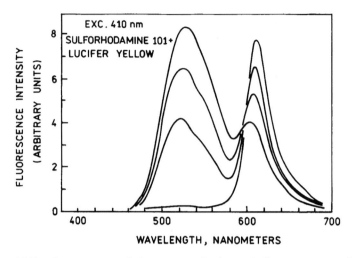

Figure 15.5 Fluorescence emission spectra of ratiometric fluorescence standards.

Table 15.1 Ratiometric standards

Curve at 520 nm	[Sulforhodamine 101] (μM)	[Lucifer Yellow] (μM)
Highest	4.48	17.5
High	10	15
Low	15	10
Lowest	25	1

convenient aqueous buffer, aliquotted in 1 ml volumes in multiple 1.5 ml microfuge tubes, and stored frozen in the dark; the standards can be frozen and thawed a few times before they deteriorate.

The lifetime standards are more complex. Quite a number of fluorescent compounds have been designated as lifetime standards (Lakowicz, 1983; Lakowicz et al., 1991; Thompson and Gratton, 1988; Thompson et al., 1992), but few of these were usable in the red to near infrared wavelength of our copper sensor. Rayleigh scatterers are commonly used as lifetime standards because scattering is very fast (femtoseconds) compared to fluorescence and therefore effectively has a zero lifetime. However, the multiple optical elements in the paths of excitation and emission all represent potential scattering sources with different path lengths, which would confound scattering's use as a standard in fiber optic fluorometers. What was required was a reproducible means of producing different phase delays and modulations, essentially by modulating the lifetime. This was done by mixing a suitable NIR fluorophore (DTDCI) at a fixed concentration with a nonfluorescent dye (Janus Green B, in this case) at relatively high concentrations, such that the dye quenches the fluorophore's fluorescence by resonance energy transfer and reduces its lifetime. DTDCI was chosen because its fluorescence closely mimics that of the Alexa Fluor 660 label used in copper sensing, such that the standard can be used under the same instrumental conditions as the sensor. Frequency-dependent phases and modulations for a set of standards are shown in Fig. 15.6 (McCranor and Thompson, submitted for publication) and Table 15.2. Evidently we can produce a range of phases of 35° and 22% modulation by varying the concentration of the dye. More important, the principle can be extended to other wavelengths and lifetime ranges depending on the choice of fluorophore and dye.

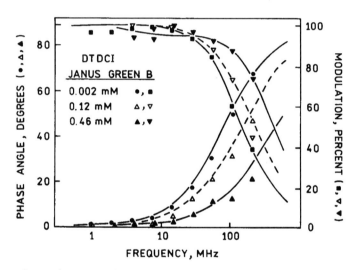

Figure 15.6 Fluorescence lifetime standards. Frequency-dependent phase angles and modulations for DTDCI together with 0.002 mM (●, ■), 0.12 mM (▲, ▼), and 0.46 mM (△, ▽) Janus Green B.

Table 15.2 Lifetime standards

Standard No.	[Janus Green B] (mM)	$\Delta\varphi$ (°)	Modulation (%)
1	0.0	48.01	0.645
2	0.06	41.25	0.673
3	0.12	31.88	0.726
4	0.46	12.43	0.861

6. Construction of Fiber Optic Probes

The fiber optic probe itself is essentially a pigtail with the fluorescent transducer somehow immobilized or entrapped at the distal end, and a suitable connector (usually an FC) at the proximal end. The fluorescent transducer (in our case a variant of carbonic anhydrase) must be polymeric or otherwise immobilized by attachment to a polymer or particle since small molecules will tend to diffuse away, at the same time small molecule or ionic analytes can diffuse in. At present we make two types, neither of which is particularly novel: dialysis probes and gel probes. The dialysis probes (C) essentially have a small ($\sim$1 μl) chamber at the end of the fiber with the fluorescent transducer inside and an opening covered with a dialysis membrane (see Fig. 15.7). The gel probe (B) has the transducer material entrapped in a porous hydrogel at the distal end. The advantage of the dialysis probe is that it is relatively easy to assemble and fill, and it can be refilled if the protein no longer responds or becomes photobleached. However, its response time is likely to be slow due to the relatively long path for the analyte to diffuse through to the distal end face of the fiber itself, and the diameter of the probe (limited by the ring which holds the dialysis tubing in place) is inconveniently large for *in vivo* use in blood vessels or the brain. By comparison, the gel probe is more difficult to make and cannot be emptied and refilled, but by use of a thin gel layer, can equilibrate much faster; Barnard and Walt have described means to make very thin ($\sim$5 μm thick) gel layers for rapid response (Barnard and Walt, 1991). We had

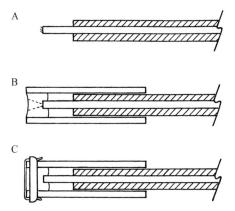

Figure 15.7 Fiber optic sensor probes: (A) bare fiber, with the jacket or buffer cross-hatched; (B) gel probe, with the bare end of the fiber inserted in a small volume of gel polymerized *in situ*; and (C) a dialysis probe, with the dialysis membrane secured over the end by an O-ring made of Tygon tubing.

previously described an approach for covalently attaching what amounts to a monolayer of fluorescent protein transducer to the distal end face of the fiber, which can provide a very fast response (Bhatia *et al.*, 1989; Zeng *et al.*, 2003), but the signal is correspondingly weak and the transducer is directly exposed to the medium, which can result in rapid breakdown of the sensing layer.

Construction of the dialysis probes is straightforward. Construction and filling should be done in a HEPA-filtered laminar flow hood to minimize contamination (Envirco makes a convenient one for benchtop use) and dust-free gloves (Oak Technical "Long-Length" Class M1.5 compatible gloves cat. No. 96–337). A 2-m length of fiber optic cable connectorized at one end (Newport Corp., F-MLD-C) has a 150 mm length of the outer jacket and Kevlar strength member stripped, leaving the buffered fiber exposed; the OD of the buffer is 0.034 in.; the fiber should be cleaved so that a flat end face is exposed with a few centimeters of fiber exposed. An acid-cleaned and air-dried (in the laminar flow hood) 50 μl micropipette (VWR cat. No. 53432–783) is slid over the end of the exposed buffer, leaving perhaps 1 mm or less along the optical axis between the distal fiber end face and the (fire-polished) end of the capillary. A droplet of UV-curable epoxy (e.g., Norland Optical Adhesive type 68, Norland Products, Cranbury, NJ 08512, or available through Newport Corp.) is placed near the proximal end of the capillary tube and readily wicks into the thin space between buffer and the inside of the glass capillary. The epoxy cures in a few minutes under illumination with a UV hand lamp; a jig like that shown in Fig. 15.8 is convenient for holding probes while they are being glued. The end of the probe is filled with the fluorescent transducer

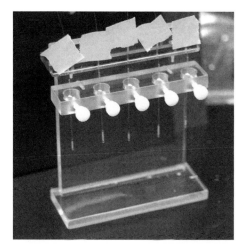

Figure 15.8 Jig for holding fiber optic probes while being glued, etc.

solution using a nonmetallic World Precision Instruments (Sarasota, FL) Microfil 34AWG syringe needle for filling micropipettes on an all-plastic 5 ml syringe (Norm-Ject Henke-Sass Wolf GmbH Tuttlingen, Germany, sold in US by Sigma-Aldrich cat. No. Z248010). Because the filling needle is long and narrow, there is substantial hysteresis between when the syringe plunger is depressed and the transducer solution emerges from the needle. When filling it is important to avoid introducing bubbles, and leave a positive meniscus; it is worthwhile to practice the manipulations of filling with some colored sham solution before working with precious real sensor solutions. Finally, the open end of the filled capillary is covered with a clean, wet 5×5 mm^2 of Spectrapor 3 dialysis tubing (10,000 molecular weight cutoff: Spectrum Medical Industries) and secured with a 2 mm ring of 1/16th inch ID, 1/8th OD Tygon tubing (Fisher cat No. 14–169–1B, formula R-3603); the corners of the secured dialysis tubing may be trimmed. Different capillary tubes and Tygon tubing may be necessary for different diameter fiber buffers.

Preparation of the gel probes is similar, but involves mixing the (polymeric) fluorescent transducer in a solution of gel monomers, loading the probe, and having the gel polymerize in the glass capillary. The empty probe is prepared as above, except that the fiber is not glued inside the capillary before filling; rather, the fiber and capillary are held in the jig (Fig. 15.8) and the end of the capillary is quickly filled with the gel solution by capillary action in a glove box under nitrogen and allowed to polymerize, then the two are glued using the UV-curable epoxy at the proximal end as above. The gel solution is a classical polyacrylamide gel formulation, except no SDS is present and the aqueous component contains the fluorescent sensor transducer: for zinc (see below) this might be wild-type apo-CA II and poly-ABDN (each 10 μM final concentration), and for Cu(II) L198C-Alexa Fluor 660 apo-CA (3 μM final concentration). The concentrations of the polymerization initiators TEMED and ammonium persulfate may need to be adjusted slightly so the polymerization does not occur too quickly.

Gel Recipe

40% Acrylamide solution in water (Bio-Rad)	100 μl
Gel buffer stripped of metal with Chelex 100 (Bio-Rad) contains *apo*-carbonic anhydrase, and poly-ABDN (for Zn)	50 μl
2% N,N-methylene *bis*-acrylamide in water (Bio-Rad)	28 μl
N,N,N',N'-Tetramethylethylenediamine (TEMED; Aldrich T2, 250–0)	2 μl
Ammonium persulfate (100 mg/ml in pure water; make fresh)	2 μl

The acrylamide, *bis*-acrylamide, and buffer are mixed and have nitrogen (prepurified grade) blown over the top. The protein (and ABDN, if measuring Zn) is added and mixed, and finally the TEMED and ammonium persulfate are added to initiate polymerization. The capillaries must be rapidly filled.

Following polymerization, the probes are dialyzed versus 50 mM HEPES pH 7.5 or another suitable buffer overnight to remove unreacted acrylamide, *bis*-acrylamide, TEMED, and ammonium persulfate, which can quench fluorescence and/or damage the transducer molecules. They should be stored under refrigeration in the dark until used.

7. Zn^{2+} Probe

For dialysis or gel probes for zinc we use apocarbonic anhydrase entrapped together with a polymeric form of ABDN (Thompson *et al.*, 2000) to determine free zinc either by ratio or lifetime. The end of the dialysis probe is filled with a solution of ~1 μM apoprotein, ~5 μM poly-ABDN and secured; the probe can be calibrated using zinc buffers as described in the accompanying chapter. A ratio calibration is shown later for three different variants of CA having differing zinc affinities (Fig. 15.9). In this case the variants are the wild-type, E117A, and H94N with the apparent affinities depicted; note that the affinities of the variants (as well as the metal ion buffers themselves) change with pH and ionic strength and the buffers must be reformulated accordingly. E117A has about 10-fold weaker affinity than

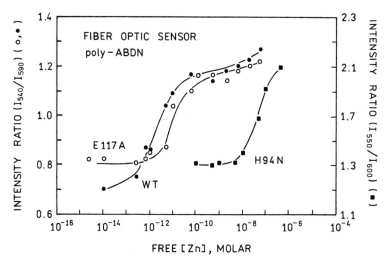

Figure 15.9 Calibration curves of fiber optic zinc sensors with carbonic anhydrase transducers with varying affinity; from left to right, wild-type, E117A, and H94N.

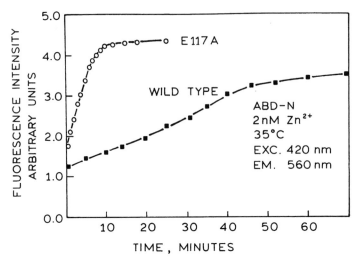

Figure 15.10 Time-dependent fluorescence response of fiber probes made with wild-type (■) and E117A (○) apocarbonic anhydrases, illustrating the more rapid response of the latter.

the wild-type but its zinc association rate constant is about 800-fold faster, providing a quicker response in the absence of catalysts. Enhanced speed is shown in Fig. 15.10 where the E117A probe responds to a nanomolar level of free Zn(II) in 10 min, whereas the wild-type protein takes over an hour.

8. Cu^{2+} Probe

For determining Cu(II) we have primarily used two forms of CA (N67C and L198C), labeled with two different fluorophores (Oregon Green and Alexa Fluor 660). L198C-labeled wild-type CA is more stable than N67C but harder to strip of zinc. Oregon Green seems to be more stable and photostable, but the shorter wavelengths at which it is excited and emits (488 nm excitation/510 nm emission) compared with Alexa Fluor 660 (660 nm excitation/680 nm emission) are more strongly attenuated in all fiber optics and thus less usable with longer fibers (25 m) than Alexa Fluor 660. The principle of the probe is simple: Cu(II) when bound to the apoprotein exhibits weak d–d absorbance bands which can serve as energy transfer acceptors for a fluorophore attached close to the binding site whose emission overlaps the absorbance, thereby reducing the fluorophore's intensity and lifetime (Thompson et al., 1996). If no Cu(II) is bound there is no absorbance and no energy transfer, and one observes the unquenched intensity and lifetime. At intermediate fractional saturations of the binding

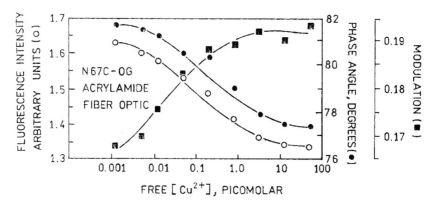

Figure 15.11 Cu(II)-dependent phase shifts (●), intensities (o), and demodulations (■) at 92 MHz of N67C-Oregon Green.

site with copper, one observes the intermediate intensity or mixture of lifetimes. Accurately measuring and calibrating intensities is difficult, particularly through optical fiber, so for some years we have used lifetimes exclusively. Measuring lifetime changes through fiber optics in the frequency domain is easier than in the time domain (Betts et al., 1990; Thompson and Lakowicz, 1993), but there seems no reason these measurements cannot be carried out in the time domain. An example of the copper-dependent phase and modulation changes observed at a single frequency is depicted in Fig. 15.11. As in the case of zinc we can also acquire the whole decay using the Fast Scan ISS software and a preexisting reference measurement, and determine the proportions of free and bound (and thereby the free zinc concentration) directly.

It is certainly desirable to measure Cu(II) and other analytes at the end of long lengths of optical fiber. Several factors limit the length of fiber which can be used. First and foremost is the wavelength-dependent attenuation of the fiber itself, as mentioned earlier. For wavelengths in the ultraviolet, attenuation is of the order of −100 dB/km, which effectively means that fiber lengths greater than a few meters are unusable. At 660 nm in the red attenuation is ~10 dB/km, which is quite usable up to hundreds of meters (see below). A second issue is modal dispersion; as discussed earlier, modal dispersion (in multimode fibers such as we use) demodulates the excitation and emission. The modal dispersion (expressed as 3 dB bandwidth) of the 50/125 μm core/cladding gradient index multimode fiber we use is specified at 400 MHz km at 830 nm; the dispersion is greater at the shorter wavelengths we use, but improved performance would require the use of a single-mode fiber. We find through relatively long lengths (250 m) of fiber that the signal is substantially demodulated (AC/DC becomes low), and ultimately the precision of phase and modulation measurements (at the relatively high modulation frequencies necessary for the long wavelength labels with their short lifetimes) becomes unacceptably poor. Manufacturers have introduced longer wavelength fluorescent labels which would probably be usable, but their likely short lifetimes would still require high frequencies wherein modal dispersion would limit fiber lengths. We have found (with adequate barrier filters and no drift in the excitation wavelength) that the photoluminescence background in the fiber in the absence of a fluorescent label at 660 nm is quite modest and essentially unmodulated. At the shorter wavelengths used for zinc (excitation 405 nm, emission 560+ nm), the background is significantly greater and readily compromises performance. We find it essential to not launch light into the cladding to minimize background; in particular, the modal quality of the launched laser light must be high to give a small spot which does not get coupled into the cladding.

Approaches which might permit very long (km) lengths of fiber would seem to require longer wavelength labels with longer lifetimes to limit the effect of modal dispersion; with microsecond lifetimes such as those exhibited by lanthanide chelates (Lovgren *et al.*, 1985) one could use multimode fibers with higher numerical apertures which would collect more fluorescence from the distal end.

9. Use of Fiber Optic Probes for Discrete Samples

One of the issues with using sensors that bind to or react with analytes at trace concentrations is that the analyte amount in a fixed sample volume is necessarily limited. For instance, consider an equilibrium binding indicator or sensor with a K_D for the analyte of one nanomolar, and an analyte concentration of 100 pM in a sample. If a finite amount of sensor were in an infinite bath of this concentration the sensor would be 10% occupied. However, if 50 nM sensor were necessary to provide enough signal for accurate measurement and this were done on a sample volume of (for instance) 1 l, the high-affinity sensor would scavenge up the limited amount of analyte, while still remaining over 99% unbound. The answer is to limit the total amount of sensor present but concentrate it in front of the detector to get a good signal. The fiber optic probe does this well by concentrating the indicator in a very small volume (less than 1 μl) at the distal end of the fiber, so that even at a concentration of 1 μM indicator only 1 pmol total binding sites are present. Thus one can use a fiber sensor with the affinity described earlier to accurately determine 100 pM analyte without significantly depleting even as little as 1 l of sample (Thompson *et al.*, in preparation). For geochemical determinations which otherwise require collection and processing of tens or hundreds of liters of sample, this may be particularly attractive.

10. Operating Issues: Noise, Background, Thermal Drift, and Mode Hopping

The advantages enumerated earlier for diode lasers argue strongly for their use, but they also have issues which must be addressed for the best performance. Diode lasers become dramatically noisy if their output is retroreflected back into the laser cavity; these are also called etaloning effects. Very good alignment of the exterior reflector is required for this to occur, and the most likely source of reflection is Fresnel reflection from the proximal face of the fiber optic. These effects are easily detected by blocking the beam after it exits the objective but measuring its intensity using an intervening detector (or the reference channel on the ISS instruments); the laser noise will jump from fluctuations of ~1% to tens of percent over seconds. This reflection can be eliminated in principle by antireflection coating of the fiber end face, polishing it at a slight angle, or polishing it into a convex surface so it no longer retroreflects. In practice even slight misalignment of the system prevents retroreflection, with little compromise of sensitivity.

Unlike ion lasers or other solid state lasers, diode lasers can change their lasing mode, which changes their emission wavelength unpredictably; this is termed mode hopping (Thompson, 1994). This phenomenon not only creates intensity fluctuations and fluctuations in the phase of the emission, but may also redshift the laser emission enough to leak past the emission filters, all of which are unsatisfactory. To prevent mode hopping in diode lasers one may use an external cavity laser which permits only a single mode of this type (but this is costly), or one can minimize it less expensively by thermostatting the laser using a Peltier junction (preferred) or an external jacket connected to a circulating bath. Attention must also be paid to how the synthesizers in the phase fluorometer are synchronized. The effects of thermostatting on phase stability are depicted in Fig. 15.12; the periodic fluctuations in phase angle of the unthermostatted laser of nearly $15°$ (upper panel) are reduced to about $0.5°$ RMS with temperature control in the lower panel. Note the ordinary short term phase fluctuations of the instrument are of the order of $0.3°$.

For lifetime-based sensors one can monitor changes in phase and modulation at a single frequency in comparison to some original measurement (corresponding to zero analyte) and compute the analyte level from the changes in modulation and phase when the sensor is put in contact with the sample. Evidently this requires excellent stability of the system, since any drift in the phase angle could be interpreted as a change in analyte concentration; this is less of an issue with the modulation. Preferred is the FastScan approach in the ISS phase fluorometers, wherein a complete frequency scan of a known lifetime sample (like those earlier) is collected and used to correct a complete unknown sample frequency scan. Note that the phase shift at the detector is the sum of the phase delay due to the lifetime and an

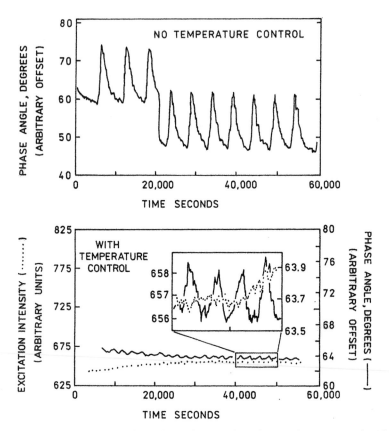

Figure 15.12 Upper: time-dependent phase angles of excitation measured without thermostatting. Lower: time-dependent phase angles (—) and intensities (...); the inset shows an expanded scale.

arbitrary phase delay imposed by the length of the fiber. We have not attempted lifetime-based sensing through fiber optics in the time domain by, for example, time-correlated single photon counting. In all cases we attempt to maximize the stability of the system by leaving lasers and synthesizers on at all times. If the laser is thermostatted using a circulating bath one must consider the risk of a leak of the circulating fluid on the experiment and those of one's colleagues.

11. Field and Shipboard Use

We have deployed the instrument on ships and in field laboratories, and these venues offer particular challenges. On ships the instrument must be vibration-isolated from the "thrumming" of the ship's engines which is poorly damped within the steel superstructure; we found that the diesel-electric propulsion of the *RV Knorr* and ships like her offered much reduced vibration. In addition, all components of the instrument must be mounted on the lab bench to prevent movement in response to the ship's motion. This conveniently done by having the benchtop be a piece of plywood to which the instrument and other components (computer, monitor, power supply, etc.) can be bolted or tied fast to eyebolts, etc. Consideration should be given to operation of the instrument in high sea states; under very rough conditions the instrument is unlikely to be operated but securing it against violent motion, 90° rolls, etc., becomes a concern. Most research vessels have air conditioned laboratory space and conditioned ("clean") power, but these must be checked in advance. It is helpful in aligning the instrument (see above) to be able to make the space quite dark; we have found that the "blackout cloth" (a type of fabric comprising a layer of thin, opaque rubber sandwiched with cloth used to make curtains) is more convenient and durable than black felt for this purpose. Care should be taken to avoid drafts of moist, salty air over the instrument, which can corrode the electronics. For shipping the instrument we have found the Zarges boxes to be satisfactory and much better than the cardboard boxes that computers, etc., are shipped in from manufacturers. In some cases, several boxes can be conveniently shrink-wrapped on a palette to minimize loss or theft. It is very helpful to have a compact laminar flow hood available on shipboard for manipulating and refilling probes while on shipboard.

Field operation in extemporized laboratories involves different challenges. We have modified standard 20 ft. shipping containers into portable laboratories with success. Clean power is an issue, as well as environmental control if the lab is not air conditioned. In most field laboratories security from thieves and vandals is a concern when the laboratory is unoccupied; in our case it has helped when the laboratory is on (posted) private property or a patrolled nature reserve. Consideration must be given also to the security of deployed fiber optics to inadvertent destruction by transiting vessels, fishermen, clam diggers, fish, birds, tides, waves, and powerboat wakes. In the polar regions, ice movement, heating, and condensation become concerns. All these consideration make field experiments challenging and interesting to perform.

ACKNOWLEDGMENTS

We would like to thank Carol Fierke and Andrea Stoddard for carbonic anhydrase variants and much wise counsel; Carl Villarruel of the Naval Research Laboratory for flawless advice on all things fiber optic; and Gabe Sinclair for working tireless 4-h days in manufacturing myriad obscure devices for us. This work sponsored by the National Science Foundation, the Cooperative Institute for Coastal, and Estuarine Environmental Technology, and the Office of Naval Research.

REFERENCES

Barnard, S. M., and Walt, D. M. (1991). A fiber-optic chemical sensor with discrete sensing sites. *Nature* **353,** 338–340.

Betts, T. A., Bright, F. V., Catena, G. C., Huang, J., Litwiler, K. S., and Paterniti, D. P. (1990). Laser-based approaches to increase fiber optic-based selectivity: Dynamic fluorescence spectroscopy. *In* "Laser Techniques in Luminescence Spectroscopy, ASTM STP-1006" (T. Vo-Dinh and D. Eastwood, eds.), pp. 88–95. American Society for Testing and Materials, Philadelphia.

Bhatia, S. K., Shriver-Lake, L. C., Prior, K. J., Georger, J., Calvert, J. M., Bredehorst, R., and Ligler, F. S. (1989). Use of thiol-terminal silanes and heterobifunctional cross-linkers for immobilization of antibodies on silica surfaces. *Anal. Biochem.* **178,** 408–413.

Golden, J. P., Shriver-Lake, L. C., Anderson, G. P., Thompson, R. B., and Ligler, F. S. (1992). Fluorometer and tapered fiber optic probes for sensing in the evanescent wave. *Opt. Eng.* **31,** 1458–1462.

Grattan, K. T. V., and Zhang, Z. Y. (1994). Fiber optic fluorescence thermometry. *In* "Topics in Fluorescence Spectroscopy, Vol. 4: Probe Design and Chemical Sensing" (J. R. Lakowicz, ed.), pp. 335–376. Plenum Press, New York.

Harrick, N. J., and Loeb, G. I. (1973). Multiple internal reflection fluorescence spectrometry. *Anal. Chem.* **45,** 687–691.

Herron, J. N., Wang, H.-K., Terry, A. H., Durtsch, J. D., Tan, L., Astill, M. E., Smith, R. S., and Christensen, D. A. (1998). Rapid clinical diagnostics assays using injection-molded planar waveguides. *In* "SPIE Conference on Systems and Technologies for Clinical Diagnostics and Drug Discovery" (G. E. Cohn, ed.), Vol. 3259, pp. 54–64. SPIE, San Jose, CA.

Hirschfeld, T. E. (1984). Fluorescent Immunoassay Employing Optical Fiber in Capillary Tube US Patent No. 4447546.

Kronick, M. N., and Little, W. A. (1975). A new immunoassay based on fluorescence excitation by internal reflection spectroscopy. *J. Immunol. Methods* **8,** 235–240.

Lakowicz, J. R. (1983). "Principles of Fluorescence Spectroscopy." Plenum Press, New York.

Lakowicz, J. R., Gryczynski, I., Laczko, G., and Gloyna, D. (1991). Picosecond fluorescence lifetime standards for frequency- and time-domain fluorescence. *J. Fluorescence* **1,** 87.

Lippitsch, M. E., Pusterhofer, J., Leiner, M. J. P., and Wolfbeis, O. S. (1988). Fibre-optic oxygen sensor with the fluorescence decay time as the information carrier. *Anal. Chim. Acta* **205,** 1–6.

Lovgren, T., Hemmila, I., Pettersson, K., and Halonen, P. (1985). Time-resolved fluorometry in immunoassay. *In* "Alternative Immunoassays" (W. P. Collins, ed.), pp. 203–217. John Wiley and Sons, New York.

Marcuse, D. (1991). "Theory of Dielectric Optical Waveguides." Academic Press, New York.

Papaioannou, T., Preyer, N., Fang, Q., Kurt, H., Carnohan, M., Ross, R., Brightwell, A., Cottone, G., Jones, L., and Marcu, L. (2003). Performance evaluation of fiber optic probes for tissue lifetime fluorescence spectroscopy. *In* "Advanced Biomedical and Clinical Diagnostic Systems" (T. Vo-Dinh, W. S. Grundfest, D. A. Benaron, and G. E. Cohn, eds.) Vol. 4958, pp. 43–50. Society of Photooptical Instrumentation Engineers, San Jose, CA.

Peterson, J. I., Goldstein, S. R., Fitzgerald, R. V., and Buckhold, D. K. (1980). Fiber optic pH probe for physiological use. *Anal. Chem.* **52,** 864–869.

Sipior, J., Carter, G. M., Lakowicz, J. R., and Rao, G. (1996). Single quantum well light emitting diodes demonstrated as excitation sources for nanosecond phase-modulation fluorescence lifetime measurements. *Rev. Sci. Instrum.* **67,** 3795–3798.

Snyder, A. W., and Love, J. D. (1983). "Optical Waveguide Theory." Chapman & Hall, New York.

Thompson, R. B. (1991). Fluorescence-based fiber optic sensors. In "Topics in Fluorescence Spectroscopy, Vol. 2: Principles" (J. R. Lakowicz, ed.), pp. 345–365. Plenum Press, New York.

Thompson, R. B. (1994). Red and near-infrared fluorometry. In "Topics in Fluorescence Spectroscopy, Vol. 4: Probe Design and Chemical Sensing" (J. R. Lakowicz, ed.), pp. 151–181. Plenum Press, New York.

Thompson, R. B. (2006). In "Fluorescence Sensors and Biosensors," p. 394. CRC Press, Boca Raton, FL.

Thompson, R. B., and Gratton, E. (1988). Phase fluorimetric measurement for determination of standard lifetimes. Anal. Chem. **60,** 670–674.

Thompson, R. B., and Lakowicz, J. R. (1993). Fiber optic pH sensor based on phase fluorescence lifetimes. Anal. Chem. **65,** 853–856.

Thompson, R. B., Levine, M., and Kondracki, L. (1990). Component selection for fiber optic fluorometry. Appl. Spectrosc. **44,** 117–122.

Thompson, R. B., Frisoli, J. K., and Lakowicz, J. R. (1992). Phase fluorometry using a continuously modulated laser diode. Anal. Chem. **64,** 2075–2078.

Thompson, R. B., Ge, Z., Patchan, M. W., Huang, C.-C., and Fierke, C. A. (1996). Fiber optic biosensor for Co(II) and Cu(II) based on fluorescence energy transfer with an enzyme transducer. Biosens. Bioelectron. **11,** 557–564.

Thompson, R. B., Jr., Maliwal, B. P., Fierke, C. A., and Frederickson, C. J. (2000). Fluorescence microscopy of stimulated Zn(II) release from organotypic cultures of mammalian hippocampus using a carbonic anhydrase-based biosensor system. J. Neurosci. Methods **96,** 35–45.

Thompson, R. B., Zeng, H. H., Fierke, C. A., Fones, G., and Moffett, J. (2002). Real-time in situ determination of free Cu(II) at picomolar levels in sea water using a fluorescence lifetime-based fiber optic biosensor. In "Clinical Diagnostic Systems: Technologies and Instrumentation" (G. E. Cohn, ed.), Vol. 4625, pp. 137–143. Society of Photooptical Instrumentation Engineers, San Jose, CA.

Wolfbeis, O. S. (1991). "Fiber Optic Chemical Sensors and Biosensors." CRC Press, Boca Raton.

Zeng, H. H., Thompson, R. B., Maliwal, B. P., Fones, G. R., Moffett, J. W., and Fierke, C. A. (2003). Real-time determination of picomolar free Cu(II) in seawater using a fluorescence-based fiber optic biosensor. Anal. Chem. **75,** 6807–6812.

Author Index

A

Abbyad, P., 175
Abduragimov, A. R., 24, 32
Abraham, M. H., 39
Adair, B. D., 109
Ahearn, G. A., 297
Ahmed, M., 289
Ahmed, R., 8, 13
Ahuja, P., 24, 32
Ala-aho, R., 14
Albertini, R. A., 162, 173
Alcala, J. R., 171
Alexiev, U., 160
Al-Hassan, K. A., 74
Allain, F. H. T., 186, 191
Altman, J. D., 8, 12
Ambrose, W. P., 147
Amirgoulova, E. V., 138
Anderson, B. J., 253, 262, 265
Anderson, C. F., 190
Anderson, G. P., 312
Andersson, S., 216
Andreatte, D., 175
Andrews, D. L., 160
Angel, P., 98
Angulo, G., 24, 32
Anselmetti, D., 197
Aota, S., 3
Archambault, J., 258, 262
Arden-Jacob, J., 261
Ariese, F., 72
Ariga, T., 131
Arnaut, L. G., 41
Arold, S. T., 97
Artym, V. V., 15
Arzhantsev, S., 161
Ashkin, A., 131
Aslamkhan, A., 297
Aslamkhan, A. G., 297
Astill, M. E., 312
Auguin, D., 97
Augustyn, K. E., 206
Austin, R. H., 208
Avilov, S. V., 52, 53
Axelman, K., 11
Axelrod, D., 136, 240, 291

B

Babcock, H. P., 131, 132, 236
Bacia, K., 94, 95, 97, 98
Bader, A. N., 72
Baer, T., 167
Bagchi, B., 33
Baird, T. T., 289
Baker, J. O., 297
Bakhshiev, N. G., 38, 39, 44, 53, 69, 71
Balaji, R. V., 302
Balis, F. M., 199, 201, 202, 204, 205, 206, 207, 208, 210, 215, 219, 220, 221
Banyasz, A., 175, 176
Barbara, P. F., 161, 162
Barbeau, K. A., 296
Barber, D. L., 8, 13
Barbic, A., 191
Barch, M., 198
Barkley, M. D., 160, 171
Barnard, S. M., 315, 326
Barone, V., 175, 176
Barrick, J. E., 233
Barthe, P., 97
Bartley, L. E., 236
Basche, T., 147
Bastiaens, P. I., 98
Baud, S., 92
Baudendistel, N., 98
Becker, P. C., 167
Becker, W., 161
Beechem, J. M., 81, 160, 171, 214, 222
Been, M. D., 233
Behlke, M. A., 266
Bell, C. A., 125
Bell, J. K., 3
Benkovic, S. J., 131
Berg, M., 175
Berger, D. H., 14
Berkow, R., 296
Berland, K. M., 97
Bernard, J., 147
Berry, D. A., 198
Betts, T. A., 331
Betzler, K., 168

Bevington, P. R., 278
Bhatia, S. K., 327
Bhattacharya, A. A., 50
Bhattacharya, K., 33
Biasutti, M. A., 24, 32
Bieschke, J., 98
Bird, P. I., 3
Biswas, R., 24
Blab, G. A., 132
Blanchard, S. C., 131, 234, 236
Blanco, F. G., 39, 53
Bliss, D. E., 218
Blumen, A., 60
Böhnlein, S., 247
Boise, L. H., 8
Bokinsky, G., 236, 249
Bonaventure, J., 123
Bortner, C. D., 10
Bossy-Wetzel, E., 8
Boukobza, E., 137
Boulahtouf, A., 95, 97
Bourdoncle, A., 95, 97
Bourguet, W., 88, 89, 90
Bowater, R. P., 237
Boxer, S., 175
Boyer, M., 92, 93, 94
Bozhkov, P. V., 8
Bozym, R. A., 287, 288, 291, 294, 301, 302, 304, 306
Bradley, D. J., 161
Bradrick, T. D., 222
Brand, L., 3, 4, 24, 26, 31, 32, 69, 70, 71, 81, 160, 162, 171, 173, 174, 214, 255, 258, 259, 260, 261, 263, 264, 265
Brand, L. C. E. O., 220
Brand, L. E., 296
Brault, K., 262
Breaker, R. R., 233
Bredehorst, R., 327
Bredesen, D., 11
Brenowitz, M., 235
Breslauer, K. J., 198, 208
Brewer, G. J., 289, 297, 303
Brich, D. J., 161
Bright, F. V., 331
Brightwell, A., 316
Brini, M., 306
Brito, C. H., 167
Brochon, J. C., 24
Brock, R., 97
Brotz, T. M., 7, 11
Brown, M. J., 5, 7, 8, 10, 12
Brown, M. P., 89, 92
Brown, R., 147
Brun, F., 91
Brus, L. E., 161
Bryant, S. F., 161
Buckhold, D. K., 312

Bullard, D. R., 237
Burghardt, T. P., 240
Burke, J. M., 243
Bustamante, C., 131
Bykov, A. B., 161

C

Callis, P. R., 74, 160, 162, 173, 175, 179
Calvert, J. M., 327
Cambi, A., 130, 131
Cao, J., 146, 154
Carillo, M. O. M., 186, 191
Carnohan, M., 316
Carranza, M. A., 32
Carrell, R. W., 3
Carter, G. M., 318
Cary, P. D., 186, 191
Casas, R., 112
Case, D. A., 186, 191, 189
Castanho, M. A., 130
Castner, E. W., Jr., 198
Catalan, J., 38, 39, 53
Cate, J. H., 235
Catena, G. C., 331
Cavailles, V., 95, 97
Cepero, E., 8
Chachisvilis, M., 160
Chadha, H. S., 39
Chahroudi, A., 8, 12
Chai, G., 249
Chan, S. S., 208
Chance, M. R., 235
Chang, C. P., 73, 74
Chang, S. H., 8
Changenet, P., 162
Changenet-Barret, P., 160
Chapman, K. T., 5
Chattopadhyay, A., 60
Chauvet, J.-P., 24
Chemla, D. S., 238
Chen, C. P., 17
Chen, L., 109, 116, 118, 120, 124
Chen, R. F., 171, 172, 290, 293
Chen, S. S.-Y., 101
Chen, Y., 95, 97
Cheng, W., 131, 132
Cheng, Y., 269
Cheng, Y.-M., 41
Childs, W., 175
Chin, A. S., 200
Cho, J. Y., 124
Chou, P.-T., 41, 42, 43, 73, 74
Choudhury, S. D., 24, 32
Chowdhury, P., 171
Christensen, D. A., 312
Christensen, T., 303

Christianson, D. W., 288, 289, 290, 291, 303
Chu, S., 131, 132, 236
Chun, T. W., 15
Church, F. C., 3
Churchill, M. E. A., 186, 189, 191, 189, 190
Cidlowski, J. A., 10
Clamme, J. P., 98
Clayton, A., 171
Clegg, R. M., 81, 189, 190, 236, 259, 262
Clements, J. H., 41, 42, 43, 73, 74
Clore, G. M., 186, 191, 189
Cochrane, A. W., 247
Cognet, L., 132
Cohen, B., 175
Coker, G., 101
Cole, J. L., 205, 212, 213
Cole, T. D., 198
Coleman, R., 175
Coleman, R. G., 160
Collet, A., 24
Collins, C. A., 233
Collins, K., 236, 249
Conn, G. L., 235
Cornilescu, G., 186, 191, 189
Correa, N. M., 24, 32
Corrie, J. E., 131
Cotter, T. G., 8
Cotto-Cumba, C., 288, 291
Cotton, F. A., 297
Cottone, G., 316
Coughlin, P. B., 3
Cova, S., 131
Craig, D. B., 133, 138
Craik, C. S., 3, 5
Cramer, M. L., 288, 289, 291, 294, 301, 304, 305, 311
Crane-Robinson, C., 186, 189, 191, 189, 190, 193
Crawford, F., 11
Cremazy, F. G., 98
Croce, R., 161
Croot, P. L., 296
Crothers, D. M., 191, 206
Cullen, B. R., 247
Curry, S., 50
Cusack, S., 235
Cvetanovic, M., 8

D

Dabatin, K. M., 11
Dahan, M., 238
Dahl, K., 24
Dale, R. E., 214
Darras, B., 306
Dash, C., 198
Dashnau, J. L., 160

Datta, A., 24, 32
Datta, S., 265
Davenport, L., 32, 214
Davis, S. P., 198
De Bakker, B., 130
Dedonder, C., 160
de Foresta, B., 33, 50
De Grooth, B. G., 131
Dekker, J. P., 162
de La Hoz Ayuso, A., 160
de Lange, F., 130
de Lera, A. R., 88, 89, 90
Deltau, G., 261
Demchenko, A. P., 37, 40, 41, 42, 45, 46, 47, 50, 51, 52, 53, 54, 55, 59, 60, 64, 70, 71, 72, 73, 74, 75, 76, 171
Demidov, A. A., 160
Deng, C. X., 124
Deniz, A. A., 22, 236, 238, 243, 245
Dennis, W. M., 161
Deras, M., 237
Dertouzos, J., 129, 131, 133, 137, 142, 145
DeToma, R. P., 24
Deveraux, Q., 11
de Vries, A., 123
Diekman, S., 190
Diekmann, S., 187, 191
Dillon, P. J., 247
DiMauro, S., 306
Dimicoli, J. L., 91
Dittrich, P. S., 220
DiTusa, C. A., 303
Dix, J. A., 291
Dixit, V. M., 11, 14
Dobek, K., 24
Donoghue, D. J., 124, 125
Doroshenko, A. O., 42
Doudna, J. A., 235
Douhal, A., 32, 160
Dowdy, S. F., 305
Downey, C. D., 236
Dragan, A. I., 185, 186, 187, 189, 191, 189, 190, 192, 193
Draper, D. E., 235
Dreskin, S. C., 215
Drews, G., 161
Drexhage, K. H., 261
Driscoll, J. S., 211
Driscoll, P. C., 186, 191
Driscoll, S. L., 220
Dubruil, T. P., 162
Duguay, M. A., 161
Duportail, G., 42, 45, 52, 54, 55
Durtsch, J. D., 312
Dwyer, M., 131
Dyck, A. C., 133, 138

E

Easter, J. H., 24
Eaton, W. A., 131
Ebbs, M. L., 8
Ebie, A., 265
Eckel, R., 197
Eckstein, F., 204
Edman, L., 131, 259
Eftink, M. R., 302
Eggeling, C., 259
Eggeling, C. J., 220
Ehler, L. A., 15
Eigen, M., 98
Eimerl, D., 164, 168
Ekert, P. G., 8
Elbaum, D., 290, 291
El-Byoumi, M. A., 74
Ellerby, H. M., 11
Ellerby, L. M., 11
Elson, E. L., 82, 131
Enderlein, J., 220
Engelman, D. M., 109
England, T. E., 237
Ercelen, S., 42, 51, 55
Erickson, S., 14
Eriksson, A. E., 288
Ernsting, N., 175
Ernsting, N. P., 160, 177
Erpelding, T., 175
Espagne, A., 160
Evans, W. S., 274

F

Fabrizi, G. M., 306
Fang, Q., 316
Farhy, L. S., 273, 274
Farris, F. J., 86
Farztdinov, V., 177
Fasshauer, D., 131
Fauci, A. S., 15
Faure, J., 24
Fedor, M. J., 233, 243
Feigon, J., 186, 191
Feinberg, M. B., 8, 12
Felekyan, S., 131
Feliccia, V. L., 290, 292
Felsenstein, K. M., 11
Feorino, P., 211
Ferré-D'Amaré, A. R., 243
Fiebig, T., 160
Fierke, C. A., 287, 288, 289, 290, 291,
    292, 293, 294, 301, 302, 303, 304,
    306, 313, 316, 323, 327, 329, 330
Figdor, C. G., 130, 131
Filonova, L. H., 8
Finucane, D. M., 8
Fiore, J. L., 236

Fiskum, G., 288, 291
Fite, B., 175
Fitzgerald, R. V., 312
Flannery, B. P., 275, 276
Fleming, G. R., 32, 71, 160, 162, 171, 178
Fletterick, R. J., 3
Fleury, L., 147
Foldes-Papp, Z., 131
Fones, G. R., 288, 313, 327
Fork, R. L., 167
Forkey, J. N., 131
Formosinho, S. J., 41
Forster, A. C., 233
Forster, T., 189, 293
Francke, U., 306
Frank, H.-G., 8
Frank, L., 193
Franks, N. P., 50
Frasier-Cadoret, S. G., 276
Frederickson, C. J., 287, 288, 289, 290,
    291, 292, 293, 301, 302, 304,
    305, 316, 323, 329
Frederiks, W. M., 14
French, D. M., 14
Fridland, A., 211
Fries, J. R., 259
Friesen, C., 11
Frisch, M., 175, 176
Frisoli, J. K., 324
Froelich, C. J., 8
Fronczek, F. T., 171
Fujimoto, B. S., 197
Fujiwara, T., 132
Fukunaga, Y., 171
Fukushima, P., 8
Fulda, S., 11
Funatsu, T., 22

G

Gadella, T. J., Jr., 98
Gafni, A., 129, 131, 133, 137,
    142, 145
Gallay, J., 33, 70, 71, 171
Gamsjager, R., 132
Garcia-Calvo, M., 5
Garcia-Ochoa, I., 160
Garcia-Parajo, M., 130
Gasymov, M. K., 24, 32
Gauss, G. H., 131, 132
Ge, Z., 330
Genet, R., 171
Gensch, T., 161
Georger, J., 327
Georghiou, S., 222
Germain, P., 88, 89, 90
Gettins, P. G. W., 3
Ghoneim, N., 38, 39, 43, 49, 70

Ghosh, G., 164, 168
Ghosh, P., 116, 123
Giblin, L. J., 302
Giese, K., 186, 191, 189
Gitsov, I., 111
Glasgow, B. J., 24, 32
Glegg, R. M., 197
Gloyna, D., 324
Gobets, B., 161, 162, 198
Goetz, D. H., 3
Gohlike, J. R., 69, 70, 71
Gohlke, C., 197, 259, 262
Gokulrangan, G., 197, 215, 216
Golden, J. P., 312
Goldman, Y. E., 131
Goldner, L. S., 210, 211, 220
Goldstein, L. S. B., 131
Goldstein, S. R., 312
Gondry, M., 171
Gonzalez, R. L., Jr., 131, 236
Gooijer, C., 72
Gopich, I., 245
Gordus, A., 236
Grainger, R. J., 266
Grampp, G., 24
Graslund, A., 216
Grattan, K. T. V., 312
Gratton, E., 95, 97, 171, 324
Graver, K. J., 162, 173
Green, D. R., 8, 11
Green, M. R., 247
Green, R., 234
Gregoire, G., 160
Greve, J., 131
Grigorovich, A. V., 42
Grillo, A. O., 89, 92
Grinvald, A., 24, 172
Gronemeyer, H., 88, 89, 90
Gronenborn, A. M., 160, 171
Grosschedl, R., 186, 191, 189
Grove, A., 109
Grubmüller, H., 131, 247
Gryczynski, I., 160, 324
Gryczynski, Z., 160
Gu, J. L., 161
Gubmuller, H., 160
Guedez, L., 8
Guichou, J. F., 88, 89, 90
Guja, K., 253
Günther, R., 259
Guo, C. S., 124
Gussakovsky, E., 131
Gustavsson, T., 175, 176
Guthrie, C., 233

H

Ha, T., 131, 132, 134, 236, 241, 242, 243
Haft, R. J. F., 267

Hager, G. L., 98
Haley, B., 233
Hallahan, C. W., 15
Halliday, L. A., 162
Halonen, P., 332
Hammes, G. G., 131
Han, X., 116, 123
Hansen, J. W., 161
Haran, G., 131, 137
Hardman, S. J., 198
Hargreaves, V., 265
Harley, M. J., 259, 262, 265, 267
Harms, G. S., 132
Harpur, A. G., 98
Harrick, N. J., 312
Harris, J. L., 3, 5
Hart, K. C., 125
Harvey, K. J., 8
Hashimoto, H., 98
Hassanali, A. A., 171
Hauber, J., 247
Haugland, R. P., 73, 235
Haustein, E., 83, 131
Hawkins, M., 205, 212, 213
Hawkins, M. E., 195, 199, 200, 201, 202, 204, 205, 206, 207, 208, 210, 211, 215, 219, 220, 221
Hayashi, Y., 301
Heikal, A. A., 219
Heinze, K. G., 94, 95, 98
Helene, C., 91
Hellec, F., 50
Heller, C. A., 218
Helmersson, A., 8
Hemmila, I., 332
Henderson, L. E., 288
Henkart, P. A., 5, 7, 10, 11, 13
Henry, R. A., 218
Herron, J. N., 312
Herschlag, D., 235, 236
Hershfinkel, M., 301, 304
Hess, S. T., 220
Heyduk, T., 91
Heyes, C. D., 138
Hill, J. J., 197, 213
Hillisch, A., 187, 190, 191
Hirata, H., 8
Hirch, M. D., 162
Hirschfeld, T. E., 312, 316
Hisahara, S., 8
Ho, A., 305
Hodak, J. H., 236
Hof, M., 33
Hogan, M. E., 208
Hohng, S., 236, 243
Holten, D., 160
Holub, D. F., 215
Holzwarth, A., 161

Hong, J. W., 138
Hong, M., 249
Hope, T. J., 247
Horton, W. A., 124
Houlden, H., 11
Houtzager, V. M., 5
Hristova, K., 107, 108, 109, 110, 111, 112, 114, 115, 116, 118, 120, 121, 123, 124
Hu, D., 141
Hu, W.-P., 41
Huang, C.-C., 289, 303, 330
Huang, J., 331
Huang, S., 219
Hudig, D., 5
Hudson, B. S., 114
Hui, F. K., 211
Huijbens, R., 130
Humpolickova, J., 33
Hung, F.-T., 41
Hunihan, L. W., 11
Hunt, J. A., 289
Hunt, J. B., 303
Huppertz, B., 8
Hurst, T., 287
Hutchins, C. J., 233

I

Ihalainen, J. A., 161
Iino, R., 132
Imhof, R. E., 161
Improta, R., 175, 176
Inchauspe, C. M. G., 167
Ingram, B., 215
Ippen, E. P., 160
Ippolito, J. A., 289
Ira, 24, 31, 32
Irving, J. A., 3
Ishii, Y., 22
Ito, N., 24
Itoh, R. E., 132
Iwamoto, T., 108, 109, 110
Iwane, A. H., 141
Iwata, T., 124

J

Jabs, E. W., 123
Jahn, R., 131
Jan, L., 175
Jan, Y., 175
Jarvie, K., 236
Jarzeba, W., 162
Jaworski, J., 301
Jean, J., 175
Jensen, K. F., 131, 133, 137, 142, 145
Jensen, O. N., 18
Jimenez, R., 178

Jin, W., 42
Johnson, B. W., 8
Johnson, C. K., 197, 215, 216
Johnson, M. L., 273, 274, 276
Johnson, R. C., 186, 191
Jones, E. R., 290
Jones, L., 316
Jones, T. A., 288
Joo, C., 242
Jouvet, C., 160
Jouvin, M.-H., 215
Jovin, T. M., 97
Jovine, L., 235
Jung, K.-Y., 198
Jungmann, O., 199, 202, 206, 207, 208, 219
Jurgensmeier, J. M., 11
Justement, J. S., 15

K

Kadenbach, B., 306
Kafka, J. D., 167
Kahari, V. M., 14
Kahlow, M. A., 162
Kahr, H., 132
Kahya, N., 97
Kaiser, W. J., 8, 12
Kammerer, S., 88, 89, 90
Kannan, K. K., 288
Kanuka, H., 8
Kao, Y., 171
Kapanidis, A., 240
Karnchanaphanurach, P., 131
Kasha, M., 41, 45
Katunuma, N., 14
Kaufmann, P. P., 8
Kavlick, M. F., 211
Kelley, F. D., 41, 42, 43
Kennis, J. T. M., 162
Kernohan, J., 290, 293
Kerppola, T. K., 186
Kessler, D. J., 208
Kettling, U., 98
Kharlanov, V., 160
Kiefer, L. L., 289, 303
Kim, H. D., 131, 236
Kim, J., 171
Kim, J. H., 8
Kim, S. A., 94, 95
Kinet, J.-P., 215
Kingdom, J., 8
Kitamura, K., 141
Klass, J., 186, 189, 191, 189
Klein, W. L., 132
Klostermeier, D., 235
Klymchenko, A. S., 37, 41, 42, 45, 46, 47, 51, 52, 53, 54, 55
Knutson, J. R., 32, 81, 159, 160, 162, 171, 172, 173, 199, 204, 206, 210, 214, 215, 221

Kobayashi, S., 8
Kobayashi, T., 132
Kobitski, A. Yu., 138
Koda, M., 14
Kohl, T., 98
Koivunen, E., 18
Koltermann, A., 98
Komoriya, A., 1, 3, 4, 5, 7, 8, 9, 10, 11, 12, 13, 15
Kondracki, L., 317
Koner, A. L., 24, 32
Kong, X., 240
Konig, M., 131
Konopasek, I., 31
Korr, H., 8
Koti, A. S. R., 24, 26, 27, 28, 29, 30, 31, 32
Kovacs, C., 15
Kovalenko, S. S., 160, 175
Kowalczyk, A. A., 214
Kragh-Hansen, U., 50
Kramer, G., 98
Krammer, P. H., 11
Krebs, J. W., 211
Krezel, A., 302
Krishna, M. M. G., 24, 26, 27, 28, 29, 31, 32
Krishnamoorthy, G., 24, 31, 32, 98, 198
Kronick, M. N., 312
Kubata, T., 38, 53
Kubicki, J., 24
Kuglstatter, A., 235
Kuhlemann, R., 98
Kühn-Hölsken, E., 247
Kulzer, F., 22
Kumar, P. V., 178
Kumbhakar, M., 24, 32
Kunzelmann, C., 42
Kuo, M. S., 73, 74
Kurokawa, K., 132
Kurt, H., 316
Kurz, L. C., 175
Kusumi, A., 132
Kwok, S., 211
Kwon, K., 3

L

Lacaze, P.-A., 210, 211, 220
Laczko, G., 324
Ladohin, A. S., 60, 75
Laguitton-Pasquier, H., 24
Lahue, E. E., 261
Lakowicz, J. R., 24, 32, 60, 132, 171, 207, 254, 273, 293, 318, 324, 331
Lampa-Pastrk, S., 171
Landgraf, S., 24
Langowski, J., 98, 261
Lannfelt, L., 11

Lapidus, L. J., 236
Larkin, C., 253, 265, 267
Larsen, O. F. A., 198
Laurence, T. A., 238, 240
Laws, W. R., 26, 171, 200, 220
Lazzarotto, E., 175, 176
Lee, B. J., 198
Lee, J. C., 91
Lee, K. N., 3
Lee, N. K., 240
Lee, T., 236
Lee, W. P., 14
Lee, Y. J., 8
Le Grice, S. F., 198
Leiner, M. J. P., 316, 318
le Maire, M., 50
Lemke, E. A., 22
Lesburg, C. A., 289, 303
Levine, M., 317
Li, E., 108, 109, 110, 111, 112, 114, 115, 116, 120, 121, 123, 124
Li, F., 11
Li, T., 171
Li, X., 186, 191, 189
Ligler, F. S., 312, 327
Lilius, L., 11
Liljas, A., 288
Lilley, D. M., 236, 266
Lindskog, S., 288
Linn, S., 198
Liphardt, J., 131
Lipman, E. A., 131
Lippard, S. J., 301
Lippert, E. L., 38, 39, 44
Lippitsch, M. E., 316, 318
Little, W. A., 312
Litwiler, K. S., 331
Liu, C., 198
Liu, L., 8, 12
Liu, L. Q., 164, 168
Liu, S., 15, 249
Liu, W., 42
Liu, Y., 187, 192, 193
Livesey, A. K., 24
Loeb, G. I., 312
Lohman, T. M., 131, 132, 190
Lomas, D. A., 3
Lommerse, P. H., 132
Lorenz, M., 187, 191
Louie, T., 131
Louis, J. M., 186, 191, 189
Love, J. D., 313
Love, J. J., 186, 191, 189
Lovgren, T., 332
Lowe, B., 261
Lu, H. P., 131, 133, 141, 146, 147, 148, 149
Lu, W., 171
Lührmann, R., 247

M

Luke, C. J., 3
Lukovic, D., 8, 10
Lunstrum, G. P., 124
Luo, G., 131
Lurkin, I., 123
Lushington, G. H., 197, 216
Lustres, J., 175, 177

Maciejewski, A., 24
Mack, D. H., 211
MacKenzie, K. R., 108
Mackie, H., 198
Magde, D., 131
Mahapatro, M., 303
Mahoney, W., 289, 305
Mahr, H., 162
Majewski, J., 262
Majoul, I. V., 98
Makeyeva, E. N., 187, 189, 190, 192, 193
Malim, M. H., 247
Maliwal, B. P., 288, 289, 290, 291, 292, 293, 294, 302, 305, 316, 323, 327, 329
Mandal, P. K., 55
Manders, E. M., 98
Manger, M., 160
Mann, M., 18
Manning, G. S., 190
Mansoor, A., 8
Maravei, D. V., 8
Marcu, L., 316
Marcus, R., 72
Marcuse, D., 313
Maren, T. H., 290
Margeat, E., 92, 93, 94, 95, 97, 240
Margittai, M., 131
Margueron, R., 95, 97
Mariathasan, S., 14
Marino, J. P., 198
Markovitsi, D., 175, 176
Maroncelli, M., 24, 71, 161, 178
Martin, C. T., 198
Martin, J. L., 171
Martin, M. M., 160
Martinez, M. L., 41, 42, 43
Martinez, O. E., 167
Marx, N. J., 261
Masaike, T., 131
Masalha, R., 302
Masse, J. E., 186, 191
Mataga, N., 38, 53
Matson, S. W., 261
Matsuda, M., 132
Matsumura, M., 198
May, J. M., 160
Mazumder, A., 199, 201, 205

Mazurenko, Y. T., 69, 71
McAdoo, D. J., 302
McAnaney, T., 175
McCall, K. A., 289, 290, 292, 297, 303
McCourt, L., 131
McCranor, B., 311
McGee, S. A., 289
Mcinerney, J., 161
McInnis, J., 71
Mckinney, S., 242, 243
McLaughlin, B. A., 218
McLaughlin, L. W., 216
McLaughlin, M., 15
McLaughlin, M. L., 171
Mcmahon, L. P., 171
Mead, R., 276
Meadow, D., 171, 174
Meadow, N. D., 162, 173
Mely, Y., 46, 52, 54, 55, 98
Mély, Y., 42, 45, 54, 55
Menez, A., 171
Mengel, T., 306
Merzlyakov, M., 107, 109, 110, 111, 112, 114, 116, 118, 120, 124
Metkar, S. S., 8
Mets, U., 259
Metzler, D. E., 25
Meyer, A. N., 125
Meyers, G. A., 123
Miannay, F., 176
Michaels, C. A., 171
Midwinter, J. E., 163, 165, 166
Mihalusova, M., 236, 249
Mihrus, S., 3
Milgontina, E. I., 189, 190
Millar, D. P., 198, 233, 235, 236, 243, 245, 247, 248, 249
Minoguchi, S., 132
Mirzabekov, A., 243
Misra, V. K., 236
Mitchell, R. C., 39
Mitchell, S. W., 211
Mitra, R. K., 33
Mitsuya, H., 211
Miura, H., 8
Mochizuki, N., 132
Moerner, W. E., 131, 147, 236
Moffett, J. W., 288, 296, 311, 313, 327
Moller, J. V., 50
Mollova, E. T., 235
Monack, D. M., 14
Montal, M., 109
Montal, M. O., 109
Mook, O. R., 14
Moore, M. J., 237
Moore, P. B., 235
Moore, S. A., 205
Morales, G. A., 171

Moreira, B. G., 266
Morgan, J. J., 301
Morosiuotto, T., 161
Morton, B. S., 261
Moyer, R. W., 3
Muelle, R. N., 302
Muino, P. L., 179
Mukerji, I., 205, 206, 208, 212, 213
Mukherjee, T. K., 24, 32
Mukhopadhyay, S., 22
Mulkherjee, S., 60
Mullan, M., 11
Muller, C. D., 42
Muller, G., 98
Muller, J. D., 95, 97
Muller, M. G., 161
Müller, R., 261
Munro, I. R., 123
Munsen, V. A., 258
Murakoshi, H., 132
Murao, T., 161
Murgia, M., 306
Murphy, C., 175
Murphy, E. C., 186, 191, 189
Myers, J. C., 205

N

Nagai, K., 235
Nagashima, K., 164, 168
Nahas, M. K., 236
Nahum, R., 8
Nair, R., 18
Nair, S. K., 290, 291
Nakamura, H., 256
Nakase, H., 306
Nakazawa, N., 124
Nalin, C. M., 247
Nanda, V., 4
Narlikar, G. J., 235
Nass, R., 274
Nath, S., 24, 32
Nazarenko, I., 261
Negrerie, M., 171
Neher, E., 130
Nelder, J. A., 276
Nelson, W. C., 261
Nemkovich, N. A., 60, 61, 64, 68, 71, 72
Nesbitt, D. J., 236
Newton, K., 14
Nicholson, D. W., 5
Nie, S., 131, 134, 236
Nienhaus, G. U., 138, 236
Nilsson, L., 216
Nilsson, T., 98
Nishimoto, E., 171
Nivón, L. G., 249
Noguchi, M., 97

Noji, H., 131
Nolan, E. M., 301
Nomizu, M., 3
Nomura, T., 14
Nord, S., 261
Nordlund, T. M., 198, 216
Nordstrom, P. A., 5
Norman, D. G., 186, 191, 266
Nottrott, S., 247
Novaira, M., 24, 32
Nuccitelli, R., 297
Nyman, P. O., 288

O

Obaidy, M., 261
O'connor, C. M., 236, 249
O'Connor, D. V., 24, 161
O'Connor, L., 11
Oda, M., 256
Oesterhelt, F., 247
Ohba, Y., 132
Ohnemus, D., 311
Ohshima, C., 132
Ojemann, J., 208
Okamoto, K.-I., 301
Okano, H., 8
Okasaki, T., 8
Okobiah, O., 171
Olsen, H. S., 247
Omit, M., 22
Oncul, S., 54
Organero, J. A., 32
Orlow, S. J., 123
Orr, J. W., 131, 236
Orrit, M., 147, 236
Osman, R., 200
Otosu, T., 171
Otsuki, T., 124
Ou, C. Y., 211
Oubridge, C., 235
Owczarzy, R., 266
Owens, T. J., 262
Ozturk, T., 42

P

Packard, B. Z., 1, 3, 4, 5, 7, 8, 9, 10, 11, 12, 13, 15
Paik, D. H., 160
Pal, H., 24, 32
Pal, S. K., 32, 33
Palfey, B., 131, 133, 137, 142, 145
Pan, C. P., 160
Pan, J., 235
Pan, T., 234, 236
Panda, D., 24, 32
Pandya, P., 15
Pansu, R., 24

Pansu, R. B., 24
Papaioannou, T., 316
Pardi, A., 235, 236
Park, E., 175
Park, J., 175
Park, Y.-W., 208
Parslow, T. G., 247
Passner, J. M., 208
Patchan, M. W., 290, 291, 293, 330
Paterniti, D. P., 331
Paterno, S. A., 289
Patten, L. C., 14
Pearson, W. H., 198
Pemberton, P. A., 3
Pereira, M. J., 131, 236
Perez, E., 88, 89, 90
Perez, G. L., 8
Perez, P., 39, 53
Periasamy, N., 21, 24, 26, 27, 28, 29, 30, 31, 32
Pernot, P., 24
Peter, M. E., 11
Peterman, E. J. G., 131
Peterson, D., 289, 305
Peterson, E. P., 5
Peterson, J. I., 312
Petrich, J. W., 171
Pettersson, K., 332
Pfleiderer, W., 199, 200, 201, 202, 204, 205, 206, 207, 208, 210, 215, 219, 220, 221
Philippsetis, A., 222
Phillips, D., 24, 161
Piemont, E., 45, 46, 54
Piestert, O., 236
Pipes, L., 274
Pires, R., 261
Piston, D. W., 136
Pivovarenko, V. G., 40, 42, 55, 72
Pizzo, P., 306
Pleiss, J. A., 237
Pljevaljcic, G., 233, 236, 243, 245
Pogenberg, V., 88, 89, 90
Pollard, V. W., 247
Pommier, Y. G., 199, 201, 205
Pond, S. J., 247, 248, 249
Porter, D., 171, 172, 199, 204, 210, 215, 221
Porter, R. A., 171
Poujol, N., 92, 93, 94, 95, 97
Pozzan, T., 306
Prathapam, R., 236, 249
Prendergast, F. G., 171
Press, W. H., 275, 276
Preyer, N., 316
Price, P. J., 289, 297, 303
Prior, K. J., 327
Privalov, P. L., 185, 186, 187, 189, 191, 189, 190, 192, 193

Procheka, K., 33
Proudnikov, D., 243
Prough, D. S., 302
Prusoff, W. H., 269
Przylepa, K. A., 123
Pu, S.-C., 41
Puglisi, J. D., 131, 234, 236
Pusterhofer, J., 316, 318

Q

Qiu, T., 161
Qiu, W., 171, 171
Quake, S. R., 138
Query, C. C., 237
Quinlan, M. E., 131

R

Rader, S. D., 237
Radvanyi, F., 123
Rai, P., 198
Raja, S. M., 8
Rajagopalan, P. T. R., 131
Ramakrishnan, V., 235
Ramreddy, T., 198
Randolph, J. B., 198
Rano, T. A., 5
Rao, B. J., 198
Rao, G., 318
Rarbach, M., 98
Rashtchian, A., 261
Rasnik, I., 131, 132, 242, 243
Rasper, D. M., 5
Rathjen, P. D., 233
Rausch, J. W., 198
Rayner, D. M., 171
Read, C., 186, 189, 191, 189
Read, C. M., 189, 190
Rech, I., 131
Record, M. T., Jr., 190
Reed, J. C., 9, 11
Reichardt, C., 38, 44
Reichert, A., 168
Reister, F., 8
Remold-O'Donnell, E., 3
Rensen, W., 130
Reutzepis, P. P., 161
Rhee, M. J., 303
Rhodes, E., 131
Rhodes, M. M., 236
Richardson, P. L., 3
Richert, R., 60
Ridgeway, W., 247, 248, 249
Ridgler, R., 131
Rigler, R., 98, 216, 220, 259
Rippe, K., 98, 261
Rist, M. J., 198
Rizzuto, R., 306

Roberts, S. B., 11
Robertson, R., 247, 248, 249
Robertson, S. C., 125
Robles, R., 8
Rodgers, M. E., 258
Romanin, C., 132
Roose-Flrma, M., 14
Ros, R., 197
Roseman, S., 162, 173
Rosen, C. A., 247
Rosenthal, R. E., 288, 291
Roshal, A. D., 42
Ross, J. A., 171
Ross, J. B. A., 200
Ross, R., 316
Rosspeintner, A., 24
Rost, B., 17, 18
Roumestand, C., 97
Roy, R., 243
Roy, S., 5
Royer, C. A., 79, 88, 89, 90, 91, 92, 93, 94, 95, 97, 197, 213
Rubinov, A. N., 60, 61, 64, 68, 71, 72
Rubtsov, I. V., 160
Rueda, D., 236
Ruggiero, A. J., 171
Runnels, L. W., 97, 101, 102
Ruoslahti, E., 18
Rupert, P. B., 243
Rusch, R. M., 91
Russell, R., 236
Rust, M., 236

S

Sachl, R., 33
Sackmann, E., 113
Safrit, J. T., 8, 12
Saito, K., 98, 141
Sakaguchi, H., 124
Sakmann, B., 130
Sako, Y., 132
Salvesen, G. S., 9
Samanta, A., 55
Sanabia, J. E., 210, 211, 220
Sandison, D. R., 136, 140
Santos, N. C., 130
Sarkar, N., 175, 176
Sarvey, J. M., 302
Sasada, M., 8
Sauer, M., 261
Savtchenko, R., 162, 171, 173, 174
Sawai, H., 8
Sawamoto, K., 8
Sawyer, W., 171
Scaffidi, C. S., 11
Scalmani, G., 175, 176
Scaringe, S. A., 237

Scarlata, S. F., 79, 91, 97, 101, 102
Schanz, R., 160
Schenter, G. K., 146
Scherer, N. F., 236
Scherfeld, D., 97
Schildbach, J. F., 253, 258, 259, 260, 261, 262, 263, 264, 265, 267
Schlarb, U., 168
Schmidt, T., 132
Schochetman, G., 211
Schon, E. A., 306
Schroder, G. F., 131, 160, 247
Schuler, B., 131
Schultz, P. G., 238
Schulz, A., 261
Schurr, J. M., 197
Schwalb, N., 176
Schwartz, G. P., 171
Schwarze, S. R., 305
Schweinberger, E., 131
Schwille, P., 83, 94, 95, 97, 98, 220
Sclavi, B., 235
Scozzafava, A., 290
Seeger, S., 261
Seibert, E., 200
Seidel, C. A., 131, 220, 247, 259, 261
Sen, S., 175
Sercel, A. D., 198
Sevenich, F. W., 261
Sewald, N., 197
Shakked, Z., 206
Shamoo, Y., 205
Shank, C. V., 160, 167
Shaw, A. K., 32
Shaw, M. A., 131
Shen, G., 42
Shen, X., 160, 171, 172
Shen, Y. R., 160, 163, 164, 166, 171, 172
Sheng, M., 301
Shi, J., 129, 131, 133, 137, 142, 145
Shi, J. L., 161
Shi, X., 175
Shih, I. H., 233
Shin, H., 11
Shinitzky, M., 72
Shoji, S., 8
Shriver-Lake, L. C., 312, 327
Shui, L., 171
Shynkar, V. V., 42, 45, 46, 54, 55
Siano, D. B., 25
Sibbett, W., 161
Silber, J. J., 24, 32
Silke, J., 8
Silverman, G. A., 3
Silvestri, G., 8, 12
Singer, S. J., 171
Sinha, S. S., 33
Sipior, J., 318

Sischka, A., 197
Skrzydlewska, E., 14
Smertenko, A. P., 8
Smith, D., 131
Smith, R. S., 312
Smith, S. B., 131
Smolarczyk, G., 54
Snel, M. M., 131
Sninsky, J. J., 211
Snyder, A. W., 313
So, P. T., 97
Sobolewski, A., 160
Soldatov, N. M., 132
Somers, R. L., 198
Song, H. K., 3
Sonnenfeld, A., 137
Sosa, H., 131
Sosnick, T. R., 234, 236
Spaink, H. P., 132
Spangler, J., 108, 109, 110, 116, 123
Squire, A., 98
Srinivas, R. V., 211
Srinivasan, A., 11
Srividya, N., 236
Stanley, R. J., 217, 218
Stark, M. R., 237
Steel, D., 129, 131, 133, 137, 142, 145
Steinberg, I. Z., 24, 172
Stennicke, H. R., 11
Stepanek, M., 33
Stern, J. C., 258, 259, 260, 261, 262, 263, 264, 265
Stern, M. H., 97
Stetler-Stevenson, W. G., 8
Stine, W. B., Jr., 132
Stoddard, A., 287, 288, 291
Stoddard, A. K., 288, 302, 306
Stoeckel, H., 42
Stone, M. D., 236, 249
Storm, C. B., 303
Strandberg, B., 288
Strasser, A., 11
Straume, M., 276
Strickler, J., 136, 140
Strub, K., 235
Stryer, L., 235
Stuhmeier, F., 190
Stumm, W., 301
Suarez, M. F., 8
Sugiyama, H., 198
Suh, S. W., 3, 289, 305
Sulkowska, M., 14
Sulkowski, S., 14
Sullivan, M., 235
Suppan, P., 38, 39, 43, 49, 70
Supuran, C. T., 290
Sutin, N., 72
Svobodova, J., 31

Swiney, T. C., 41, 42, 43
Symons, R. H., 233
Sytnik, A. I., 72, 73
Szabo, A. G., 171, 245
Szmacinski, H., 293

T

Takahashi, A., 8
Takahashi, R., 9
Takeda, K., 42
Tan, E., 236
Tan, K. T., 261
Tan, L., 312
Tan, W., 133, 138
Tanaka, M., 113, 211
Taniwaki, M., 124
Tao, X.-J., 8
Tasset, D., 215
Taylor, J. R., 161
Telford, W. G., 7, 9, 13
Temps, F., 176
Teplow, D. B., 132
Terry, A. H., 312
Teukolsky, S. A., 275, 276
Thiery, J. P., 123
Thirumalai, D., 235
Thompson, E., 198
Thompson, K. C., 198
Thompson, N. L., 220, 240
Thompson, R. B., 287, 288, 289, 290, 291, 292, 293, 294, 301, 302, 304, 305, 306, 311, 312, 313, 317, 324, 327, 330, 331, 333
Thompson, R. B., Jr., 316, 323, 329
Thornberry, N. A., 5
Thorner, M. O., 274
Tilly, J. L., 8
Tilly, K. L., 8
Timkey, T., 5
Titolo, S., 258, 262
Todd, D. C., 171
Toensing, K., 197
Tokunaga, M., 141
Tomaselli, K. J., 11
Tomich, J. M., 108, 109, 110
Tomin, V. I., 54, 60, 61, 64, 68, 71, 72
Toone, E. J., 303
Topp, M. R., 160, 162
Toptygin, D. D., 3, 4, 31, 160, 162, 171, 173, 174, 255, 258, 259, 260, 261, 262, 263, 264, 265
Torello, M., 124
Travers, K. J., 236
Traxler, B., 267
Trbovich, A. M., 8
Treiber, D. K., 234
Troxler, T., 259, 262
Tynan, J. A., 125

U

Ucker, D. S., 8, 10
Ueki, A., 124
Ueno, T., 211
Uhlenbeck, O. C., 237
Uhrin, D., 266
Unruh, J. R., 197, 215, 216

V

Vaillancourt, J. P., 5
Valadkhan, S., 233
Valdenebro, M. d., 302
Vamosi, G., 197, 259, 262
van Amerongen, H., 198
van den Berg, C., 296
Vanderkooi, J. M., 160
van der Kwast, T. H., 123
van der Meer, M. J., 162
Van der Meer, W., 101
van Driel, R., 98
van Grondelle, R., 162, 198
van Hulst, N., 130
van Munster, E. B., 98
Van Noorden, C. J., 14
van Rhijin, B., 123
van Stokkum, I. H. M., 161, 162, 198
van Tilborg, A., 123
Vaux, D. L., 8
Vela, M. A., 171
Veldhuis, P. P., 273, 274, 276
Verkman, A. S., 291
Verschure, P. J., 98
Vettering, W. T., 275, 276
Vihinen, P., 14
Vincent, M., 33, 70, 71, 171
Vivat-Hannah, V., 88, 89, 90
Vocero-Akbani, A., 305
Volker, J., 198
von Arnold, S., 8
Vucic, D., 14

W

Wada, H., 124
Walbridge, D. G., 81
Waldeck, W., 98
Walsh, F., 306
Walt, D. M., 315, 326
Walter, N., 236
Walter, N. G., 131, 243
Wang, B., 8, 18
Wang, H.-K., 312
Wang, J., 247, 248, 249
Wang, K., 258
Wang, L., 171
Wang, Y., 42
Wang, Z. X., 270

Warfield, D., 211
Waterhouse, N. J., 8
Waxham, M. N., 94, 95
Wazawa, T., 22
Webb, W., 136, 140
Webb, W. W., 136, 217, 218, 219, 220
Weber, G., 72, 80, 86, 291, 302
Webster, M. K., 124
Weeks, K. M., 249
Wei, L. N., 97
Weichenrieder, O., 235
Weiglhofer, M., 24
Weiss, M., 98
Weiss, S., 131, 238, 240
Weiss, V., 261
Welchner, E., 258
Wells, K. S., 136, 140
Wennmalm, S., 131
Wernett, M. E., 8, 12
Westerberg, N., 287, 288, 291
Westerberg, N. M., 301, 304
Wherry, E. J., 8, 13
Whetsell, W. O., Jr., 290, 292, 293, 302
White, P. W., 258, 262
White, S. H., 108, 114
Whiting, G. S., 39
Widengren, J., 131
Wiegand, T. W., 215
Wild, K., 235
Wiles, N. C., 208
Wilking, S. D., 197
Wilkinson, G., 297
Willemsen, O. H., 131
Williams, A., 198
Williams, A. P., 208
Williams, P. B., 215
Williams, R. M., 136, 140, 219
Williamson, J. R., 131, 234, 236
Wilson, G. S., 197, 215, 216
Wilson, T. J., 236
Wimley, W. C., 108, 109, 110, 111, 114, 116, 121, 123, 124
Winblad, B., 11
Windsor, M. W., 160
Winkler, T., 98
Winkler, W. C., 233
Wise, D. S., 198
Witkoskie, J. B., 154
Wojtuszewski, K., 205, 206, 212, 213
Wolber, P. K., 114
Wolf, B. B., 11
Wolfbeis, O. S., 312, 316, 318
Wolff, L., 8
Wolfrum, J., 261
Wong, B., 186, 191
Woodson, S. A., 235
Wouters, F. S., 98
Wozniak, A. K., 247

Wright, P. E., 186, 191, 189
Wu, M.-L., 5, 7, 10
Wyssbrod, H. R., 171

X

Xie, X. S., 131, 133, 141, 146, 147, 148, 149
Xie, Z., 236
Xu, C., 217, 218, 219
Xu, J., 159, 162, 172, 173
Xu, Y., 198
Xun, L., 131, 133, 141, 147, 148, 149

Y

Yamada, K. M., 3, 15
Yamada, O., 124
Yamamoto, K., 8
Yamashita, K., 171
Yamashita, S., 171
Yamazaki, I., 161
Yanagida, T., 22, 132, 141
Yang, A., 217, 218
Yang, H., 131, 141
Yang, S., 146, 154
Yang, X., 11
Yang, Y., 171
Yaniv, M., 91
Yawata, Y., 124
Yen, Y. M., 186, 191
Yesylevskyy, S. O., 41
Yeung, E. S., 133, 138
Yonehara, S., 8
Yoshida, M., 131
Yoshihara, K., 24, 160, 161
Yoshimura, A., 132
Yoshizaki, H., 132
You, M., 108, 109, 110, 111, 112, 115, 116, 120, 121, 123, 124
You, Y., 266

Yu, B. L., 161
Yu, H. T., 171
Yu, M., 3
Yu, R., 42
Yu, W.-S., 41
Yu, Y.-C., 41

Z

Zamore, P. D., 233
Zander, C., 220, 261
Zander, C. K. H. D., 220
Zapata, J. M., 9
Zapp, M. L., 247
Zare, R. N., 131, 134, 236
Zelent, B., 160
Zeng, H.-H., 288, 290, 291, 292, 311, 313, 327
Zeng, Y., 302
Zernicke, F., 163, 165, 166
Zewail, A. H., 160
Zhang, H., 162
Zhang, L., 171
Zhang, Z., 131
Zhang, Z. Y., 312
Zhao, L. J., 177
Zhao, W., 236
Zhivotovsky, B., 8
Zhong, D., 171, 171
Zhou, Q., 11
Zhuang, X., 131, 236, 249
Zhurkin, V. B., 186, 191, 189
Ziffer, H., 171, 172
Zimmer, D. P., 191
Zipfel, W., 219
Zornow, M. H., 302
Zumbusch, A., 147
Zwarthoff, E., 123
Zwicker, H. R., 161

Subject Index

A

AFM, *see* Atomic force microscopy
Area-normalized time-resolved emission spectroscopy, *see also* Time-resolved emission spectroscopy
  fluorophores in microheterogeneous and biological media
    dyes in surfactant micelles or lipid membranes, 30–31
    special cases
      anisotropy decay, 32
      site heterogeneity, 32
      solvation dynamics, 33
    isoemissive point, 29–30
    overview of steps, 24–25
    simple case examples
      fluorophore in dilute solution
        with excited-state reaction, 26
        without excited-state reaction, 26
      fluorophore in dilute, viscous solution
        without excited-state reaction, 28
      fluorophore mixture in dilute solution
        without excited-state reaction, 27
    spectra interpretation, 29–30
Atomic force microscopy, resolution, 130
A-tract, pteridine-containing oligonucleotide studies
  characterization with 6MAP, 206–208
  temperature-dependent behavior of duplexes, 208–210
Autocorrelation function, single-molecule spectroscopy, 147–148, 154
AutoDecon
  concordant secretion events, 279–280
  event detection criteria, 274–275
  mathematical model of hormone concentration time-series data, 276
  modules
    combined modules, 278–279
    fitting module, 276
    insertion module, 277
    overview, 275–276
    triage module, 277, 286
  pulsatile signals in noisy data, 274
  simulated data and results, 275, 280–284
  steps, 284

B

Biosensors, *see* Carbonic anhydrase fluorescence biosensors; Fluorescence-based fiber optic biosensors

C

Carbonic anhydrase fluorescence biosensors
  advantages, 288
  apoenzyme preparation for biosensor generation, 303–304
  buffer preparation with varying zinc concentrations, 297–301
  distinction between free and bound zinc ion, 295–297
  fluorescence resonance energy transfer, 293–295
  intracellular sensing
    expressible biosensor, 305–306
    TAT tagging, 304–305
  kinetics of metal-binding reactions, 301–302
  metal-binding specificity, 289
  principles of zinc sensing, 288–289
  ratiometric determination of free zinc in solution, 302–303
  zinc binding transduction to fluorescence change, 290–295
6-Carboxyfluorescein, labeled oligonucleotide studies of DNA-binding proteins, 259–260, 262
Carboxytetramethylrhodamine, labeled oligonucleotide studies of DNA-binding proteins, 259–262, 264–265
Caspase, intramolecular H-type excitonic homodimer probe assays
  activation studies
    cascade, 10–11
    macrophage caspase-1 in inflammation, 13–14
  specificity studies, 5–7
  subcellular localization of activity, 11
Cathepsin D, subcellular localization of activity with intramolecular H-type excitonic homodimer probe, 11
Cholesterol oxidase, single-molecule spectroscopy, 147–149
Circular permutation assay, DNA bending, 186

353

Cladding, fiber optic sensors, 313
Copper, fluorescence-based fiber optic biosensors, 330–332

D

DAS, *see* Decay-associated spectra
Decay-associated spectra, upconversion fluorescence spectroscopy, 171–172
4′-(Diethylamino)-3-hydroxyflavone, *see* Solvatochromic fluorescent dyes
Dihydroorotate dehydrogenase, single-molecule spectroscopy, 142–148
DNA bending, *see* Fluorescence resonance energy transfer
DNA-binding protein
 fluorescence anisotropy, 91–95
 fluorescence correlation spectroscopy, 91, 93, 95–96
 fluorescence resonance energy transfer of protein-induced DNA bending
  DNA duplex labeling, 187–189
  efficiency equations, 189–190
  electrostatic component of Gibbs binding energy, 190
  large bend studies of high-mobility group protein domains, 191–189
  overview, 186–187
  salt effects, 192–189
  small bend studies
   DNA U-shaped construct and labeling, 191–193
   titration, 193
 fluorophore-labeled oligonucleotide binding studies
  advantages, 256–257
  anisotropy, 259–260, 263–264
  competition assays for specificity determination, 269–270
  disadvantages, 257–258
  experimental design, 266–268
  oligonucleotide design, 258–259
  pH and salt effects, 262
DNA hairpin, structure probing with pteridine-containing oligonucleotides, 215–216

E

Electrophoretic mobility shift assay, DNA-binding protein studies, 254, 257–258
EmEx-FRET, *see* Fluorescence resonance energy transfer
EMSA, *see* Electrophoretic mobility shift assay
ESIPT, *see* Excited state intramolecular proton transfer
Excited state intramolecular proton transfer, solvatochromic fluorescent dyes, 40–41, 45, 47, 55

F

6-FAM, *see*, 6-Carboxyfluorescein
FCS, *see* Fluorescence correlation spectroscopy
FGFR3, *see* Fibroblast growth factor receptor-3
Fiber optic biosensors, *see* Fluorescence-based fiber optic biosensors
Fibroblast growth factor receptor-3, transmembrane helix dimerization energetics measurements with fluorescence resonance energy transfer, 123–125
Fluorescence anisotropy
 calculation, 254
 DNA-binding protein studies, 259–260, 263–264
 lifetime relationship with Perrin equation, 255
 magnitude, 255
 principles, 84–85
 protein–nucleic acid interaction studies, 91–95
 protein–protein interaction studies
  membrane proteins, 99
  solution, 97–98
 pteridine-containing oligonucleotide studies
  protein binding, 212–213
  steady-state and time-resolved anisotropy, 213, 215
 time-resolved emission spectroscopy, 32
 titration of fluorescent peptide ligands, 88–89
Fluorescence-based fiber optic biosensors
 advantages, 312–313
 alignment of sensors, 316–318
 construction, 326–329
 copper probe, 330–332
 discrete samples, 332
 field applications, 334–335
 instrument mounting and alignment, 319–323
 mode hopping, 333–334
 noise, 333
 overview, 312
 principles, 313–316
 standards
  lifetime standards, 324–325
  ratiometric standards, 323–324
 thermal drift, 333
 zinc probe, 329–330
Fluorescence correlation spectroscopy
 principles, 82–84
 protein–nucleic acid interaction studies, 91, 93, 95–96
 protein–protein interaction studies
  membrane proteins, 99
  solution, 96–97
 resolution, 131
 titration of fluorescent peptide ligands, 89–90
Fluorescence intensity
 decay measurements, 81
 emission energy in binding profiles, 84

Subject Index

protein–protein interaction studies in membranes, 99–101
pteridine oligonucleotide studies, 204–206
Fluorescence lifetime
  anisotropy relationship with Perrin equation, 255
  pteridine-containing oligonucleotide studies, 214
Fluorescence resonance energy transfer
  carbonic anhydrase fluorescence biosensors for zinc measurement, 293–295
  principles, 81–82, 132
  protein-induced DNA bending studies
    DNA duplex labeling, 187–189
    efficiency equations, 189–190
    electrostatic component of Gibbs binding energy, 190
    large bend studies of high-mobility group protein domains, 191–189
    overview, 186–187
    salt effects, 192–189
    small bend studies
      DNA U-shaped construct and labeling, 191–193
      titration, 193
  protein–nucleic acid interaction studies, 91
  protein–protein interaction studies
    membrane proteins, 99, 101–103
    solution, 97–98
  Red-Edge effects, 59, 72–73
  RNA folding studies
    diffusion single-pair fluorescence resonance energy transfer
      advantages, 240
      efficiency, 238, 239
      hairpin ribozyme folding analysis, 243–247
      instrumentation, 238, 239
      overview, 235
    transmembrane helix dimerization energetics measurements
      bilayer platforms
        large unilamellar vesicles, 111
        multilamellar vesicles, 110–111
        small unilamellar vesicles, 111
        surface-supported bailers, 111–114
      challenges, 108–110
      efficiency of transfer
        direct calculation from donor quenching, 115–116
        equation, 114
      EmEx-FRET
        overview, 116–117
        steps, 119–120
        theory, 117–118
      fibroblast growth factor receptor-3 transmembrane domain studies, 123–125
      proximity fluorescence resonance energy transfer, 114–115
      thermodynamics and calculations, 120–123
Fluorescence upconversion, see Upconversion fluorescence spectroscopy
FRET, see Fluorescence resonance energy transfer

G

G protein, reconstitution for fluorescence studies, 100

H

Hairpin ribozyme, see RNA folding
Hydrogen bonding, see Solvatochromic fluorescent dyes
3-Hydroxyflavone dyes, see Solvatochromic fluorescent dyes

I

ICT, see Intramolecular charge transfer
Intramolecular charge transfer, solvatochromic fluorescent dyes, 41, 53
Intramolecular H-type excitonic homodimer (IHED)
  cell-permeable probes
    applications, 8–9
    caspase activation studies
      cascade, 10–11
      macrophage caspase-1 in inflammation, 13–14
    cell-mediated cytotoxicity studies, 12–13
    metastatic cancer cell migration and invasiveness studies, 14–15
    nuclease activity and hybridization assays, 15
    overview, 7
    subcellular localization of protease activities, 11
    toxicity, 8
    viral protease studies, 15
  prospects, 16–17
  protease substrates
    caspase specificity studies, 5–7
    design, 2–5
    nomenclature, 2
    NorFES properties, 3–4
Isoemissive point, area-normalized time-resolved emission spectroscopy spectrum, 29–30
Isothermal calorimetry, DNA-binding protein studies, 254, 256, 258
ITC, see Isothermal calorimetry

L

Laser-induced fluorescence, see Single-molecule spectroscopy

M

6MAP, *see* Pteridine nucleoside analogs
3MI, *see* Pteridine nucleoside analogs
6MI, *see* Pteridine nucleoside analogs
Multiparametric probing, *see* Solvatochromic fluorescent dyes

N

Near-field scanning optical microscopy, resolution, 130
Noisy fluorescent data, *see* AutoDecon
NorFES, *see* Intramolecular H-type excitonic homodimer
NSOM, *see* Near-field scanning optical microscopy
Numerical aperture, fiber optic sensors, 314–315

O

Optical Kerr shutter, ultrafast fluorescence spectroscopy, 161

P

Perrin equation, 255
Photomultiplier tube, ultrafast fluorescence spectroscopy, 160–161
Photoselection, *see* Red-Edge effects
PMT, *see* Photomultiplier tube
PPI, *see* Probable Position Index
Probable Position Index, AutoDecon, 277
Protease fluorescence assay, *see* Intramolecular H-type excitonic homodimer
Pteridine nucleoside analogs
  advantages, 197–198
  applications
    A-tracts
      characterization with 6MAP, 206–208
      temperature-dependent behavior of duplexes, 208–210
    bulge hybridization, 210–212
    fluorescence anisotropy studies
      protein binding, 212–213
      steady-state and time-resolved anisotropy, 213, 215
    fluorescence intensity changes, 204
    fluorescence lifetime studies, 214
    hairpin structure probing of aptamer with 6MI, 215–216
    single molecule detection with 3MI, 220–221
    two-photon excitation of 6MAP, 216–219
  characterization of oligonucleotides, 204
  DNA fluorescence labeling approaches, 197
  oligonucleotide synthesis
    deprotection, 203
    phosphoramidite conservation, 202
    purification, 203–204
  properties
    base pairing, 201–202
    fluorescence intensity and spectral shifts, 199–200
    stability, 200–201
  prospects for study, 221–222
  structures, 198
Pulsatile signal analysis, *see* AutoDecon

Q

Quasistatic self-quenching, tryptophan in dipeptides, 172

R

Ratiometric fluorescence
  fluorescence-based fiber optic biosensor standards, 323–324
  zinc in solution with carbonic anhydrase fluorescence biosensors, 302–303
Red-Edge effects
  excitation fluorescence shift magnitude and dielectric relaxations, 66–70
  excited state reactions, 71–73
  fluorescent probes, 73–74
  ground-state heterogeneity, 65–66
  high-resolution time-resolved spectroscopy, 70–71, 75
  origins, 60–61
  overview, 59–60
  photoselection principles, 61–64
  prospects, 75–76
  tryptophan, 74–75
Rev–Rev response element complex
  overview of assembly, 247
  total internal reflection fluorescence of assembly
    advantages, 242
    instrumentation, 240–242
    kinetic analysis, 248–249
    principles, 240–241, 247–248
    RNA immobilization, 242–243
Ribounucleoprotein assembly, *see* Rev–Rev response element complex
RNA folding
  fluorescence resonance energy transfer, 235
  overview, 234
  single molecule fluorescence
    applications, 236
    labeling, 236–237
    detection, 238
    diffusion single-pair fluorescence resonance energy transfer
      advantages, 240
      efficiency, 238, 239

Subject Index

hairpin ribozyme folding analysis, 243–247
instrumentation, 238, 239
total internal reflection fluorescence
   advantages, 242
   instrumentation, 240–242
   principles, 240–241
   RNA immobilization, 242–243

S

Single-molecule spectroscopy
   applications, 131–132
   ensemble studies, 132–133
   enzyme kinetics and mechanism studies
      comparison with steady-state and stopped-flow measurements, 151, 153
      experimental considerations, 134–136
      instrumentation, 138–141
      sample preparation, 136–138
      simulated data, 153–154
      trajectory analysis
         autocorrelation function, 147–148, 154
         cholesterol oxidase, 147–149
         dihydroorotate dehydrogenase, 142–148
         dynamic heterogeneity, 146–147
         kinetics of single oligomeric enzyme, 147–151
         static heterogeneity, 145–146
   flavoenzymes, 133–134
   prospects, 154
SMS, see Single-molecule spectroscopy
Solvatochromic fluorescent dyes
   4′-(diethylamino)–3-hydroxyflavone, 41–42, 44, 46
   multiparametric probing
      algorithm, 46–49
      applications
         binary solvent mixture probing, 49–50
         protein-binding site probing, 50–51
      limitations
         hydrogen bonding heterogeneity within sites of same polarity, 51
         site heterogeneity, 52–53
      rationale, 38
   noncovalent interactions, universal versus specific, 38–40
   prospects, 55
   spectroscopic data correlation with solvatochromic variables
      absorption band position, 43
      fluorescence bands N★ and T★
         positions, 43
         ratio of intensities, 44–45
         Stokes shift, 44
   types, 53–54
Spot radius, equation for fiber optic sensors, 321

Streak camera, ultrafast fluorescence spectroscopy, 161
Sum frequency generation, see Upconversion fluorescence spectroscopy

T

TAMRA, see Carboxytetramethylrhodamine
TAT, tagging of carbonic anhydrase fluorescence biosensors, 304–305
TCSPC, see Time-correlated single photon counting
Time-correlated single photon counting, ultrafast fluorescence spectroscopy, 161
Time-resolved emission spectroscopy, see also Area-normalized time-resolved emission spectroscopy
   fluorophores in microheterogeneous and biological media
      dyes in surfactant micelles or lipid membranes, 30–31
      special cases
         fluorescence anisotropy decay, 32
         site heterogeneity, 32
         solvation dynamics, 33
      homogeneity in ground state, 22–23
      overview of steps, 24–25
      relaxation, 23–24
      simple case examples
         fluorophore in dilute solution
            with excited-state reaction, 26
            without excited-state reaction, 26
         fluorophore in dilute, viscous solution without excited-state reaction, 28
         fluorophore mixture in dilute solution without excited-state reaction, 27
      single species in ground state, 22–23
TIRF, see Total internal reflection fluorescence
Total internal reflection fluorescence
   advantage in RNA studies, 242
   principles, 240–241, 247–248
   Rev–Rev response element complex assembly
      instrumentation, 240–242
      kinetic analysis, 248–249
      RNA immobilization, 242–243
TraI36, DNA-binding studies with fluorescence, 259–265, 267
TRANES, see Area-normalized time-resolved emission spectroscopy
Transmembrane helix dimerization energetics, see Fluorescence resonance energy transfer
TRES, see Time-resolved emission spectroscopy
Tryptophan
   protein–ligand interaction studies, 86–87
   quantum yield, 85
   Red-Edge effects, 74–75
   upconversion fluorescence spectroscopy
      protein dynamics, 173–175

Tryptophan (*cont.*)
    quasistatic self-quenching in dipeptides, 172
    solvent relaxation in water, 171–172
Two-photon excitation, pteridine-containing oligonucleotide studies, 216–219

U

Upconversion fluorescence spectroscopy
    acceptance angle, 165–166
    crystal choice, 168
    group velocity mismatch, 166–167
    historical perspective, 162
    instrumentation, 168–170
    phase-matching angle, 163–164
    polarization, 167
    prospects, 177
    quantum efficiency, 166
    sample handling, 167
    spectral bandwidth, 164–165
    sum frequency generation principles, 162–163
    ultrafast photophysics
        decay-associated spectra, 171–172
        DNA dynamics, 175–177
        tryptophan
            protein dynamics, 173–175
            quasistatic self-quenching in dipeptides, 172
            solvent relaxation in water, 171–172

Z

Zinc biosensors, *see* Carbonic anhydrase fluorescence biosensors; Fluorescence-based fiber optic biosensors

Beverly Z. Packard and Akira Komoriya, Figure 1.4 Comparison of the effect of two probes on the signal of a metabolic indicator, $DiOC_6$ in viable cells. Top panel: the blue shows strong mitochondrial polarization in each control NT-2 cell. Since two IHED caspase probes were present throughout the 18-h incubation period, the strong blue signal indicates a lack of toxicity of IHED probes. Bottom panel: in cells induced to undergo apoptosis the green due to caspase 9 activity followed by red due to caspase 8 and decreased level of blue illustrate the loss of mitochondrial membrane potential as cells undergo apoptosis.

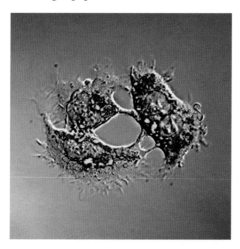

Beverly Z. Packard and Akira Komoriya, Figure 1.5 The fluorescence signals of the cleaved Cathepsin D substrate (red) and LysoTracker (green) in MCF-7 cells only partially colocalize (yellow) suggesting the intracellular site of Cathepsin D to be of a higher pH than previously believed.

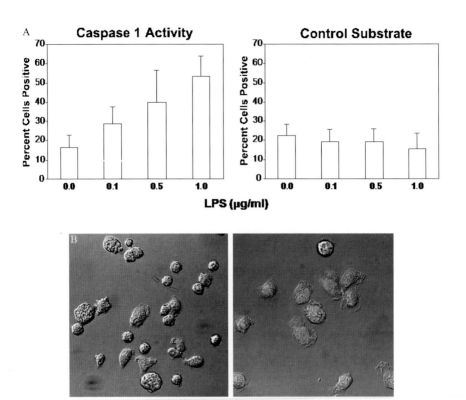

Beverly Z. Packard and Akira Komoriya, Figure 1.7 Caspase 1 activity was induced and measured in live peritoneal exudate cells (PEC) from mice by LPS in a dose-dependent manner. (A) High-throughput analysis of live cells. (B) Confocal imaging in the presence of a caspase 1 activity probe inside live cells induced by LPS (left) compared with a control population (right).

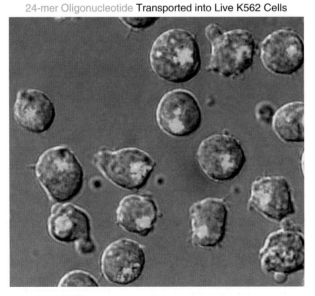

Beverly Z. Packard and Akira Komoriya, Figure 1.8 Confocal imaging of a field of live (PI-negative) cells following addition of a homodoubly labeled 24-mer oligonucleotide (turquoise) containing the complementary sequence to β-actin.

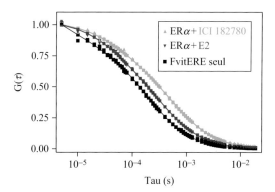

Catherine A. Royer and Suzanne F. Scarlata, Figure 5.7 Normalized two-photon FCS autocorrelation profiles of the fluorescein-labeled F-vitERE from Fig. 5.5 in alone and in presence of saturating human estrogen receptor β and saturating agonist estradiol (E2) and in presence of the saturating receptor and saturating antagonist ICI 182780 (A. Bourdoncle and C. A. Royer, unpublished results).

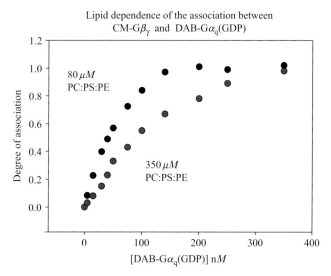

Catherine A. Royer and Suzanne F. Scarlata, Figure 5.9 Shift in the binding curves between CM-G$\beta\gamma$ and DAB-Gα(q) as the concentration of lipid in which the proteins are bound is raised. Binding is assessed by FRET.

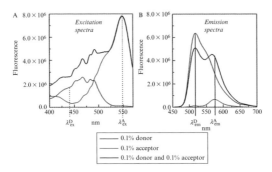

Mikhail Merzlyakov and Kalina Hristova, Figure 6.3 Emission and excitation spectra of fluorescein (Fl) and rhodamine (Rhod), a common FRET pair, conjugated to wild-type FGFR3 TM domain. POPC concentration was 0.25 mg/ml; the concentration of the protein is reported in moles of protein per mole of lipid. (A) Excitation spectra, collected by recording emission at 595 nm while scanning the excitation from 400 to 570 nm. (B) Emission spectra, with excitation fixed at 439 nm, and emission scanned from 450 to 700 nm. In (A) and (B), the blue lines correspond to 0.1 mol% Fl-TM$_{WT}$, while the red lines are for 0.1 mol% Rhod-TM$_{WT}$. The blue and red spectra serve as standard excitation and emission spectra, to be used with the EmEx-FRET method. These spectra are averages, derived from measurements of multiple samples. Also shown are the excitation and emission spectra of 0.1 mol% fluorescein-labeled wild-type and 0.1 mol% rhodamine-labeled wild-type (the donor/acceptor sample, or "FRET sample", black lines). The FRET spectrum (black) is the sum of three contributions: direct donor emission in the presence of the acceptor F_{DA}, direct acceptor emission F_A, and sensitized acceptor emission F_{sen}. Inspection of the excitation spectra in (A) reveals that at excitation wavelength λ_{ex}^A the acceptor excitation reaches its maximum, while the donor excitation is negligible. As a result, the excitation of the donor/acceptor sample (black) at λ_{ex}^A is contributed by the acceptor only (red). $\lambda_{ex}^D = 439$ nm is the excitation wavelength used in the acquisition of the emission spectra. In (B), the emissions of the donor and the acceptor reach their maxima at λ_{em}^D and λ_{em}^A, respectively. Note that the emission of the acceptor is low, but not negligible, at λ_{em}^D. Reprinted from (Merzlyakov et al., 2007), with permission from Springer Science and Business Media.

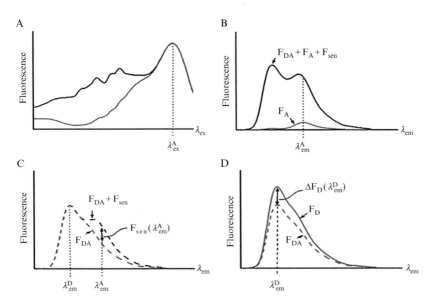

Mikhail Merzlyakov and Kalina Hristova, Figure 6.4 The EmEx-FRET method. (A) An acquired FRET excitation spectrum (black line) is compared to the excitation standard spectrum

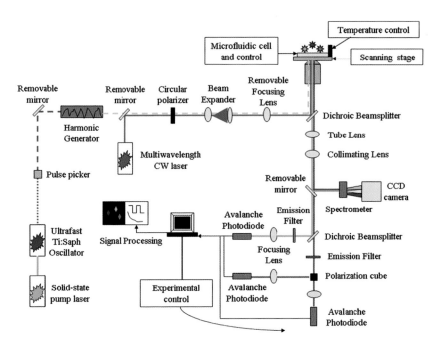

Jue Shi et al., Figure 7.2 Diagram of a single-molecule spectrometer system with a microfluidic reaction cell, which is capable at measuring single-molecule intensity, emission spectrum, lifetime, anisotropy, and FRET.

---

of the acceptor (Fig. 3A, red line.) This step gives the concentration of the acceptor in the sample. (B) The emission standard of the acceptor (Fig. 3B, red line) is scaled according to the acceptor concentration determined in (A) (red line). The difference between the red line and the FRET emission spectrum (black line) is the sum $F_{DA} + F_{sen}$. (C) Dashed black line: $F_{DA} + F_{sen}$, sum of the direct donor (Bell et al., 2000) emission in the presence of the acceptor, F_{DA}, and the sensitized acceptor emission, F_{sen}. The standard emission spectrum of the donor (Fig. 3B, blue line) is scaled, such that the amplitude of the scaled standard (blue dashed line) is identical to the amplitude of the dashed black line at λ_{em}^A. The difference between the black dashed and the blue dashed line is the sensitized acceptor emission, F_{sen}. The value of the sensitized emission at λ_{em}^A, $F_{sen}(\lambda_{em}^A)$, is related to the decrease in donor emission at λ_{em}^D, $\Delta F_D(\lambda_{em}^D)$ (Merzlyakov et al., 2007). (D) The value $\Delta F_D(\lambda_{em}^D)$, determined in (C), is used to determine the donor emission in the absence of the acceptor F_D (blue line), and the donor concentration. Reprinted from (Merzlyakov et al., 2007), with permission from Springer Science and Business Media.

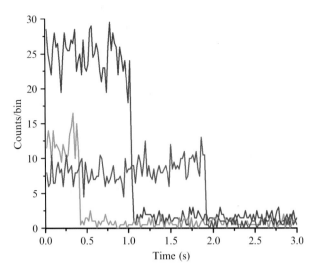

Jue Shi et al., Figure 7.3 Fluorescence trajectories of single molecules of a dihydroorotate dehydrogenase (DHOD) mutant. DHOD is a monomeric flavoenzyme with FMN bound as the prosthetic group. The fluorescence loss is the result of FMN dissociating from the enzyme and rapidly diffusing away from the field of observation. The one-step fluorescence loss indicates that a single DHOD molecule is being observed.

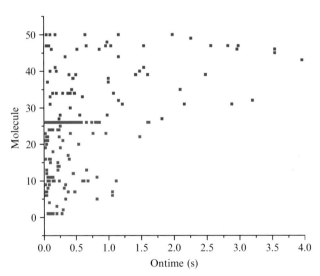

Jue Shi et al., Figure 7.6 Individual on-times of 50 different single DHOD molecules in turnovers. Slow molecules, shown in blue, have less than 10% on-times shorter than 0.1 s, while fast molecules, shown in red, have few on-time longer than 1 s. Different molecules were grouped according to their kinetic behavior and then numbered.